The Role of Nonliving Organic Matter in the Earth's Carbon Cycle

Goal of this Dahlem Workshop:

to devise experimental and modeling strategies for assessment of the sensitivity of the global carbon cycle to changes in nonliving organic pools.

Environmental Sciences Research Report ES 16

Held and published on behalf of the
Freie Universität Berlin

Sponsored by:
Funds from the Stiftung Unilever im
Stifterverband für die Deutsche Wissenschaft

The Role of Nonliving Organic Matter in the Earth's Carbon Cycle

Edited by

R.G. ZEPP and Ch. SONNTAG

Report of the Dahlem Workshop on
The Role of Nonliving Organic Matter in the Earth's Carbon Cycle
Berlin 1993, September 12–17

Program Advisory Committee:
R.G. Zepp and Ch. Sonntag, Chairpersons
A.F. Bouwman, R.F. Christman, K.-C. Emeis,
A.-Y. Huc, S.E. Trumbore

JOHN WILEY & SONS
Chichester • New York • Brisbane • Toronto • Singapore

British Library Cataloguing-in-Publication Data

A catalogue record of this book is available from the British Library

ISBN 0-471-95463-2

Dahlem Editorial Staff: J. Lupp, C. Rued-Engel, G. Custance
Typeset in 10/12pt Times by Alden Multimedia, Northampton
Printed and bound in Great Britain by Biddles Ltd, Guildford, Surrey

Contents

The Dahlem Konferenzen

In 1974, the Stifterverband für die Deutsche Wissenschaft[1] in cooperation with the Deutsche Forschungsgemeinschaft[2] founded the *Dahlem Konferenzen*. It was created to promote an interdisciplinary exchange of scientific ideas as well as to stimulate cooperation in research among international scientists. Dahlem Konferenzen proved itself to be an invaluable tool for communication in science, and so, to secure its long-term future, it was integrated into the Freie Universität Berlin in January, 1990.

As has been evident over recent years, scientific research has become highly interdisciplinary. Now, before real progress can be made in any one field, the concepts, methods, and strategies of related fields must be understood and able to be applied. Coordinated research efforts, scientific cooperation, and basic communication between the disciplines and the scientists themselves must be promoted in order for science to advance.

To meet these demands, Dahlem Konferenzen created a special type of forum for communication, now internationally recognized as the *Dahlem Workshop Model*. These workshops are the framework in which coherent discussions between the disciplines take place and are focused around a topic of high priority interest to the disciplines concerned. At a Dahlem Workshop, scientists are able to pose questions and solicit alternative opinions on contentious issues from colleagues of related fields. The overall goal of a workshop is not necessarily to reach a consensus, but rather to identify gaps in knowledge, to find new ways of approaching controversial issues, and to define priorities for future research. This philosophy is implemented at every stage of a workshop: from the selection of the theme to its breakdown in the discussion groups, from the writing of the background papers to the composition of the group reports.

Workshop topics are proposed by leading scientists and are approved by a scientific board, which is advised by qualified referees. Once a topic has been

---

[1]The Donors Association for the Promotion of Sciences and Humanities, a foundation created in 1921, in Berlin, and supported by German trade and industry to fund basic research in the sciences.
[2]German Science Foundation.

approved, a Program Advisory Committee of scientists meets approximately one year before the workshop to delineate the scientific parameters of the meeting, select participants, and assign them their tasks. Participants are invited on the basis of their scientific standing alone.

Each workshop is organized around four key questions, and these are addressed by four discussion groups of approximately ten participants. Lectures or formal presentations are tabu at Dahlem. Instead, concentrated discussion—within a group and between groups—is the means by which maximum communication is achieved. To facilitate this discussion, participants prepare the workshop theme prior to the meeting through the "background papers," the themes and authors of which are chosen by the Program Advisory Committee. These papers specifically review a particular aspect of the group's discussion topic as well as function as a springboard to the group discussion, by introducing controversies or unresolved problem areas.

During the workshop week, each group sets its own agenda to cover the discussion topic. Cross-fertilization between groups is both stressed and encouraged. By the end of the week, in a collective effort, each group has prepared a report reflecting the ideas, opinions, and contentious issues of the group as well as identifying directions for future research and problem areas still in need of resolution.

A Dahlem Workshop initiates and facilitates discussion between a certain number—necessarily restricted—of scientists. Because it is imperative that the discussion and communication should continue after a workshop, we present the results to the scientific community at large in the form of this published volume. In it you will find the revised background papers and group reports, as well as an introduction to the workshop theme itself.

The difference between proceedings of many conventional meetings and this workshop report will be easily discernable. Here, the background papers have not only been reviewed by formal referees, they have been revised according to the many comments and suggestions made by *all* participants. In this sense, they are reviewed more thoroughly than scientific articles in most archival journals. In addition, an extensive editorial procedure ensures a coherent volume. I am sure that you, too, will appreciate the tireless efforts of the many reviewers, authors, and editors.

On their behalf, I sincerely hope that the spirit of this workshop as well as the ideas and controversies raised will stimulate you in your work and future endeavors.

Prof. Dr. Klaus Roth, Director
Dahlem Konferenzen der Freien Universität Berlin
Rothenburgstr. 33, D–12165 Berlin, Germany

List of Participants with Fields of Research

GÖRAN ÅGREN Department of Ecology and Environmental Research, Swedish University of Agricultural Sciences, Box 7072, S–750 07 Uppsala, Sweden

Theories and models of carbon and nutrient cycling in terrestrial ecosystems

MARC J. ALPERIN 12–8 Venable Hall, Marine Sciences Program, University of North Carolina, Chapel Hill, NC 27599–3300, U.S.A.

Biogeochemistry of dissloved organic carbon in anoxic marine sediments; factors that control organic carbon decomposition and burial in coastal sediments

DARWIN W. ANDERSON Department of Soil Science, University of Saskatchewan, Saskatoon, SK S7N 0W0, Canada

Organic carbon cycling in soils; soil formation

JÉROME BALESDENT Institut National de la Recherche Agronomique, Unité de Science du Sol, Route de Saint-Cyr, F–78026 Versailles Cedex, France

Use of ^{13}C and ^{14}C natural abundances in measuring soil C turnover

RONALD H. BENNER Marine Science Institute, University of Texas at Austin, P.O. Box 1267, Port Aransas, TX 8373–1267, U.S.A.

Biogeochemistry; carbon and nitrogen cycles in aquatic ecosystems

NEIL V. BLOUGH Department of Chemistry and Biochemistry, University of Maryland, College Park, MD 20742, U.S.A.

Photochemistry and photophysica of colored organic matter

LEX BOUWMAN National Institute of Public Health and Environmental Protection, P.O. Box 1, NL–3720 BA Bilthoven, The Netherlands

Soil science and biogenic trace gas production

RUSS F. CHRISTMAN Department of Environmental Sciences, School of Public Health, CB 7400, University of North Carolina, Chapel Hill, NC 27514, U.S.A.

Chemical properties of dissolved organic matter in terrestrial streams

ANDREAS DAHMKE Fachbereich Geowissenschaften, Universität Bremen, Klagenfurter Straße, D–28359 Bremen, Germany

Rates of organic matter degradation in oxic and anoxic environments (especially deep sea sediments from the South Atlantic)

OTTO DANIEL Institut für Terrestrische Ökologie, Fachbereich Bodenbiologie, ETH Zürich, Grabenstr. 3, CH–8952 Schlieren, Switzerland

Interactions between soil microorganisms and animals

ERIC DAVIDSON The Woods Hole Research Center, P.O. Box 296, Woods Hole, MA 02543, U.S.A.

Soil microbial ecology; emissions of C and N trace gases from soils

MARGARET DELANEY Institute of Marine Sciences, University of California, Santa Cruz, CA 95064, U.S.A.

Cenozoic chemical mass balances; trace elements in Great Barrier Reef corals; phosphorus in marine sediments

ELLEN R.M. DRUFFEL Earth System Science, University of California, Irvine, CA 92717, U.S.A.

Use of carbon isotopes (^{14}C, ^{13}C) to study cycling of carbon in the oceans, both organic carbon and CO_2

KAY-CHRISTIAN EMEIS Geologisch-Paläontologisches Institut, Universität Kiel, Olshausenstr. 40–60, D–24118 Kiel, Germany

Marine geology; marine geochemistry; upwelling sediments

Fritz H. Frimmel Engler-Bunte-Institut, Universität Karlsruhe (TH), Richard-Willstätter-Allee 5, D–76131 Karlsruhe, Germany

Chemistry of aquatic systems; analytical characterization of dissolved organic matter; advanced water treatment (oxidation; disinfection; membrane technology)

Georg Guggenberger Institute of Soil Science and Soil Geography, University of Bayreuth, D–95440 Bayreuth, Germany

Organic matter chemistry; dissolved organic carbon in soils; organic carbon dynamics in soils

Richard A. Houghton Code YSE, NASA Headquarters, Washington, D.C. 20546, U.S.A.

Terrestrial ecosystems in the global carbon cycle; effects of land-use change and climatic change on the storage of carbon in vegetation and soil

Alain-Yves Huc Institut Français du Pétrole, 1 & 4, av. de Bois-Préau, F–92506 Rueil-Malmaison, France

Organic geochemistry applied to petroleum sedimentology of organic matter

Claude Largeau Laboratoire de Chimie Bioorganique, Ecole Nationale Supérieure de Chimie, CNRS/UA 1381, 11, rue Pierre et Marie Curie, F–75231 Paris Cedex 05, France

Organic geochemistry; natural product chemistry; microbial lipids; microbial cell walls; macromolecular constituents of microalgae and bacteria

Ray Lassiter Environmental Research Laboratory–Athens, 960 College Station Rd., Athens, GA 30605–2720, U.S.A.

Modeling biogeochemical cycling in soils, including the role of organic matter and the exchange of greenhouse gases with the atmosphere

Cindy Lee Max-Planck-Institute for Marine Microbiology, Fahrenheitstraße 1, D–28359 Bremen, Germany

Organic geochemistry

Jerry M. Melillo The Ecosystems Center, Marine Biological Laboratory, Woods Hole, MA 02543, U.S.A.

Biogeochemistry; ecosystem ecology

KENNETH MOPPER Chemistry Department, Washington State University, Pullman, WA 99164–4630, U.S.A.

Carbon cycling in the oceans; photochemical degradation and photoprocesses involving marine organic matter; characterization of marine organic matter

KNUTE J. NADELHOFFER The Ecosystems Center, Marine Biological Laboratory, Woods Hole, MA 02543, U.S.A.

Controls on primary production and decomposition in forest and tundra ecosystems; stable isotopes in ecosystem research

RAYMOND G. NAJJAR The Pennsylvania State University, Department of Meteorology, Walker Building, University Park, PA 16802, U.S.A.

The interaction between marine biogeochemistry and climate; modeling and data analysis studies relating to the air-sea fluxes of gases important in climate and atmospheric chemistry

HEINZ-ULRICH NEUE International Rice Research Institute (IRRI), Soil and Water Sciences Division, P.O. Box 933, 1099 Manila, Philippines

CH_4 and N_2O fluxes in rice fields; organic matter turnover in rice soils; sustainability indices for fertility of rice soils

J. MALCOLM OADES Department of Soil Science, Waite Institute, Glen Osmond, South Australia 5064

Organic matter chemistry – ^{13}C NMR; physical fractionation of organic matter; organic matter and soil aggregation

THOMAS F. PEDERSEN Department of Oceanography, University of British Columbia, Vancouver, BC V6T 1Z4, Canada

History of organic carbon burial during the quaternary in equatorial and margin areas; use of stable C and N isotopes to assess pCO_2 and nutrient history; chemical diagenesis of sediments

WILFRED M. POST Environmental Sciences Division, Oak Ridge National Laboratory, P.O. Box 2008, Bldg. 1000, Oak Ridge, TN 37831–6335, U.S.A.

Soil organic matter dynamics; global carbon cycling; forest ecosystem ecology

WILLIAM S. REEBURGH Earth System Science, Physical Sciences Research Unit, University of California Irvine, Irvine, CA 92717, U.S.A.

Methane geochemistry

DAN J. REPETA Marine Chemistry and Geochemistry, Woods Hole Oceanographic Institution, Woods Hole, MA 02539, U.S.A.

Marine organic chemistry; structural analysis using NMR, mass spectrometry; bacterial metabolism of high molecular weight carbohydrates

HANS HERMANN RICHNOW Institut für Biogeochemie und Meereschemie, Universität Hamburg, Bundesstraße 55, D–20146 Hamburg, Germany

Origin, structure, and fate of organic macromolecules in the geosphere (humic substances, asphaltenes, kerogens, and resin)

WOLFGANG SCHÄFER Institut für Umweltphysik, Universität Heidelberg, Im Neuenheimer Feld 366, D–69120 Heidelberg, Germany

Numerical modeling of biochemical processes in groundwater

DAVID S. SCHIMEL National Center for Atmospheric Research (NCAR), P.O. Box 3000, Boulder, CO 80307–3000, U.S.A.

Soil carbon storage and turnover; remote sensing of terrestrial carboon dynamics

MARY C. SCHOLES Department of Botany, University of Witwatersrand, Private Bag 3, 2050 Wits, Johannesburg, South Africa

Soil biological processes; nitrogen, phosphorus, and carbon cycling in semi-arid savannas; biomass burning and trace gas emissions

ROBERT J. SCHOLES Division of Forest Science and Technology, CSIR, P.O. Box 395, Pretoria 0001, South Africa

Plant and soil interactions in African savannas, particularly in relation to global land use, climate, and atmospheric change

CHRISTIAN SONNTAG Institut für Umweltphysik, Universität Heidelberg, Im Neuenheimer Feld 366, D–69120 Heidelberg, Germany

Isotope hydrology and paleoclimatology; isotopes in organic substances

WILLIAM G. SUNDA Beaufort Laboratory, NMFS/NOAA, 101 Pivers Island Road, Beaufort, NC 28516, U.S.A.

Marine trace metal chemistry; trace metal/organic interactions in seawater; trace metal/biological interactions in seawater

ROGER S. SWIFT CSIRO Division of Soils, PMB 2, Glen Osmond, South Australia 5064

Composition, structure, and properties of soil organic matter and soil humic substances, including their age and stability; the role of soil organic matter in nutrient cycling

SUSAN E. TRUMBORE Earth System Science, University of California, Irvine, CA 92717–3100, U.S.A.

Dynamics of soil, sedimentary, and dissolved organic carbon from ^{14}C measurements

REIN VAIKMÄE Institute of Geology, Estonian Academy of Sciences, Estonia Avenue 7, Tallinn EE0101, Estonia

Stable isotopes in ice cores, permafrost, and ground ice; lake sediments and groundwater as palaeoclimate evidence

EDSO VELDKAMP OTS–La Selva, INTERLINK 341, P.O. Box 02–5635, Miami, FL 33152, U.S.A.

Carbon/nitrogen dynamics in tropical soils; nitrous oxide gas emissions from soils

EUGENE B. WELCH Department of Civil Engineering, University of Washington, Seattle, WA 98195, U.S.A.

Lake and stream entrophication; cycling of phosphorus; cost-effectiveness of lake management techniques

RICHARD G. ZEPP Environmental Research Laboratory, U.S. Environmental Protection Agency, 960 College Station Rd, Athens, GA 30605–2700, U.S.A.

Biogeochemical processes affecting carbon cycle

1

Introduction

R.G. Zepp[1] and Ch. Sonntag[2]

[1]Environmental Research Laboratory, U.S. Environmental Protection Agency,
960 College Station Rd., Athens, Georgia 30605–2700, U.S.A.
[2]Institut für Umweltphysik, Universität Heidelberg,
Im Neuenheimer Feld 366, D–69120 Heidelberg, Germany

Nonliving organic matter (NLOM) comprises the bulk of the organic carbon stored in the terrestrial biosphere and a major part of the organic carbon in the sea. These organic substances, which include litter and soil organic matter, aquatic detritus, dissolved organic matter, and sedimentary organic matter, have diverse effects on the Earth's biogeochemical processes. For example, they have important effects on soil water retention, pH, redox, and nutrient levels; they provide reservoirs of nitrogen, phosphorus, and other important nutrients; they play a key role in the transport of organic carbon from the upper ocean to the deep ocean and bottom sediments; they affect the penetration of harmful ultraviolet and photosynthetically active solar radiation into the sea and fresh waters; and they regulate bioavailability, transport, and transformations of essential trace metals and pollutants in soils and the oceans. Given this broad spectrum of effects, efforts to adapt to or perhaps ameliorate adverse impacts of global change require that we better understand and predict the role of NLOM in the global environment.

NLOM also serves as a major reservoir of biospheric carbon that can be transformed to carbon dioxide (CO_2), methane (CH_4), and other "greenhouse gases." Based on analyses of historical data, it is generally assumed that these biospheric sources of atmospheric CO_2 are approximately matched by biospheric sinks involving photosynthesis and that the current buildup in atmospheric CO_2 is attributable to anthropogenic emissions. However, although more or less balanced in the long term, the fluxes of CO_2 into and out of the biosphere are much larger than anthropogenic emissions. For example, the approximately 100 Pg (1 Pg $= 10^{15}$ g) of CO_2 that are exchanged between

Role of Nonliving Organic Matter in the Earth's Carbon Cycle
Edited by R.G. Zepp and Ch. Sonntag © 1995 John Wiley & Sons Ltd.

the atmosphere and terrestrial biosphere annually are over an order of magnitude larger than the emissions derived from fossil-fuel burning. Thus, small fractional changes in the biospheric fluxes would have major effects on the rate of CO_2 buildup in the atmosphere. Such changes in the balance of the global carbon cycle may be induced by anthropogenic or climatic perturbations.

The most recent evidence of the potential impacts of such changes is the unique "stalling" of atmospheric CO_2 buildup (Sarmiento 1993), the slowing of atmospheric CH_4 increases (Kerr 1994), and apparent changes in the CO trend from positive to negative (Novelli et al. 1994). These surprising phenomena are apparently related, at least in part, to climatic effects of aerosols injected into the atmosphere by the recent eruption of Mt. Pinatubo in the Philippines. Some scientists have suggested that the changes in trace gas trends can partly be ascribed to aerosol-related global cooling that has reduced the conversion of NLOM pools to CO_2, CH_4, and other greenhouse gases.

Progress in evaluating and predicting the role of NLOM in biogeochemical cycles has been hindered not only by incomplete data bases and insufficient process understanding and models, but also by the fact that NLOM research pertaining to one environmental compartment, e.g., the oceans, may be poorly known to scientists working on another compartment, e.g., terrestrial soils. The closing of such gaps in awareness and the resulting enhancement of scientific perception is at the heart of the objectives of the Dahlem Konferenzen, which features a workshop format wholly designed to promote interdisciplinary discussion. Thus, to address these hindrances, a Planning Committee was convened in Berlin in October, 1992, to define the topics for a workshop on the subject of nonliving organic matter, focusing on NLOM's role in the carbon cycle and its relationship to global environmental processes. The committee decided that the primary goal of the workshop should be to devise experimental and modeling strategies for the assessment of the sensitivity of the global carbon cycle to changes in nonliving organic pools, and to accomplish this task, an interdisciplinary group of international scientists with expertise related to both marine and terrestrial systems was invited to participate.

One of the initial tasks of the workshop participants involved the development of meaningful classifications for NLOM. NLOM consists of biomolecules that exhibit a wide range of chemical and physical properties, with residence times in the environment that range from minutes to thousands of years. For terrestrial soils, terms such as labile, active, slow, passive, refractory, and inert have been used to define NLOM in terms of the turnover times of various fractions. On the other hand, perhaps in part because it is much more mobile than soil NLOM, aquatic NLOM is usually classified in terms of its form and location, e.g., dissolved organic carbon, particulate organic carbon, and sedimentary organic carbon. Although workshop participants had to deal with a certain tyranny of phrases that have historically been attached to this subject, the focal point was on NLOM that persists on time scales ranging from centuries to millenia.

The persistence of NLOM depends partly upon (a) its intrinsic reactivity (substrate quality), i.e., the degree to which it resists microbially mediated turnover, which is a function of its chemical structures, and partly on (b) matrix effects related to the local environment in which the NLOM resides. Resistant NLOM can be generated directly by primary producers (or heterotrophs); it can result from polymerization reactions of complex biopolymers; and it can be formed by fire and photochemical processes. Matrix effects that affect persistence include a variety of phenomena such as "physical protection" through occlusion and formation of clay-organic complexes in soils and sediments, supply of oxidants, nutrient levels, microbial populations, ultraviolet radiation, and temperature. Interactive feedbacks can affect persistence as well. For example, low inputs of nutrients to bogs from vegetation serve to slow microbial activity, which results in greater persistence of NLOM. Slower decomposition of NLOM, in turn, further sequesters nutrients thus reinforcing conditions that favor growth of plants that conserve nutrients.

To discuss these and other related issues, four interdisciplinary groups considered key research questions related to the primary goal. All workshop participants had the opportunity to contribute to the discussion of all groups. Background papers by leading scientists provided thorough discussions of various aspects of NLOM research and were used to stimulate group discussion. These papers are included in the volume.

The first group (see Alperin et al., this volume) focused on the characterization and quantification of NLOM. This group considered methods for measuring NLOM pools and fluxes on a global basis, including classifications based on vegetation types or climatic regimes as well as new remote sensing techniques and time series measurements of atmospheric composition that provide regional to global integration of changes in transfer rates among NLOM, biomass, and atmospheric carbon pools. The group also focused on methods for characterizing the reactivity, sources, and fates of NLOM, such as biological techniques, spectroscopic methods, and measurements of isotopic composition. Possible alterations of NLOM composition by various analytical procedures were evaluated as well.

The role of NLOM cycling on a global scale was considered by the second group (see Post et al., this volume). Using geochemical models of the current global carbon cycle as a framework to examine carbon cycle dynamics, the group discussed the possibility that changes in some fluxes of carbon into and back out of NLOM reservoirs may account for the "missing carbon sink". The group delved into the Quaternary and geologic past to help discern how changes in the present carbon cycle may sequester additional carbon. Finally, this group discussed new approaches to advancing our knowledge of NLOM at the global scale. Inverse modeling approaches that utilize observations of the geographic distribution of atmospheric CO_2 or oxygen concentrations and isotopic composition to estimate carbon fluxes indirectly were considered as one particularly useful approach. Critical components that must be captured in models of NLOM dynamics in terrestrial and aquatic systems were defined.

The biogeochemical cycles of various elements are tightly linked. Thus, changes in the global carbon cycle, including the parts that involve NLOM, are accompanied by changes in cycles of the growth-limiting nutrients, including nitrogen, phosphorus, potassium, sulfur, and transition metals (especially iron). Human and climatic perturbations of these interactions between NLOM and nutrients were discussed by the third group (see Nadelhoffer et al., this volume). Using "minimum" models (terrestrial and aquatic) as a contextual framework, the group considered feedback interactions in terrestrial ecosystems between plant, soil NLOM, and microbial processes, including effects of nutrient availability on NLOM amounts and quality and consequences of NLOM element ratios for carbon sequestration. Discussions of aquatic ecosystems examined key interactions that affect sedimentation and NLOM burial in marine and freshwater ecosystems, including the role of NLOM in the oceanic biological CO_2 pump. The wide-ranging discussions of this group also considered perturbations of these interactions by various human activities: nutrient redistributions, land-use changes, fire, nutrient and sediment loadings in continental margins, and increased deposition of eolian iron in Fe-limited regions of the sea.

Our fourth discussion group considered the physical, chemical, and biological processes that control the production and degradation of NLOM (see Davidson et al., this volume). This group focused most of its discussion on identifying the main sources and processes of formation and decomposition of persistent NLOM. Effects of substrate quality on resistance to decomposition in aquatic and terrestrial ecosystems were discussed, including direct production of persistent NLOM by primary producers and production of persistent NLOM such as charcoal by fire. The important role played by chemical and physical protection in soils (and possibly sediments) and mechanisms for such protection were also major topics of discussion. Finally, the group considered relationships between processes that affect NLOM persistence and key environmental variables, such as oxidant supply, nutrients, transport in water, acidity, fauna, and ultraviolet radiation.

This workshop was an outstanding experience for all of the participants. We are indebted to Klaus Roth and the Dahlem staff for their highly successful efforts in establishing logistics and organization for the workshop. We also express our thanks to Russell Christman and Jennifer Altman for their encouragement and assistance during the early stages of preparing the proposal for the workshop. It is impossible to describe adequately the intensity and excitement of the scientific discussions that took place at the Clubhaus of the Freie

Universität Berlin (and other parts of Berlin) during September, 1993. The group reports included in this volume at least capture the essence of our far-ranging discussions. A unique aspect of this workshop was the opportunity for extensive exchanges between leading international scientists from both marine and terrestrial backgrounds. These interactions will no doubt lead to future fruitful collaborations among the workshop participants and, through this volume, the broader scientific community. We hope that this brief introduction to the critical roles of NLOM in the Earth's biogeochemical cycles is sufficient to convey the spirit and thrust of our discussions.

REFERENCES

Kerr, R.A. 1994. Methane increase put on pause. *Science* **263**:751.
Novelli, P., K.A. Masarie, P.P. Tans, and P.M. Lang. 1994. Recent changes in atmospheric carbon monoxide. *Science* **263**:1587–1589.
Sarmiento, J. 1993. Atmospheric CO_2 stalled. *Nature* **365**:697–698.

2

Carbon Isotopes for Characterizing Sources and Turnover of Nonliving Organic Matter

S.E. TRUMBORE and E.R.M. DRUFFEL
Earth System Science, University of California, Irvine,
California 92717–3100, U.S.A.

ABSTRACT

Carbon isotope ratios within nonliving organic matter (NLOM) reflect the processes and rates associated with its production, internal transformations, and ultimate fate. In this chapter, examples from marine and terrestrial systems that demonstrate the power of using isotopic tools to decipher NLOM sources and turnover rates are presented, similarities and differences in a variety of NLOM pools are explored, and new approaches for future study are suggested. The carbon isotopes ^{13}C and ^{14}C are emphasized, as they have been most widely applied to NLOM studies.

GENERAL PROPERTIES OF CARBON ISOTOPES

Stable isotopes, such as ^{13}C and ^{15}N, are generally used in studies of nonliving organic matter (NLOM) to understand the balance of sources as well as the processes operating to transform living organic matter (LOM) into NLOM (Williams et al. 1992; Druffel and Williams 1992). Stable isotope abundances are commonly expressed as the deviation (in per mil) in the ratio of the heavy (rare) isotope to the light (abundant) isotope in a sample (e.g., $[^{13}C:^{12}C]_{sam}$) from that of a commonly accepted standard ($[^{13}C:^{12}C]_{std}$):

Role of Nonliving Organic Matter in the Earth's Carbon Cycle
Edited by R.G. Zepp and Ch. Sonntag © 1995 John Wiley & Sons Ltd.

$$\delta^{13}C = \left\{ \frac{\left(\frac{^{13}C}{^{12}C}\right)_{sam} - \left(\frac{^{13}C}{^{12}C}\right)_{std}}{\left(\frac{^{13}C}{^{12}C}\right)_{std}} \right\} \times 1000. \qquad (2.1)$$

The standard used for reporting ^{13}C data ($^{13}C = 0‰$) is Pee Dee Belemnite limestone (PDB). The standard for ^{15}N data is atmospheric N_2 ($\delta^{15}N = 0‰$).

Figure 2.1 shows the range of $\delta^{13}C$ values found in some of the natural carbon reservoirs. The systematic differences seen in ^{13}C result from both equilibrium and kinetic fractionation processes. Equilibrium effects arise from differences in the relative strengths of chemical bonds between heavy and light isotope species; at equilibrium, the heavier isotope will be concentrated in the phase or compound with stronger chemical bonds. An example of equilibrium fractionation is the difference in ^{13}C content between atmospheric CO_2 (about $-7.8‰$ in 1990) and marine dissolved inorganic carbon (DIC) in surface ocean waters ($\delta^{13}C = -2$ to $+2‰$). Kinetic fractionation,

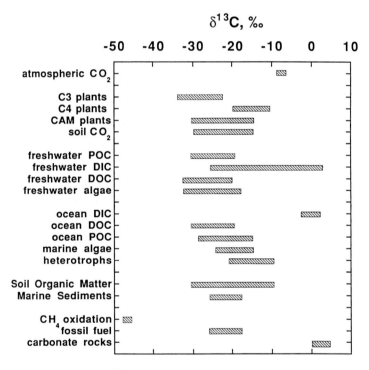

Figure 2.1   The ranges of $\delta^{13}C$ values for common carbon reservoirs associated with the formation or decomposition of NLOM.

on the other hand, results from the faster diffusion or reaction of compounds that contain lighter isotopes over those that contain heavier isotopes. For example, the ^{13}C content of terrestrial plants is dominated by two kinetic processes: the resistance to gas transfer of CO_2 through the plant stoma ($^{12}CO_2$ diffuses faster than $^{13}CO_2$) and the tendency of the RuDP enzyme (which converts dissolved CO_2 to a sugar) to use ^{12}C preferentially (Farquhar et al. 1982). An intermediate step, dissolution of the CO_2 in intracellular leaf water, involves an equilibrium fractionation. Both equilibrium and kinetic fractionations are temperature dependent.

The ^{13}C content of atmospheric CO_2 reflects a balance between the opposing influences of exchange with terrestrial organic (more ^{13}C-depleted) and marine inorganic (more ^{13}C-enriched) carbon reservoirs. Changes in the apportionment of carbon among these three reservoirs will be reflected as a change in the atmospheric CO_2 ^{13}C value. ^{13}C measurements of ice core CO_2 (Leuenberger et al. 1992) and C_4 plant remains (Marino et al. 1992) show that atmospheric $^{13}CO_2$ values were lower during the last glacial period than during the subsequent interglacial. Also, the ^{13}C content of the atmosphere has decreased during the past century, due to accumulation of fossil fuel and biosphere-derived carbon in the atmosphere (Keeling et al. 1989).

Radiocarbon (^{14}C; half-life = 5730 years) is produced by the interaction of cosmic-ray-produced neutrons with ^{14}N in the upper atmosphere. The ^{14}C produced is quickly oxidized to $^{14}CO_2$ and mixed into the troposphere, where it exchanges with carbon in nonatmospheric reservoirs. ^{14}C data related to geochemical problems are reported as the ratio of heavy to light isotope compared to that of a standard (19th century wood or 95% of NBS oxalic acid-I). The distribution of ^{14}C within a carbon reservoir is determined both by mass-dependent fractionation, which affects the initial isotope ratio, and by subsequent radioactive decay. To isolate the effects of radioactive decay, measurements of ^{14}C are corrected for isotope fractionation. Because the mass difference between ^{14}C and ^{12}C is twice as great as that found between ^{13}C and ^{12}C, ^{14}C is depleted or enriched twice as much as ^{13}C. The correction of all ^{14}C measurements to a constant $\delta^{13}C$ of –25‰ eliminates any influence of fractionation on the resultant $\Delta^{14}C$ values. ^{14}C data are thus reported not only with respect to the $^{14}C:^{12}C$ value of the standard (Equation 2.2), but are also corrected to a ^{13}C value of –25‰ (Equation 2.3) (Stuiver and Polach 1977):

$$\delta^{14}C = \left\{ \frac{\left(\frac{^{14}C}{^{12}C}\right)_{sam}}{\left(\frac{^{14}C}{^{12}C}\right)_{std}} - 1 \right\} \times 1000, \qquad (2.2)$$

$$\Delta^{14}C = \delta^{14}C - 2(25 + \delta^{13}C)\left(1 - \frac{\delta^{14}C}{1000}\right). \qquad (2.3)$$

The natural production rate of ^{14}C has remained approximately constant ($\pm 10\%$) for the past 10,000 years, as recorded in the (decay-corrected) ^{14}C content of known-aged tree rings. Isolation of a carbon reservoir (such as a plant seed) from exchange with atmospheric $^{14}CO_2$ results in a decrease in the $^{14}C{:}^{12}C$ ratio in the reservoir over time due to ^{14}C decay. In closed, homogeneous systems, such as a preserved seed, the $^{14}C{:}^{12}C$ value is a measure of the time since death of the organism. In heterogeneous reservoirs that are continuously exchanging part of their carbon with the atmosphere, such as soil or sedimentary organic matter, the ^{14}C content reflects a mixture of materials with both long and short residence times. In these heterogeneous systems, it is particularly important to distinguish between the average age of a carbon atom in the reservoir (given by the decrease in $^{14}C{:}^{12}C$ in the reservoir below that of fresh carbon inputs) and the residence time of a carbon atom in the reservoir (given by the reservoir size divided by the input flux). For reservoirs such as NLOM in soils and sediments, in which a majority of the carbon added annually is rapidly decomposed but a small portion may accumulate slowly over long time scales, the age of carbon derived from ^{14}C will be greater than the turnover time for carbon in the reservoir.

Atmospheric testing of thermonuclear weapons approximately doubled the atmospheric content of ^{14}C by 1963. This "bomb ^{14}C" is now present as a contaminant in all living and recently dead organic matter, thereby complicating, but sometimes aiding in, the interpretation of ^{14}C in contemporary NLOM reservoirs. The subsequent distribution of bomb ^{14}C has provided a global tracer that has been particularly useful for following the exchange of carbon between atmospheric, oceanic, and terrestrial reservoirs on annual to decadal time scales (Broecker and Peng 1982).

USE OF CARBON ISOTOPES IN THE STUDY OF NLOM: COMMON THEMES

Simple measurement of NLOM isotope ratios will not reveal the processes involved in its formation and destruction. An adequate interpretation of isotope data in NLOM requires an understanding of its sources and ultimate decomposition products. This section uses specific examples from marine and terrestrial systems to illustrate several of the approaches using isotopes to study NLOM:

- Determination of the relative importance of different NLOM sources by comparing bulk isotopic composition of NLOM with that of various LOM precursors.
- Comparison of stable isotope composition of NLOM with its (known) LOM precursor to determine biochemical fractionation pathways leading to NLOM formation.
- Use of ^{14}C to determine the turnover time of NLOM. This includes measurement of the distribution of ^{14}C in chemically and physically fractionated NLOM and assessment of the rate of infiltration of bomb ^{14}C into the NLOM pool.
- Comparison of NLOM substrates with DIC, CO_2, and CH_4 produced during its decomposition to understand roles of chemical reactivity and "age" of stored NLOM.

Isotopes as Source Indicators

Traditionally, $\delta^{13}C$ has been used as a source indicator for a variety of carbon pools. For example, $\delta^{13}C$ values of –20‰ to –21‰ are found for DOC_{UV} (UV-oxidized dissolved organic carbon) in the oceanic water column (Williams and Gordon 1970; Williams and Druffel 1987). This range is similar to that for marine plankton in nonpolar oceans (–18‰ to –20‰), as opposed to a riverine source whose $\delta^{13}C$ range is about 5‰ to 7‰ lower. Thus, the ^{13}C data support a marine origin for DOC. It remains a mystery as to why the large riverine flux of organic matter to the ocean leaves such a small imprint on the isotopic and chemical (i.e., lignin-content) signatures (Meyers-Schulte and Hedges 1986; J. Ertel, pers. comm., 1992). It is suspected that photochemical degradation of the riverine organic matter is a significant sink (Kieber et al. 1990), though more work in this area is crucial to quantify its global significance and the major chemical pathways of destruction.

Jasper and Hayes (1990) measured the $\delta^{13}C$ values in a plankton-specific alkenone from several intervals of sediment in the northern Gulf of Mexico. Based on the relationship between the concentration of dissolved CO_2 and the $\delta^{13}C$ of phytoplankton in surface seawater (Popp et al. 1989), Jasper and Hayes (1990) reconstructed a pCO_2 record for the Pleistocene atmosphere that showed good correspondence with the ice core pCO_2 record of Barnola et al. (1987). As phytoplankton is the presumed major precursor to DOC in the oceans, it is conceivable that DOC $\delta^{13}C$ could have been significantly different during the last glacial due to changes in phytoplankton $\delta^{13}C$.

Terrestrial plant tissues, which are the ultimate source of NLOM produced in terrestrial environments, fall into two groups with distinct stable carbon isotope characteristics (see Figure 2.1). Plants utilizing a C_3 photosynthetic pathway (most trees) have $\delta^{13}C$ values between –22‰ to –32‰, while C_4

plants (mostly tropical grasses) have $\delta^{13}C$ values between $-9‰$ and $-15‰$ (Figure 2.1). The stable carbon isotope content of NLOM in soils developed under these two vegetation types reflects this difference in input materials. This information has been exploited by researchers in two ways: observed changes in the $\delta^{13}C$ content of soil organic matter following conversion of vegetation from C_3 to C_4 vegetation have been used to infer the dynamics of NLOM in soils (e.g., Balesdent et al. 1987; Cerri et al. 1991), and changes in vegetation type (C_3- or C_4-dominated) in the past have been inferred from the vertical distribution of NLOM $\delta^{13}C$ in soil profiles.

Fractionation as an Indicator of Geochemical Processes

Fractionation of carbon and other isotopes (^{15}N, ^{34}S, ^{18}O) can provide important clues as to the biogeochemical processes that take place within pools of NLOM. Differences in $\delta^{13}C$ and $\delta^{15}N$ values are observed between soil organic matter and the fresh plant material from which it is originally derived. These differences may be due to one or several of the following causes (Nadelhoffer and Fry 1988): (a) discrimination between heavy and light N and C isotopes during decomposition; (b) differential preservation of compounds with lesser or greater $\delta^{13}C$ and $\delta^{15}N$ values; (c) changes in the $\delta^{13}C$ and $\delta^{15}N$ values of plant inputs with time; or (d) selective translocation of depleted or enriched organic matter from litter to soil layers. In their study of $\delta^{13}C$ and $\delta^{15}N$ in organic matter from an upland forest, Nadelhoffer and Fry (1988) concluded that the generally higher $\delta^{13}C$ and $\delta^{15}N$ values observed in soils over litter input values are due to discrimination during decomposition of organic matter. They point out that selective preservation of lignin components, which are generally depleted in $\delta^{13}C$ relative to bulk plant tissues, would cause a decrease of $\delta^{13}C$ as LOM is degraded, rather than the observed increase. The decrease in $\delta^{13}C$ in atmospheric CO_2 caused by combustion of fossil fuels (Keeling et al. 1989) may contribute to some of the differences observed recently between the $\delta^{13}C$ values of soil organic matter and fresh plant litter.

Use of ^{14}C to Determine the Turnover Time of NLOM

^{14}C may reflect both the isotopic content of NLOM sources and the rate of turnover of the NLOM itself. Most NLOM pools consist of a mixture of carbon from a variety of sources that are likely of different ^{14}C "ages". When trying to interpret the meaning of a bulk ^{14}C measurement in terms of NLOM dynamics, it is important to consider all possible combinations of ages from the precursors of NLOM as well as the dynamics of the NLOM carbon pool. It is also important to define clearly those terms used to refer to NLOM dynamics. Here, "turnover time" is the quantity obtained when the carbon inventory of a given NLOM reservoir is divided by the annual rate of addition or loss of carbon from the reservoir. The "^{14}C apparent age" is the age determined from the ^{14}C:^{12}C ratio (see Equation 2.4, below) measured in the NLOM pool. For a steady-state, homogeneous NLOM pool with input rate I, decomposition rate k, and carbon content C, these terms will be related:

$$\text{Age} = -\frac{1}{\lambda_{14C}} \times ln\left(\frac{k}{k + \lambda_{14C}}\right), \tag{2.4}$$

$$\text{Turnover time} = \frac{C}{I} = \frac{1}{k}, \tag{2.5}$$

where λ_{14C} is the radioactive decay constant for ^{14}C (Trumbore et al. 1992). Some investigators report a "^{14}C mean residence time", which is the same as our ^{14}C apparent age. The importance of differentiating these terms when referring to nonhomogeneous reservoirs, such as most NLOM, is revealed using the following example.

Figure 2.2 shows the dynamic makeup of two hypothetical NLOM reservoirs, 1 and 2. In the figure, NLOM is broken down as the amount of the total organic matter in ten homogeneous pools with turnover times ranging from 1 year to 10,000 years. The distribution of NLOM among components with different turnover times has been chosen such that the bulk soil ^{14}C content will correspond to an apparent ^{14}C age of 650 years for both soils (see figure caption for details of the calculations). The organic carbon in NLOM-1 is more or less evenly distributed about a mean turnover time of between 500 and 1000 years, while NLOM-2 is a mixture of rapidly cycling (1–50 years) material with a smaller amount of very refractory material (turnover times greater than 1000 years). The average (bulk) turnover time of carbon in NLOM-1 and NLOM-2 is determined by assuming both reservoirs are at steady state (inputs of new material equal losses by decomposition). Decomposition rates are calculated by dividing the amount of carbon in a pool by the turnover time for that pool. The average turnover time of carbon in NLOM-1 calculated in this way is 62.5 years, as opposed to 8.3 years for NLOM-2. Both of these turnover times are surprisingly much less than the ^{14}C-derived apparent age of 650 years. In the past, models of NLOM dynamics have sometimes relied on a ^{14}C apparent age to estimate the "mean residence time" of carbon in NLOM reservoirs (thus equating ^{14}C apparent age with turnover time). The above discussion shows that such treatment can seriously underestimate the dynamic nature of NLOM, especially if even a small fraction is considerably more ^{14}C-depleted than the bulk material.

How can we distinguish NLOM-1 from NLOM-2? First, we could try to fractionate NLOM into operationally defined compound classes. For example,

S.E. Trumbore and E.R.M. Druffel

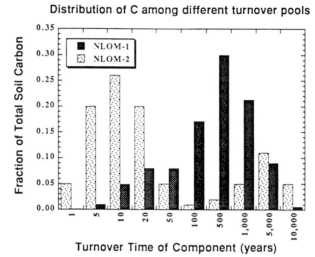

Figure 2.2 The breakdown of two hypothetical NLOM reservoirs into constituents with different turnover times. The majority of organic matter in NLOM-1 (dark shading) is distributed in pools with turnover times of 100 to 1000 years. NLOM-2 (stippled) is predominantly composed of organic matter, which cycles rapidly (less than 20 years), mixed with a smaller amount of more refractory material (5,000- to 10,000-year turnover). The steady-state $^{14}C:^{12}C$ ratio for each pool is calculated using Equation 2.5; the ratio for the whole reservoir (NLOM-1 or -2) is the sum of products of the fraction of carbon in each pool and its $^{14}C:^{12}C$ ratio. Both NLOM-1 and NLOM-2 will have the same apparent ^{14}C age (650 years) at steady state. The net annual fluxes (used to calculate bulk turnover times) of carbon into and out of NLOM-1 and NLOM-2 are calculated as the sum of the products of the amount of carbon in each pool times its decomposition rate (turnover time^{-1}).

^{14}C values for two humic fractions separated from filtered seawater from four depths using XAD macroreticular resins are lower than those for the bulk DOC (Figure 2.3). This indicates that the average apparent age of the hydrophobic humic fractions are generally older than those in total DOC in seawater, which likely contains the more labile, dissolved components (amino acids/proteins, sugars/carbohydrates and lipids) from such sources as plankton exudates and debris. Similarly, operationally defined fractionation methods have been used (with limited success) to separate soil organic matter into components with higher and lower $\Delta^{14}C$ values.

Second, Figure 2.4 shows the increase in $\Delta^{14}C$ expected for steady-state soils NLOM-1 and NLOM-2 during the bomb period. The increase is larger in NLOM-2, the pool with the shorter average turnover time. If samples are available from two time points within the past 30 years, modeling of the ^{14}C change between time points will help constrain NLOM dynamics. The use of this method to demonstrate differences in carbon cycling between temperate

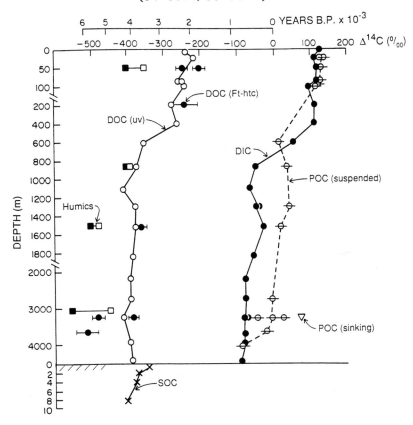

Figure 2.3 $\Delta^{14}C$ measurements of DOC_{uv} (UV-oxidizable DOC), DOC_{Ft-htc} (flow-through high temperature combustion DOC), humic materials, suspended POC (particulate organic carbon), sinking POC, and sedimentary organic carbon (SOC) measured in seawater in the Sargasso Sea (North Atlantic Ocean). Squares are isolates of DOC separated using XAD-2 (filled) and XAD-4 (open) resins. From Druffel et al. (1992).

(NLOM-1-like) and tropical (NLOM-2-like) forest soils is shown in Figure 2.5 (from Trumbore 1993).

Chemical Reactivity ("Old Sources")

Chemical reactivity, as it relates to why certain organic matter pools are preserved and others are not, is an essential concept to understand in our

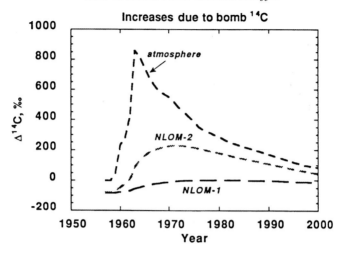

Figure 2.4 The increase in ^{14}C predicted by modeling carbon reservoirs with the characteristics of NLOM-1 and NLOM-2 (see Figure 2.2). NLOM-2, which contains more carbon in pools with turnover times of 20 years or less, takes up more bomb ^{14}C than NLOM-1. Thus the infiltration of bomb ^{14}C into a NLOM pool provides a clue to its carbon dynamics.

purview of NLOM. In the context of isotopes, a simplistic assumption often made is that the "younger" the carbon, or the higher the $\Delta^{14}C$ value, the more chemically (and biologically) labile it is. For example, in the oceans, POC has higher $\Delta^{14}C$ values than DOC does at the same depths, consonant with the 2–3 times higher relative amounts of labile components contained within POC.

Marine DIC $\delta^{13}C$ is a powerful tracer of organic matter oxidation, owing to the large difference between the $\delta^{13}C$ of organic matter and that of $CaCO_3$ and bottom water carbonate species. In a study of the remineralization processes in nearshore sediments of Buzzards Bay, Massachusetts, McNichol et al. (1991) found that the $\delta^{13}C$ of DIC added to the sediment pore waters was 5–15‰ greater than the sedimentary organic carbon (SOC). They surmised that the enrichment must be due either to fractionation during remineralization of SOC or to isotope exchange with bottom water. It remains to be determined whether the fractionation was due to the selective oxidation of specific organic compound classes (e.g., amino acids), different sources of organic carbon to coastal sediments, or whether DOC production was an important intermediate in the remineralization of SOC.

The degree to which kerogen is remineralized upon weathering of sedimentary rocks on the continents is an open question (Hedges 1992). The presence of "dead" carbon in modern, coastal sediments (e.g., Benoit et al. 1979) is evidence that significant quantities of weathered organic matter is not entirely

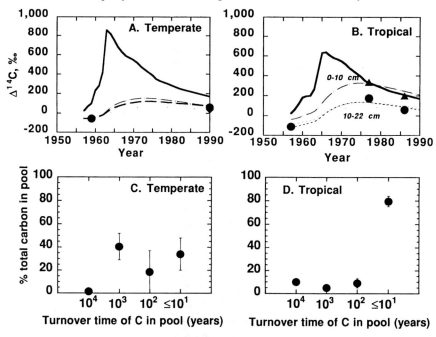

Figure 2.5 Comparison of carbon dynamics in two soils (Amazon tropical forest and dry temperate forest in California), determined by comparing the ^{14}C content in soils collected before and after the peak of atmospheric weapons testing. A and B show the increase in ^{14}C since 1959 in soil organic matter in the upper 22 cm of soil for the temperate and tropical soils (data points). The dashed lines are models that reproduce the ^{14}C increase by calculating the distribution of carbon among pools which turn over on time scales of $\leq$10, 100, 1000, and 10,000 years. C and D show the range of values (in terms of the fraction of total organic carbon in each component) which can reproduce the ^{14}C increases observed in A and B. Thus the temperate soil is more like the hypothetical NLOM-1, with more carbon in reservoirs turning over on 100- to 1000-yr time scales, while the tropical soil is more like NLOM-2, with the majority of its carbon turning over on time scales less than 10 years. Modified from Trumbore (1993).

remineralized. As kerogen constitutes 95% of the organic matter in sedimentary rocks, resolution of this issue is crucial to our understanding of the control of NLOM on the global biogeochemical cycles over geologic time.

The isotopic composition of CO_2 and CH_4 produced during the decomposition of terrestrial NLOM reflects a combination of: (a) isotopic composition of the NLOM substrate, (b) fractionation on decomposition, and (c) fractionation associated with transport from the system. Carbon dioxide and CH_4 produced in soil and wetland sediments may be depleted or enriched in Δ^{14}C relative to contemporary atmospheric CO_2 Δ^{14}C values (i.e., LOM or root-respired CO_2). These differences reflect the ^{14}C content of the organic matter being decomposed and provide valuable clues to the rates of carbon turnover in different ecosystems. CO_2 enriched in ^{14}C relative to 1992 atmospheric $^{14}CO_2$ values observed in the air-filled pore spaces of tropical forest soils (Nepstad et al. 1994) indicates the influence of decomposing organic carbon that was originally fixed from the atmosphere between 5 and 30 years ago. The $\Delta^{14}C$ of bulk soil organic matter is much lower than that of the soil atmosphere CO_2, reflecting that the SOM is a mixture of material with rapid and slow turnover times.

^{14}C values in CH_4 bubbling from peat bogs and other wetlands are often measurably less than those of contemporary LOM, indicating that, although the predominant sources of CH_4 are "young" carbon substrates, some contribution from "old" subtrates is indicated. DOC observed in freshwater streams and in soil solutions may be similarly depleted in ^{14}C (Schiff et al. 1990). Because the residence time of the water in these systems is much shorter than the time needed for significant radioactive decay of ^{14}C (as evidenced by hydrologic balances and the presence of bomb-derived tritium), the ^{14}C depletion reflects the presence of ^{14}C-depleted DOC sources (Schiff et al. 1990). Thus, organic matter that has been preserved long enough for significant radioactive decay of ^{14}C to have occurred may still be "labile" enough to be a substrate for microbial decomposition. Careful study of such systems will provide clues as to the ultimate factors controlling NLOM preservation.

UNKNOWNS AND FUTURE RESEARCH AREAS

Isotopes have not been fully utilized as a means to unravel the history of NLOM pools on Earth. There are still measurements that have not even been made, such as the $\delta^{15}N$ of DOM in the oceans, owing to the difficulty of oxidizing organic matter in dilute saline solution. As yet, the controversy surrounding the accuracy of DOC concentration measurements in seawater has not been completely resolved. The following are fertile areas of future research that require the use of isotopes for studying the sources and turnover of elements in NLOM:

1. Measurement of $\Delta^{14}C$ from separate organic compound classes and from individual organic compounds that are tracers of specific organisms or key processes. To advance our knowledge past the impasse of bulk NLOM isotope values, we need to be able to separate in sufficient quantities and to date individual organic compounds. In the same way that archaeology has been revolutionized by AMS in its ability to ^{14}C-date a single seed from an occupied site, so too shall geochemists be able to constrain the origin of widely diverse carbon pools by relating isotope measurements to organic chemistry and structure.

2. Integrated studies that include ^{14}C, ^{13}C, and ^{15}N isotopes of major organic pools in the oceans and on land.
3. Study of the preserved organic matter in coastal and slope sediments to delineate its terrestrial versus marine-derived sources.
4. Isotope labeling studies, now used in terrestrial decomposition research, have yet to be widely applied to the oceans to investigate such processes as marine snow formation and mid-water transformation reactions between soluble and particulate phases.
5. Organic chemical characterization of the NLOM using both isotopic signatures and molecular structure information (e.g., Hedges et al. 1992).
6. Exploration of the relation between the isotopic content of NLOM decomposition products and NLOM substrates.

In our pursuit of the histories of NLOM pools, we offer the following controversial issues as seeds for lively discussion.

First, to what degree are our present observations of isotopes in NLOM affected by anthropogenic changes? The increase in atmospheric CO_2 concentration due to fossil-fuel burning and land-use change has caused a decrease in the ^{13}C content of atmospheric CO_2 of about 1.2‰ to 1.5‰ since 1900. How much does this recent decrease influence the difference in ^{13}C often observed between fresh LOM and the NLOM derived from it? For example, how much of the approximately $+2‰$ shift in ^{13}C observed between detrital and mineral soil organic matter is an anthropogenic effect? ^{14}C is also being perturbed. Fossil-fuel burning dilutes ^{14}C in the atmosphere (though this effect has been overwhelmed by the bomb ^{14}C over the last 30 years). In the late 1950s, NLOM may reflect this dilution of the Suess effect, a factor not always accounted for when interpreting ^{14}C measurements in pre-bomb NLOM. Soot, or elemental carbon derived from fossil-fuel burning, is ubiquitous and often inseparable from NLOM. Estimates of the global production of soot from fossil-fuel burning are very uncertain. One estimate by Penner et al. (1993) suggests that the global source is $25\,TgC\,yr^{-1}$, equivalent to a flux of $0.05\,gC\,m^{-2}\,yr^{-1}$. Assuming that most of this falls out in the Northern Hemisphere where it is produced, a flux of this magnitude would have added approximately $5\,g\,m^{-2}$ of ^{14}C-free carbon to soils and sediments over the past 50 years. The A horizons of soils typically contain $100–500\,gC\,m^{-2}$; the bioturbated layers of deep marine sediments contain roughly $500\,gC\,m^{-2}$. Given the uncertainty of the soot production estimates, we cannot rule out the possibility that this anthropogenic influence is adding a "dead" carbon pool to soils measured today. Similarly, organic carbon derived from fossil-fuel sources is often found in sediments, identified by organic geochemists as "unresolved complex mixtures." How much does the presence of this radiocarbon-free material influence our (^{14}C-based) interpretation of NLOM dynamics in such systems?

Second, the apparent ^{14}C age of some NLOM that is very refractory and accumulates over long time scales is several thousand years. Is there a chance that reservoirs we observe with ^{14}C apparent ages in this range could be nonsteady-state (i.e., slowly accumulating) reservoirs? This is almost certainly true of refractory organic matter in soils. Could it be true of ocean DOC as well? The carbon content, C, of a carbon reservoir accumulating at a constant annual rate, I, in any given year will be $I \times t$, where t is time in years since the start of carbon accumulation. Assuming the new carbon added each year is Modern (i.e., Δ^{14}C of 0‰), the ^{14}C content of the reservoir with time will vary (as the "old" organic matter in which ^{14}C has decayed begins to dilute the Modern carbon signal). The radiocarbon content for a given year, $R(t)$ (where R is the ratio of ^{14}C:^{12}C in the NLOM reservoir compared to that of the oxalic acid standard) can be calculated from that of the previous year $R(t-1)$:

$$R(t) = \frac{R(t-1) \times C(t-1) \times (1 - \lambda_{14C}) + I \times 1.0}{C(t)} \qquad (2.6)$$

Figure 2.6 shows the evolution of the apparent ^{14}C age (or $-\frac{1}{\lambda_{14C}} \times \ln[R(t)]$) with time for a constantly accumulating reservoir. This relation, which is independent of I (since $C(t-1) = I \times (t-1)$ and $C(t) = I \times t$ in Equation 2.6), demonstrates that we should not necessarily assume refractory material in NLOM pools which accumulate over long times scales to be radiocarbon free.

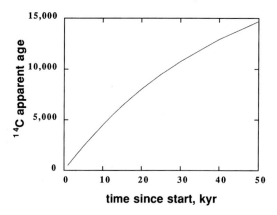

Figure 2.6   The predicted evolution of the apparent ^{14}C age for a carbon reservoir that is accumulating at a constant rate with time (see discussion in text).

REFERENCES

Balesdent, J., A. Mariotti, and B. Guillet. 1987. Natural ^{13}C abundance as a tracer for studies of soil organic matter dynamics. *Soil Biol. Biochem.* **19**:25–30.

Barnola, J.M., D. Raynaud, Y.S. Korotkevich, and L. Lorius. 1987. Vostock ice core provides 160,000 year record of atmospheric CO_2. *Nature* **329**:408–414.

Benoit, G.J., J.K. Turekian, and L.K. Benninger. 1979. Radiocarbon dating of a core from Long Island Sound. *Est. Coast. Mar. Sci.* **9**:171–180.

Broecker, W.S., and T.-H. Peng. 1982. Tracers in the Sea. L-DGO. New York: Columbia Univ. Press, 690 pp.

Cerri, C.C., B.P. Eduardo, and M.C. Piccolo. 1991. Use of stable isotopes in soil organic matter studies. In: Stable Isotopes in Plant Nutrition, Soil Fertility and Environmental Studies, Symp. Proc., pp. 237–259. Vienna: Intl. Atomic Energy Agency.

Druffel, E.R.M., and P.M. Williams. 1992. The importance of isotope and other complimentary measurements in marine organic geochemistry. *Mar. Chem.* **39**:209–215.

Druffel, E.R.M., P.M. Williams, J.E. Bauer, and J.R. Ertel. 1992. Cycling of dissolved and particulate organic matter in the open ocean. *J. Geophys. Res.* **97**:15,639–15,659.

Farquhar, G.D., M.H. O'Leary, and J.A. Berry. 1982. On the relationship between carbon isotope discrimination and the intercellular carbon dioxide concentration in leaves. *Austr. J. Plant Physiol.* **9**:121–137.

Hedges, J.I. 1992. Global biogeochemical cycles: Progress and problems. *Geochim. Cosmochim. Acta* **39**:67–93.

Hedges, J.H., P.G. Hatcher, J.R. Ertel, and K. Meyers-Schulte. 1992. A comparison of dissolved humic substances from seawater with Amazon River counterparts by ^{13}C-NMR spectrometry. *Geochim. Cosmochim. Acta* **56**:1753–1757.

Jasper, J.P., and J.M. Hayes. 1990. A carbon isotope record of CO_2 levels during the late Quaternary. *Nature* **347**:462–464.

Keeling, C.D., R.B. Bacastow, A.F. Carter, S.C. Piper, T.P. Whorf, M. Heimann, W.G. Mook, and H. Roeloffzen. 1989. A three-dimensional model of atmospheric CO_2 transport based on observed winds: 1. Analysis of observational data. *Geophys. Monogr.* **55**:165–236.

Kieber, R.J., X. Zhou, and K. Mopper. 1990. Formation of carbonyl compounds from UV-induced photodegradation of humic substances in natural waters: Fate of riverine carbon in the sea. *Limnol. Oceanogr.* **35**:1503–1515.

Leuenberger, M., U. Siegenthaler, and C.C. Langway. 1992. Carbon isotope composition of atmospheric CO_2 during the last ice age from an Antarctic ice core. *Nature* **357**:488–490.

Marino, B.D., M.B. McElroy, R.J. Salawitch, and S.W.G. Spaulding. 1992. Glacial to interglacial variations in the carbon isotopic composition of atmospheric CO_2. *Nature* **357**:461–466.

McNichol, A.P., E.R.M. Druffel, and C. Lee. 1991. Carbon cycling in coastal sediments: 2. An investigation of the sources of CO_2 to pore water using carbon isotopes. In: Organic Substances and Sediments in Water, vol. 2, Processes and Analytical, ed. R.A. Baker, pp. 249–272. Chelsea, MI: Lewis Publ.

Meyers-Schulte, K.J., and J.I. Hedges. 1986. Molecular evidence for a terrestrial component of organic matter dissolved in ocean water. *Nature* **321**:61–63.

Nadelhoffer, K.J., and B. Fry. 1988. Controls on natural N–15 and C–13 abundances in forest soil organic matter. *Soil Sci. Soc. Am. J.* **52**:1633–1640.

Nepstad, D.C., C.R. de Carvalho, E.A. Davidson, P.H. Jipp, P.A. Lefebvre, G.H. Negreiros, E.D. da Silva, T.A. Stone, S.E. Trumbore, and S. Vieira. 1994. The deep soil link between water and carbon cycles of Amazonian forests and pastures. *Nature*, in press.

Penner, J.E., H. Eddleman, and T. Novokov. 1993. Towards the development of a global inventory for black carbon emissions. *Atmos. Environ.* **27A**:1277–1295.

Popp, B.N., R. Takigiku, J.M. Hayes, J.W. Louda, and E.W. Baker. 1989. The post-Paleozoic chronology and mechanism of ^{13}C depletion in primary marine organic matter. *Am. J. Sci.* **289**:436–454.

Schiff, S.L., R. Aravena, S.E. Trumbore, and P.J. Dillon. 1990. Contribution of carbon isotopes to the understanding of the cycling of dissolved organic carbon in forested water sheds. *Water Resour. Res.* **26**:2949–2957.

Stuiver, M., and H. Polach. 1977. A discussion reporting ^{14}C data. *Radiocarbon* **19**:355–363.

Trumbore, S.E. 1993. Comparison of carbon dynamics in temperate and tropical soils using radiocarbon measurements. *Glob. Biogeochem. Cyc.* **7**:275–290.

Trumbore, S.E., S.L. Schiff, R. Aravena, and R. Elgood. 1992. Sources and transformation of dissolved organic carbon in the Harp Lake forested catchment: The role of soils. *Radiocarbon* **34**:626–635.

Williams, P.M., and E.R.M. Druffel. 1987. Radiocarbon in dissolved organic matter in the central North Pacific Ocean. *Nature* **330**:246–248.

Williams, P.M., and L.I. Gordon. 1970. Carbon-13:Carbon-12 ratios in dissolved and particulate organic matter in the sea. *Deep-Sea Res.* **17**:19–27.

Williams, P.M., K.J. Robertson, A. Souter, S.M. Griffin, and E.R.M. Druffel. 1992. Isotopic signatures (^{14}C, ^{13}C, ^{15}N) as tracers of sources and cycling of DOC and POC in the Santa Monica Basin, California. *Prog. Oceanogr.* **30**:253–290.

3

Spectroscopic Characterization and Remote Sensing of Nonliving Organic Matter

N.V. BLOUGH and S.A. GREEN
Department of Marine Chemistry and Geochemistry,
Woods Hole Oceanographic Institution,
Woods Hole, Massachusetts 02543, U.S.A.

ABSTRACT

Optical absorption and luminescence spectroscopies have played a key role in characterizing the "colored" components of dissolved organic matter (CDOM), which constitute a significant fraction of the total nonliving organic matter (NLOM) in the environment. These techniques have also been essential for establishing the ways that CDOM might be viewed by remote sensors. However, our current understanding of the spatial and temporal variability in the optical properties and their relation to CDOM structure remains limited. This article emphasizes the need to examine the relationship between the optical and molecular properties of CDOM in a more systematic fashion, and the importance of this information for both field and remote sensing studies. The optical properties of CDOM are first reviewed, followed by a brief examination of available correlations between the optical and molecular properties of this material. Possible experimental approaches to define these relationships better are then presented. Finally, the use of remote sensing to track the distribution and dynamics of CDOM in natural waters is discussed.

INTRODUCTION

The study of nonliving organic matter (NLOM) has been greatly facilitated through the use of a variety of sophisticated spectroscopic and spectrometric techniques. This has been especially true for that fraction of the NLOM classically known as humic substances, Gelbstoffe, or yellow substance. This

Role of Nonliving Organic Matter in the Earth's Carbon Cycle
Edited by R.G. Zepp and Ch. Sonntag © 1995 John Wiley & Sons Ltd.

material, referred to here as colored (or chromophoric) dissolved organic matter (CDOM), is a chemically complex mixture of anionic organic oligoelectrolytes created by the decay of plant organic matter; it represents a major, if not dominant, portion of the total dissolved NLOM in most water bodies. Because of its ubiquitous presence and complex nature, an array of techniques has been utilized to probe the chemical and structural properties of this material. These techniques have included ^{13}C and ^{1}H nuclear magnetic resonance (NMR), electron paramagnetic resonance spectroscopy, pyrolysis-mass spectrometry, gas chromatography-mass spectrometry, infrared spectroscopy, and ultraviolet/visible absorption and luminescence spectroscopies. Their use has allowed the composition of CDOM to be characterized roughly in terms of the relative amounts of common molecular substituents such as saccharide, phenolic, methoxyl, carboxyl, hydrocarbon, and aromatic moieties (Malcolm 1990). Importantly, this work has shown that the relative contributions of these substituents vary with the source and environmental history of the CDOM, thus indicating that the observed "quality" of this material is both spatially and temporally dependent (Hedges et al. 1992; Malcolm 1990).

Remote sensing from airborne or satellite platforms offers a possible way to determine the distribution and nature of CDOM over broader spatial and temporal scales than is currently possible. Because the applicable remote sensing methods are based primarily on the interaction of CDOM with the near-ultraviolet and visible portions of the electromagnetic spectrum, optical absorption and luminescence spectroscopies have played and will continue to play a critical role in establishing the ways that CDOM might be viewed by remote sensors. Unlike other techniques, optical methods are both portable and highly sensitive, allowing their use in the field for rapid analysis of CDOM and the ground-truthing of remote measurements. Unfortunately, optical methods alone are incapable of providing the level of structural information afforded by some of the other spectroscopic techniques, such as NMR. If field and remote optical measurements are ultimately to provide the means not only of determining the distribution of CDOM in surface waters rapidly, but also of identifying salient features of its structure, then *it is essential to define the relationship between the optical properties of the CDOM and the underlying structural elements responsible for producing these properties.* This goal might be achieved in part, for example, by establishing simple, empirically derived relationships between the absorption lineshapes or specific absorption coefficients of CDOM and the amount of aromatic carbon it contains as measured by ^{13}C NMR. *Ultimately, a better understanding of the molecular basis for the optical absorption and emission properties of this material is required to assess its structure in a more critical fashion.* As structure largely determines reactivity, this information is key to delineating the role that photochemical, chemical, and biological processes play in controlling the fate of this material.

In this chapter we emphasize the importance of understanding the connections between the optical and molecular properties of CDOM and the use of this information for both *in situ* and remote sensing studies. We first review the optical properties of CDOM from various sources and then briefly summarize the results pertaining to possible correlations between the optical and molecular properties of this material. Next we present an outline of experimental approaches that may provide further insight into these relationships. Finally, we discuss the use of remote sensing to identify and track CDOM in natural waters.

ABSORPTION AND LUMINESCENCE PROPERTIES OF CDOM

Absorption Properties

In most instances, CDOM exhibits broad and unstructured absorption spectra, decreasing approximately exponentially throughout the near-ultraviolet and visible wavelength regimes (Figure 3.1a). To parameterize CDOM absorption, workers have generally fit these spectra to an exponential form:

$$a(\lambda) = a(\lambda o)e^{-S(\lambda - \lambda o)}, \tag{3.1}$$

where $a(\lambda)$ and $a(\lambda o)$ are the absorption coefficients at wavelength λ and reference wavelength λo, and S is a parameter that characterizes how rapidly the absorption decreases with increasing wavelength (Figure 3.1b). $a(\lambda)$ is obtained from the relationship:

$$a(\lambda) = 2.303 \, A(\lambda)/r, \tag{3.2}$$

where A is the absorbance and r is the pathlength in meters. While Equation 3.1 usually provides a good description of the absorption spectra of CDOM in natural waters, there is no theoretical justification for its use. Moreover, there have been few attempts to account for this spectral dependence employing models that assume, for example, it arises from a simple superposition of the spectra of structures thought to be a part of this material. Nevertheless, fitting absorption data to Equation 3.1 has proved useful as a simple empirical means to characterize these spectra. This exponential form has been used to describe CDOM absorption in ocean color models (Carder et al. 1991) as well as to estimate its absorption at visible wavelengths from measurements in the ultraviolet (Bricaud et al. 1981; Davies-Colley and Vant 1987).

N.V. Blough and S.A. Green

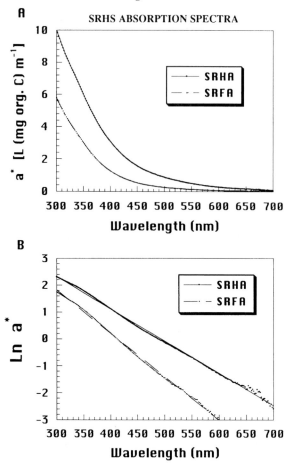

Figure 3.1 Absorption spectra of Suwanee River humic acid (SRHA) and fulvic acid (SRFA) obtained from the International Humic Substances Society. Concentrations of SRFA and SRHA were 6.6 and 6.8 mg/L, respectively, in 50 mM borate buffer, pH = 8.0. A: Wavelength dependence of the specific absorption coefficient, a^* [L (mg organic C)$^{-1}$ m^{-1}], calculated on the basis that SRFA and SRHA contain 50% C by weight. B: Wavelength dependence of the natural logarithm of a^*. The lines represent linear least squares fits to the data. Values of S acquired from these fits are provided in Table 3.1.

Another empirical index, the E4/E6 ratio (the ratio of absorbance at 465 nm to that at 665 nm), has long been employed by soil scientists to characterize the visible absorption spectra of soil humic substances (Chen et al. 1977). Assuming that the spectral dependence of these materials is precisely described by an exponential function, S and E4/E6 are related through the following expressions,

$$E4/E6 = a(465)/a(665) = e^{S \times 200} \qquad (3.3)$$

$$S = ln(E4/E6)/200 \qquad (3.4)$$

obtained by simple rearrangement of Equation 3.1. Unlike S, the E4/E6 ratio is based on measurements at only two wavelengths and does not presume a functional form for the spectral dependence. For simplicity, S is used throughout this discussion.

A variety of studies have shown that S varies with the type and source of CDOM (Table 3.1). Values for terrestrial sources range from $\sim$0.01 to 0.02 nm^{-1}, with the humic acids consistently exhibiting lower values than the fulvic acids (Zepp and Schlotzhauer 1981; Davies-Colley and Vant 1987). Similarly, Carder et al. (1989) found that S was significantly lower in isolated marine humic acids than in fulvic acids. Values for marine CDOM have also been suggested to vary between $\sim$0.01 and 0.02, with an average value of 0.014 nm^{-1} (Bricaud et al. 1981). Earlier work by Brown (1977) in Baltic waters indicated that S increases with decreasing absorption; as $a(310)$ decreased from $\approx$ 14 m^{-1} to 1.2 m^{-1} with increasing salinity, S increased from 0.015 to $\sim$0.03 nm^{-1}. More recent studies of marine waters in the eastern Caribbean (Blough et al. 1993) and off the coast of south Florida (Green 1992; Green and Blough, in prep.) have also found that S is generally larger for oligotrophic "blue" waters ($\geq$0.02 nm^{-1}) than coastal "brown" waters (0.013–0.018 nm^{-1}). These results suggest that the CDOM is modified and/or replaced by a different form during transit from coastal to offshore waters. Reflecting the rather large gradient between fresh and marine waters, values of $a(300)$ range from < 0.2 m^{-1} for "blue" sea waters to > 50 m^{-1} for some coastal and fresh waters.

Specific absorption coefficients, $a(\lambda)^*$, obtained by normalizing $a(\lambda)$ to the organic carbon concentration (c) of a CDOM sample,

$$a(\lambda)^* [\text{liter (mg org. C)}^{-1} \text{m}^{-1}] = \alpha(\lambda)/c \qquad (3.5)$$

provide an additional means of distinguishing the sources and types of this material (Blough et al. 1993; Carder et al. 1989; Zepp and Schlotzhauer 1981). This parameter furnishes a measure of how strongly CDOM absorbs light on a per unit carbon basis and is strictly defined only for isolated material stripped of nonabsorbing, noncovalently bound carbon compounds. While this parameter can be acquired for whole water samples, the values obtained must necessarily be viewed as lower bounds, since natural waters also contain nonabsorbing species. Specific absorption coefficients can be employed to estimate *in situ* CDOM concentrations using the $a(\lambda)^*$ of an isolated standard, *if* it can be demonstrated that the standard and the whole water sample contains the same type of material.

Table 3.1 Values of $a(450)$* [L (mg org. C)$^{-1}$ m^{-1}] and S [nm^{-1}] for CDOM from various sources.

Source/Type of CDOM	$a(450)$*	S	Reference
Soil Humic Acid – mean values	1.76 ± 0.21	0.0102 ± 0.0002	Zepp and Schlotzhauer (1981)
Soil Fulvic Acid – mean values	0.80 ± 0.15	0.0138 ± 0.009	Zepp and Schlotzhauer (1981)
Freshwater Aquatic Humus – mean values	0.71 ± 0.26	0.0145 ± 0.0017	Zepp and Schlotzhauer (1981)
Suwanee River Humic Acid[1]	1.57[2]	0.0120	This paper
Suwanee River Fulvic Acid[1]	0.51[2]	0.0161	This paper
Orinoco River CDOM[3]	1.23 ± 0.09	0.0141 ± 0.0008	Blough et al. (1993)
Gulf of Paria CDOM[3]	0.65 ± 0.06	0.0145 ± 0.0004	Blough et al. (1993)
Marine Fulvic Acid			
Mississippi Plume	0.007 ± 0.001	0.0194 ± 0.00044	Carder et al. (1989)
Gulf Loop Intrusion	0.005 ± 0.001	0.0184 ± 0.00166	Carder et al. (1989)
Marine Humic Acid	0.1302 ± 0.00005	0.0110 ± 0.00012	Carder et al. (1989)
Marine Aquatic Humus – mean values	0.33 ± 0.09	0.0147 ± 0.0006	Zepp and Schlotzhauer (1981)
Sedimentary Fulvic Acid Molecular Weight			
< 10,000	0.093	0.0183	Hayase and Tsubota (1985)
10,000–50,000	0.109	0.0181	Hayase and Tsubota (1985)
50,000–100,000	0.195	0.0157	Hayase and Tsubota (1985)
100,000–300,000	0.386	0.0125	Hayase and Tsubota (1985)
> 300,000	0.317	0.0130	Hayase and Tsubota (1985)

[1] International Humic Substances Society standard.
[2] Calculated assuming that this material contains 50% C by weight.
[3] Collected by solid phase extraction on C18 reversed phase cartridges.

Despite the overall lack of measurements, available results indicate that different types of CDOM show distinct variations in $a(\lambda)^*$ (Table 3.1). Generally, humic acids exhibit larger $a(\lambda)^*$ than do fulvic acids, and terrestrial CDOM displays consistently larger values than the marine material. There is also evidence that $a(\lambda)^*$ varies systematically with the molecular size of the CDOM (Hayase and Tsubota 1985; Stewart and Wetzel 1980). Several groups have reported that $a(\lambda)^*$ of fulvic acids increase with increasing molecular weight, while those of humic acids vary inversely with molecular weight (Hayase and Tsubota 1985).

Fluorescence Properties

Excitation and emission spectra, quantum yields, and lifetime measurements have all been employed to discriminate between CDOM types and to gain insight into the structural nature of the chromophores within this material. Below, we present the kinds of information provided by these measurements along with a brief survey of results for CDOM.

Excitation and Emission Spectra

Excitation and emission spectra provide information about the electronic transition energies of the light-absorbing and -emitting states of a chromophore; these energies are directly related to the structure and environment of the chromophore (e.g., the degree of extension of a π system or the presence of heteroatoms in an aromatic group). For a simple chromophore not undergoing other excited state photoprocesses, the energy and lineshape of the emission band do not change with excitation wavelength because fluorescence originates only from the first excited singlet state. Instead, the intensity of the emission band varies in direct relation to the amount of light absorbed by the chromophore at the excitation wavelength. Thus, the excitation spectrum exhibits the same appearance as the absorption spectrum, whereas the emission spectrum furnishes information about the energy of the first excited singlet state of the chromophore.

Not surprisingly, fluorescence spectra of CDOM do not show this simple behavior. Excitation spectra do not mirror the absorption spectra, and emission bands shift to longer wavelengths as the excitation wavelength is increased (Figure 3.2). The spectra are also very broad and unstructured. These results point to the presence of numerous absorbing and emitting centers, some of which are likely due to the presence of a number of different, noninteracting chromophores, while others may result from intramolecular chromophore-chromophore interactions in either the ground or excited state.

Perhaps due in part to this complexity, extreme variations in the excitation and emission maxima are not usually seen for different types of CDOM (Zepp and Schlotzhauer 1981; Senesi 1990; Green 1992). Generally, maxima in the

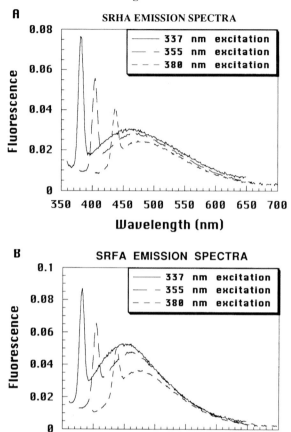

Figure 3.2 Corrected fluorescence spectra of SRHA (A) and SRFA (B) with excitation at 337, 355, and 380 nm. Concentrations of SRHA and SRFA were 1.4 and 1.3 mg l^{-1}, respectively, in 50 mM borate buffer, pH = 8.0. The water Raman scattering bands at 382, 405, and 438 nm have been retained as an internal scale to which the fluorescence can be compared.

emission and excitation spectra are observed between 400 and 500 nm and 300 and 400 nm, respectively. For most aquatic CDOM, excitation at 350 nm produces a broad emission centered at ~450–460 nm.

Because the values of these maxima are so similar, available evidence suggests that they cannot be used to discriminate cleanly between the source (soil, freshwater, marine) or the type (humic acids versus fulvic acids) of CDOM. Given the greater absorption by humic acids in the red (lower S), one might predict that their excitation and emission maxima would be red-shifted relative to the fulvic acids. Although this result has indeed been observed for sedimentary humic substances (Hayase and Tsubota 1985), other data indicate that these maxima do not differ extensively for a broad range of fulvic and humic acids originating from different sources (Zepp and Schlotzhauer 1981). Higher molecular weight subfractions of CDOM isolated from a single source do exhibit broadened and red-shifted emission bands, as well as lower fluorescence efficiencies (Senesi 1990; Green et al. 1992).

Workers have recently turned to the use of multidimensional fluorescence techniques such as three-dimensional excitation, emission matrix (3-D) spectra, in an attempt to distinguish CDOM sources more reliably as well as to acquire better spectral information on CDOM chromophores (Coble et al. 1990; Dudelzak et al. 1991; Green 1992). With the 3-D technique, emission spectra are collected at a series of successively longer excitation wavelengths and concatenated to generate a plot in which fluorescence intensity is displayed as a function of the excitation and emission wavelengths (Figures 3.3–3.6).

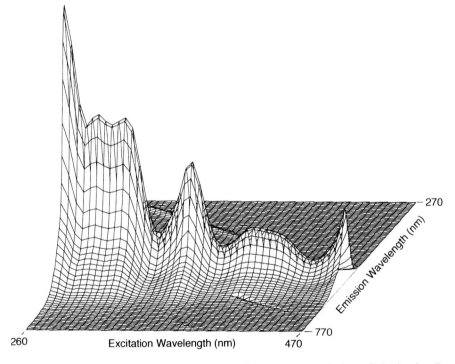

Figure 3.3 3-D fluorescence spectrum of a dilute aqueous solution of rhodamine B. The water Raman bands have been subtracted from this spectrum. Data from Green (1992).

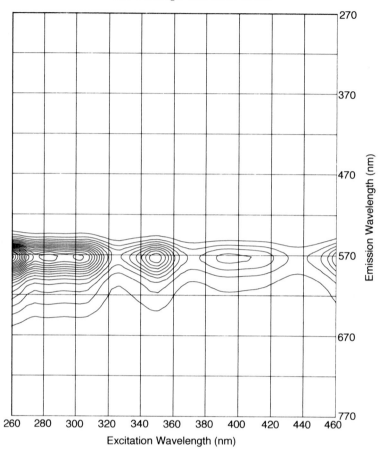

Figure 3.4 Contour plot of the data shown in Figure 3.3. Data from Green (1992).

Excepting polarization data (*vide infra*), complete (steady-state) fluorescence spectral information is contained within this plot. The 3-D spectrum of a single chromophore is quite simple and appears as a band of emission of varying intensity but constant energy and linewidth parallel to the excitation axis (Figures 3.3, 3.4). In contrast, 3-D spectra of CDOM are quite complex (Figures 3.5, 3.6), providing further evidence of multiple absorbing and emitting centers within this material.

Additional information can be obtained by dividing the individual emission scans of the 3-D plot by the absorption coefficients at the excitation wavelengths. This normalization produces a 3-D plot of relative fluorescence efficiency, which for simple chromophores eliminates the intensity variations along the excitation axis. Further, by integrating over the emission axis, the

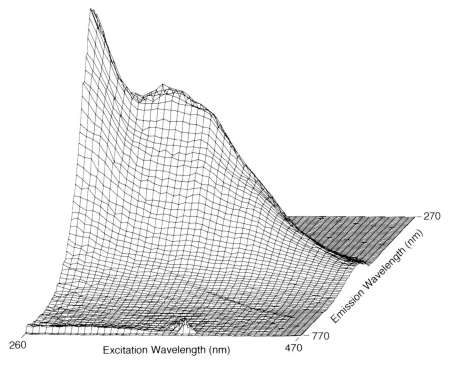

260   Excitation Wavelength (nm)   470

Figure 3.5 3-D fluorescence spectrum of Shark River water (Everglades, south Florida). The water Raman bands have been subtracted from this plot. Data from Green (1992).

wavelength dependence of the relative quantum efficiency is obtained (Figure 3.7); these relative efficiencies can be converted to quantum yields by use of an appropriate fluorescence standard (see below).

At moderate resolution (excitation and emission bandpasses, 4 or 5 nm), application of this technique to various marine and river waters has provided evidence for source-dependent spectral variations in CDOM fluorescence (Coble et al. 1990; Green 1992). In many instances, these differences are subtle and not easily displayed or characterized within the 3-D format. This technique has not been applied routinely at higher resolution to examine CDOM from more diverse sources, so an extensive data base does not yet exist. 3-D spectra have been employed to identify appropriate excitation and emission wavelengths for use in shipboard and airborne fluorescence measurements (Dudelzak et al. 1991), to estimate the impact of solar-stimulated CDOM fluorescence on remotely sensed reflectance signals (Vodacek et al. 1994; Peacock et al. 1990), as well as to assess the feasibility of using Fraunhofer line filling by solar-stimulated CDOM fluorescence to determine the levels of

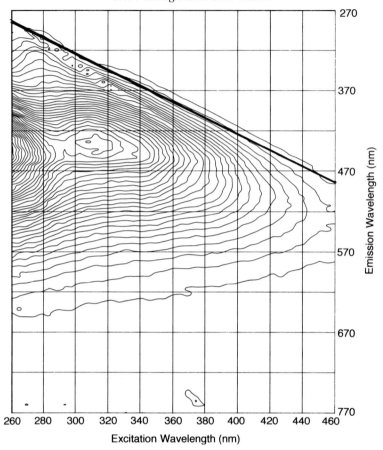

Figure 3.6 Contour plot of the data shown in Figure 3.5. The solid line represents the data that would be recorded for a synchronous scan spectrum with the excitation and emission monochromators offset by 25 nm. Data from Green (1992).

this material by *in situ* or remote sensors (Ge et al. 1992; Stoertz et al. 1969; Vodacek et al. 1994; see also below).

Some success in highlighting the often subtle differences in the fluorescence spectra of CDOM from different sources has been achieved by measuring synchronous-scan spectra (Cabaniss and Shuman 1987; Senesi 1990; Vodacek 1992). These spectra are obtained by scanning the excitation and emission monochromators simultaneously at a fixed wavelength or (preferably) frequency difference, and are equivalent to acquiring a spectral "slice" through the 3-D spectrum (Figure 3.6). Although more quickly and easily obtained than the 3-D spectra and while of possible use in the

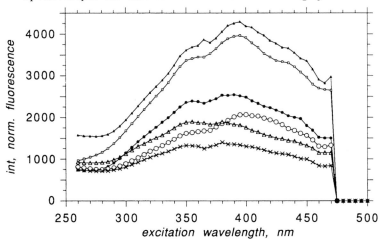

Figure 3.7 Wavelength dependence of the relative fluorescence quantum efficiencies of CDOM from different sources: isolated CDOM from the Sargasso Sea, 1500 m (□); SRFA (O); seawater from the western shelf of south Florida (△); Oyster Bay (south Florida) (●); Tiamiami River (▲); Amazon River (x). These spectra were obtained by dividing the corrected, integrated fluorescence by the absorption coefficient at each excitation wavelength. Quantum yields can be estimated from this plot by scaling to $\phi(355) = 1.51\%$ for Oyster Bay water. Data from Green (1992).

"fingerprinting" and tracking of aquatic CDOM, these spectra, as a subset of the 3-D, do not contain complete spectral information and thus cannot be used in the applications discussed above.

Quantum Yields and Absorption-Fluorescence Relationships

The fluorescence quantum yield is a well-defined photophysical quantity representing the ratio (or percentage) of the number of photons emitted to the number absorbed. This quantity is commonly measured relative to a standard compound such as quinine sulfate and calculated from the relation,

$$\phi(\lambda) = \frac{A(\lambda)_{qs} \times F_s \times \phi(\lambda)_{qs}}{A(\lambda)_s \times F_{qs}}, \tag{3.6}$$

where the subscripts S and qs refer to sample and quinine sulfate, respectively, ϕ is the quantum yield, A is the absorbance at excitation wavelength λ, and F is the integrated, corrected fluorescence.

With some exceptions (e.g., Zepp and Schlotzhauer 1981), workers have usually failed to measure this basic parameter. While numerous studies have demonstrated the existence of a strong correlation between CDOM absorption and fluorescence emission, thus implying an approximately constant quantum yield, much of this past data is unusable because workers have not employed a consistent fluorescence standardization procedure or a common set of excitation wavelengths, or worse, often have not measured the absorption coefficient at the wavelength used to excite the fluorescence.

Recent studies have shown that $\phi(355)$ for CDOM from diverse sources varies by about a factor of five, and exhibits an average (and median) value of ~1% (Green 1992; Green and Blough, in prep.). Within this limited range, the highest yields are observed for CDOM isolated from deep marine waters (2.1%), whereas the lowest yields are seen for terrestrial humic acids and some river waters (~0.4%). These yields also vary significantly with excitation wavelength; this wavelength dependence differs somewhat among CDOM sources (Figure 3.7).

The relative invariance of these yields across differing oceanic environments is also reflected in the recent work of Hoge et al. (1993). Because a major goal of this study was to calibrate fluorescence data acquired with airborne, shipboard, and laboratory instruments, fluorescence measurements were obtained at excitation wavelengths corresponding to commonly employed UV laser lines and standardized using the water Raman signal as an internal radiometric standard and quinine sulfate as an external standard. These workers showed that when this standardization was employed, the fluorescence per unit absorption (at the excitation wavelength) exhibited a maximum variability of only 36% for waters in the North Atlantic Ocean, the Gulf of Mexico, and Monterey Bay (Figure 3.8). This work demonstrates that absorption coefficients can be retrieved from fluorescence measurements with reasonable accuracy, and that airborne measurements of laser-induced CDOM fluorescence represent a very promising way of detecting and characterizing this material remotely (Hoge et al. 1993; Hoge and Swift 1981; *vide infra*).

Fluorescence Lifetimes

Time-resolved fluorescence measurements furnish the excited singlet state lifetime(s) of a chromophore, which reflects its structure and environment as well as its possible involvement in excited state photoprocesses such as energy or electron transfer. These measurements thus provide an additional way of characterizing the chromophores underlying the absorption and emission of light by CDOM.

Relatively few measurements have been performed with the requisite time (sub-nanosecond) and wavelength resolution to supply an adequate picture of the excited state properties of CDOM. Available data reveal multiexponential decay kinetics that have been interpreted in terms of a minimum of three components with lifetimes ranging from < 1 to 6 ns; fewer than 5% of these components exhibit lifetimes longer than 2.5 ns (Lochmüller and Saavedra

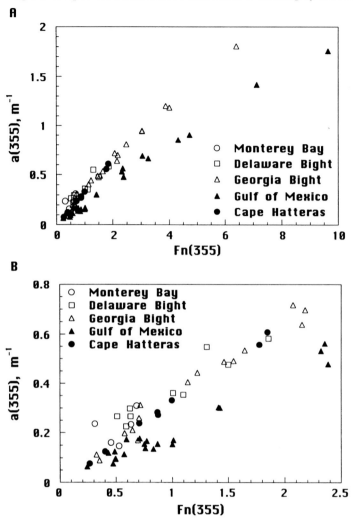

Figure 3.8 A: Relationship between the absorption coefficient at 355 nm (*a*(355)) and the fluorescence stimulated by 355 nm excitation (Fn(355)) for waters from five oceanic study sites. B: Same as Figure 3.8A, but with an expanded scale. Data from Hoge et al. (1993).

1986; Lappen and Seitz 1982). That CDOM exhibits both short fluorescence lifetimes and low quantum yields suggests that rapid nonradiative processes are limiting the singlet state lifetimes of the chromophores within this material. However, combined measurements of ϕ and fluorescence lifetimes over a wider range of excitation wavelengths are needed to establish a better understanding of the excited state processes occurring within these materials.

RELATION OF OPTICAL PROPERTIES TO MOLECULAR PROPERTIES

Empirical Relationships

Currently, quantitative comparisons between the optical and molecular properties of CDOM cannot be made for several reasons. First, there have been no systematic efforts to collect high-quality optical data on CDOM samples that have also been well characterized by other spectroscopic, chemical, and isotopic methods. Second, investigators have frequently failed to acquire defined spectroscopic parameters such as $a(\lambda)$, $a(\lambda)^*$, S, and $\phi(\lambda)$; thus the results from different laboratories often cannot be compared nor used to test for universal relationships. Third, workers generally have not taken full advantage of the suite of possible optical spectroscopic methods that could supply important information about chromophore structure, thus providing further constraints on structural models of CDOM.

Nevertheless, several general trends are evident in the literature. Results of solid-state ^{13}C NMR measurements (Malcolm 1990) have shown that terrestrial and freshwater humic substances contain the highest content of unsaturated or aromatic carbon, with the humic acids consistently exhibiting higher levels (30–40%) than the fulvic acids (20–25%). In contrast, marine humic substances show only a 15–20% aromatic carbon content (Malcolm 1990; Hedges et al. 1992). Phenolic carbon content, although much less, shows similar trends.

Observed variations in $a(\lambda)^*$ and S are compatible with these results (Table 3.1). Sources of CDOM exhibiting higher aromatic carbon content also show elevated values of $a(\lambda)^*$, consistent with the presence of a higher percentage of light-absorbing aromatic structures and/or (aromatic) structures having larger absorption cross-sections. The decrease in S with increasing aromatic carbon content most likely arises from increased levels of chromophores having extended aromatic systems (absorbing at lower energies, longer wavelengths) and/or from an increase in lower energy transitions arising from intramolecular (charge transfer) interactions between chromophores as a result of higher chromophore "densities." The small but consistent decrease in fluorescence ϕ with increasing $a(\lambda)^*$ may also reflect intramolecular chromophore interactions within the excited state leading to fluorescence quenching. Increased intramolecular interactions may also account in part for the decreases in S and ϕ and the increases in $a(\lambda)^*$ that are observed for the higher molecular weight fractions of humic substances (Table 3.1).

Simple relationships such as these could furnish the means of characterizing CDOM rapidly by both field and remote optical measurements. For example, optical methods capable of discerning shifts in S may allow CDOM to be classified roughly in terms of its aromatic content, type (humic or fulvic) or molecular weight. The inverse relationship between S and $a(\lambda)^*$, although weak, could further be used to constrain $a(\lambda)^*$, thus allowing rapid estimation of CDOM concentrations through either direct or indirect measurements of $a(\lambda)$ (Equations 3.1, 3.2). Testing and establishing these relationships in a rigorous fashion will require the collection of much better optical data for a variety of CDOM samples that have also been well characterized chemically and spectroscopically (Hedges et al. 1992; Malcolm 1990).

Experimental Approaches

Although CDOM (i.e., humic substances) has been studied for over a century, our present understanding of the structural origins of its optical properties remains rudimentary. (Conversely, however, constraints imposed on the structural nature of this material by the optical data have been ignored by some workers.) Currently, we cannot state with certainty that the absorption spectra of these materials arise from a simple sum of the absorptions of a (very large?) number of chromophores with differing structures, as opposed to a few chromophores having similar structures but undergoing a "continuum" of charge transfer (donor-acceptor) interactions (Power and Langford 1988). Similarly, we cannot state unequivocally that CDOM emission arises only from distinct chromophores (locally excited singlet states) and not in part from charge recombination from intramolecular charge transfer states (exciplex emission).

Further advances in this area will require a multidimensional approach incorporating not only the techniques discussed above, but also additional methods that can provide new and complementary information. Measurements of polarization, phosphorescence, and resonance Raman spectra represent possible examples. Further, the use of frequency domain fluorescence techniques may allow the acquisition of fluorescence lifetimes over a broader range of excitation wavelengths as well as provide time-resolved fluorescence spectra through use of phase-sensitive detection (Lakowicz 1983). A brief description of some of these methods and the information they can provide is presented below.

Polarization Spectra

The variation in the polarization (or anisotropy) of fluorescence with excitation wavelength for a chromophore embedded in a rigid medium is called a polarization spectrum. In this rigid medium (e.g., room or low-temperature glasses), the chromophore remains immobile during the lifetime of its excited state; in this situation, the value of the polarization (P) provides a measure of the angle (α) between the absorption and emission dipoles of the chromophore. P is highest when these dipoles are colinear ($\alpha = 0$, $P = 0.5$) and lowest when they are orthogonal ($\alpha = 90$, $P = -0.2$). Excitation within an absorption band corresponding to the first excited singlet state will produce $P \approx 0.5$

because emission also originates from this state ($\alpha \approx 0$). Excitation within bands corresponding to higher excited states will usually produce lower values of P because absorption and emission then take place from different states ($\alpha > 0°$).

3-D plots of polarization (or preferably anisotropy) can be constructed and used in conjunction with the absorption and 3-D fluorescence spectra to assign the absorption and emission bands of individual chromophores within complex mixtures. Further, extensive depolarization can indicate the presence of excited state photoprocesses such as energy transfer. Because of the low $\phi(\lambda)$ of CDOM, only low-resolution spectra may be achievable.

Phosphorescence Spectra

Like fluorescence spectra, phosphorescence spectra provide information about the electronic transition energies of the absorbing and emitting states of a chromophore; in this case, however, emission originates from the first excited triplet state, which is always lower in energy than the singlet. Because the inherent lifetime of this state is much longer than that of the singlet, it is highly susceptible to quenching in solution at room temperature, and thus phosphorescence is often very difficult to detect under these conditions. Instead, these spectra are usually acquired at low temperature (77°K). Because different chromophores exhibit distinct singlet and triplet state energies, phosphorescence excitation and emission spectra used in combination with 3-D fluorescence spectra can further aid in resolving individual chromophores within complex mixtures. Surprisingly, low-temperature measurements of the phosphorescence (and fluorescence) spectra of CDOM have not yet been reported.

Resonance Raman Spectroscopy

Raman spectroscopy is based on the inelastic scattering of photons by compounds containing Raman-active vibrational modes. This technique thus provides molecular vibrational spectra that are complementary to those obtained by infrared spectroscopy. Resonance Raman (RR) refers to the selective intensity enhancement of those vibrational modes associated specifically with an electronic transition. RR spectroscopy can thus be used to acquire the vibrational spectrum of a chromophore selectively. By examining the RR spectra of CDOM at a number of wavelengths across its absorption spectrum, this technique can be employed to probe the structure of the underlying chromophores. Because CDOM fluorescence could significantly interfere with the collection of the RR spectra, signal gating techniques or the use of CARS (coherent antistokes Raman scattering) might prove necessary.

REMOTE SENSING OF CDOM

Remote sensing measurements can be broadly categorized as active or passive. Passive satellite and airborne color instruments view the reflectance, $R(\lambda)$, of sunlight from the uppermost layers of the ocean at selected wavelength bands across the visible and near-infrared. Absorption and scattering of sunlight by seawater constituents determines $R(\lambda)$ through the relation,

$$R(\lambda) = C\Sigma b'_i(\lambda)/\Sigma a_i(\lambda), \tag{3.7}$$

where $b'_i(\lambda)$ and $a_i(\lambda)$ are the backscattering and absorption coefficients, respectively, of the ith constituent and C is a constant. $R(\lambda)$ can thus be roughly characterized as the inverse of the oceanic absorption spectrum weighted by the backscattering spectrum. Because the spectral dependence of CDOM absorption differs significantly from that of pigments within phytoplankton, satellite measurements can be used, in principle, to distinguish between these constituents, if sufficient spectral bands are available, particularly in the blue portion of the visible spectrum where the contribution of CDOM increases (Carder et al. 1991). The sea-viewing wide field of view sensor (Sea-WiFS), planned for launch within the next year, may fulfill these requirements with the addition of bands at 412 and 490 nm. A complicating factor is that detrital material exhibits a spectral dependence very similar to CDOM; thus these constituents cannot be determined independently. Further, because of spatial and temporal variability in the spectral properties of these constitutents (see above), the development of seasonal and regional algorithms may prove necessary (Carder et al. 1991). Additional research is needed to determine whether passive satellite measurements can ultimately provide more than crude estimates of the sum of CDOM and the detrital material.

Another potential way of detecting CDOM passively is through use of an *in situ* or airborne Fraunhofer line discriminator to measure the solar-stimulated fluorescence of this material (Ge et al. 1992; Stoertz et al. 1969; Vodacek et al. 1994). Fraunhofer lines are very sharp absorption features in the solar spectrum that result from light absorption by solar constituents. Only processes such as fluorescence or water Raman scattering, which produce a spectral redistribution of solar light, will contribute to the "filling in" of these lines. By comparing the depth of these lines with respect to the solar continuum, the line discriminator can be used to detect fluorescing species such as CDOM. This approach may be limited to coastal areas where CDOM fluorescence dominates water Raman scattering (Vodacek et al. 1994).

Active, airborne laser systems such as NASA's airborne oceanographic lidar (AOL) currently offer the best means for determining CDOM remotely. These systems employ a pulsed UV laser source to excite CDOM fluorescence; the surface return of fluorescence and water Raman scattering is collected and spectrally resolved (Hoge and Swift 1981; Hoge and Swift 1986; Figure 3.9). Ratioing the fluorescence to the water Raman scattering band provides a well-defined measure of fluorescence intensity that can be used to determine the absorption coefficient of CDOM at the excitation wavelength (Hoge et al. 1993; Figures 3.8, 3.9). Knowledge of the specific absorption coefficient of CDOM at a given locale can further provide the concentration of this material. Available evidence indicates that particulate material does not fluoresce at the same wavelengths, and thus this approach allows CDOM to be measured independently of the detrital component.

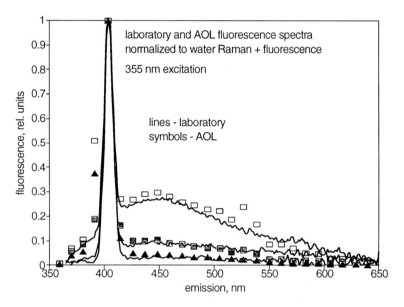

Figure 3.9 Comparison of the fluorescence spectra of surface waters collected by ship at sites in the Gulf of Mexico (solid lines) with those obtained by the airborne oceanographic lidar (AOL; NASA-Wallops Flight Facility) overflying the ship track (solid symbols). Corrected fluorescence spectra of surface waters were acquired at the Wallops Flight Facility with a Perkin-Elmer LS 50 spectrofluorometer employing 355 nm excitation and 4 nm excitation and emission bandpasses. AOL spectra were collected at an aircraft altitude of 150 m, using a pulsed, frequency-tripled Nd-YAG laser to excite CDOM fluorescence at 355 nm; the return of surface fluorescence was spectrally resolved into 32 channels (11.25 nm resolution) covering wavelengths from ~359 to 707 nm (only 27 channels shown here). Spectra were corrected by use of a NIST calibration sphere. AOL spectra were collected at the closest approach of the aircraft to the ship sample location (10 spectrum average). Temporal separation was less than 3 hours, while spatial separation was less than 2.5 km. Data courtesy of F. Hoge, A. Vodacek, and R. Swift.

If S is known, absorption coefficients in the visible can be estimated from knowledge of the absorption coefficient in the UV (Equation 3.1). Alternatively, it is possible that S could be determined remotely by using two UV laser sources; absorption by CDOM at visible wavelengths could then be estimated directly (Equation 3.1). This approach may also allow the "quality" of the CDOM to be roughly characterized (see above). Importantly, because the AOL collects both active and passive data concurrently (Hoge and Swift 1986), it can be used to check for internal consistency between algorithms developed for the active and passive detection of CDOM, as well as provide field tests of algorithms proposed for passive satellite determinations of CDOM (Hoge and Swift 1986; Hoge et al. 1993).

Ultimately, active airborne systems and other remote sensing methods may be utilized to address a wide variety of important issues including:

- the wide-scale distribution of CDOM (and in some instances, possibly dissolved organic matter) in the surface oceans and its relationship to phytoplankton blooms and their decay (Carder et al. 1991);
- the impact of terrestrial CDOM on the optical and photochemical properties of oceanic regions experiencing major freshwater inputs from rivers such as the Amazon and Orinoco (Blough et al. 1993);
- the penetration of biologically damaging UV-B radiation into the oceans (Blough and Zepp 1990; Hoge et al. 1993) as well as the influence of CDOM on the light fields that phytoplankton experience; and
- the role of photooxidation in the degradation and cycling of NLOM and in the production of atmospherically important trace gases such as CO, COS, and CO_2 (Blough and Zepp 1990; Blough and Zepp 1994).

ACKNOWLEDGEMENTS

The preparation of this article was supported by the Office of Naval Research (Grant N00014–89–J–1260), NASA through an EOS interdisciplinary investigation (NAGW–2431), and the National Science Foundation (Grant OCE–911568). We wish to thank our colleagues at NASA Wallops Flight Facility, Drs. F. Hoge, A. Vodacek, and R. Swift, for many useful discussions and for providing Figure 3.7. We also wish to thank E. Druffel and P. Williams for the Sargasso Sea isolates of CDOM and K. Ruttenberg for collecting the Amazon samples. This is contribution No. 8401 from the Woods Hole Oceanographic Institution.

REFERENCES

Blough, N.V., O.C. Zafirou, and J. Bonilla. 1993. Optical absorption spectra of waters from the Orinoco River outflow: Terrestrial input of colored organic matter to the Caribbean. *J. Geophys. Res.* **98**:2271–2278.

Blough, N.V., and R.G. Zepp, eds. 1990. Effects of solar ultraviolet radiation on biogeochemical dynamics in aquatic environments. Woods Hole Oceanographic Technical Report WHOI–90–09, Woods Hole, MA, 194 pp.

Blough, N.V., and R.G. Zepp. 1995. Reactive oxygen species (ROS) in natural waters. In: Active Oxygen: Reactive Oxygen Species in Chemistry, ed. C.S. Foote, J.S. Valentine, A. Greenberg, and J.F. Liebman. New York: Chapman and Hall, in press.

Bricaud, A., A. Morel, and L. Prieur. 1981. Absorption by dissolved organic mattter of the sea (yellow substance) in the UV and visible domains. *Limnol. Oceanogr.* **26**:43–45.

Brown, M. 1977. Transmission spectroscopy examinations of natural waters. C. Ultraviolet spectral characteristics of the transition from terrestrial humus to marine yellow substance. *Est. Coast. Mar. Sci.* **5**:309–317.

Cabaniss, S.E., and M.S. Shuman. 1987. Synchronous fluorescence spectra of natural waters: Tracing sources of dissolved organic matter. *Mar. Chem.* **21**:37–50.

Carder, K.L., S.K. Hawes, K.A. Baker, R.C. Smith, R.G. Steward, and B.G. Mitchell. 1991. Reflectance model for quantifying chlorophyll *a* in the presence of productivity degradation products. *J. Geophys. Res.* **96**:20,599–20,611.

Carder, K.L., R.G. Steward, G.R. Harvey, and P. Ortner. 1989. Marine humic and fulvic acids: Their effect on remote sensing of ocean chlorophyll. *Limnol. Oceanogr.* **34**:68–81.

Chen,Y., N. Senesi, and M. Schnitzer. 1977. Information provided on humic substances by E4/E6 ratios. *Soil Sci. Soc. Am. J.* **41**:352–358.

Coble, P.G., S.A. Green, N.V. Blough, and R.B. Gagosion. 1990. Characterization of dissolved organic matter in the Black Sea by fluorescence spectroscopy. *Nature* **348**:432–435.

Davies-Colley, R.J., and W.N. Vant. 1987. Absorption of light by yellow substance in freshwater lakes. *Limnol. Oceanogr.* **32**:416–425.

Dudelzak, A.E., S.M. Babichenko, L.V. Poryvkina, and K.J. Saar. 1991. Total luminescent spectroscopy for the remote laser diagnostics of natural waters. *Appl. Opt.* **30**:453–458.

Ge, Y., K.J. Voss, and H.R. Gordon. 1992. Measurement of oceanic inelastic scattering using solar Fraunhofer lines. *SPIE Ocean Optics XI* **1750**:161–169.

Green, S.A. 1992. Applications of fluorescence spectroscopy to environmental chemistry. Ph.D. thesis, MIT, MIT/WHOI–92–24, Woods Hole, MA, 228 pp.

Green, S.A., F.M.M. Morel, and N.V. Blough. 1992. An investigation of the electrostatic properties of humic substances by fluorescence quenching. *Environ. Sci. Technol.* **26**:294–302.

Hayase, K., and H. Tsubota. 1985. Sedimentary humic acid and fulvic acid as fluorescent organic materials. *Geochim. Cosmochim. Acta* **49**:159–163.

Hedges, J.I., P.G. Hatcher, J.R. Ertel, and K.J. Meyers-Schulte. 1992. A comparison of dissolved humic substances from seawater with Amazon River counterparts by ^{13}C-NMR spectrometry. *Geochim. Cosmochim. Acta* **56**:1753–1757.

Hoge, F.E., and R.N. Swift. 1981. Airborne simultaneous spectroscopic detection of laser-induced water Raman backscatter and fluorescence from chlorophyll *a* and other naturally occurring pigments. *Appl. Opt.* **20**:3197–3205.

Hoge, F.E., and R.N. Swift. 1986. Active-passive correlation spectroscopy: A new technique for identifying ocean color algorithm regions. *Appl. Opt.* **25**:2571–2583.

Hoge, F.E., A. Vodacek, and N.V. Blough. 1993. Inherent optical properties of the ocean: Retrieval of the absorption coefficient of chromophoric dissolved organic matter from fluorescence measurements. *Limnol. Oceanogr.* **38**:1394–1402.

Lakowicz, J.R. 1983. Principles of Fluorescence Spectroscopy. New York: Plenum.

Lappen, A.J., and W.R. Seitz. 1982. Fluorescence polarization studies of the conformation of soil fulvic acid. *Anal. Chim. Acta* **134**:31–38.

Lochmüller, C.H., and S.S. Saavedra. 1986. Conformational changes in a soil fulvic acid measured by time-dependent fluorescence depolarization. *Anal. Chem.* **58**:1978–1981.

Malcolm, R.L. 1990. The uniqueness of humic substances in each of soil, stream and marine environments. *Anal. Chem. Acta* **232**:19–30.

Peacock, T.G., K.L. Carder, C.O. Davis, and R.G. Steward. 1990. Effects of fluorescence and water Raman scattering on models of remote sensing reflectance. *SPIE Ocean Optics X* **1302**:303–319.

Power, J.F., and C.H. Langford. 1988. Optical absorbance of dissolved organic matter in natural water studies using the thermal lens effect. *Anal. Chem.* **60**:842–846.

Senesi, N. 1990. Molecular and quantitative aspects of the chemistry of fulvic acid and its interaction with metal ions and organic chemicals. II. The fluorescence spectroscopy approach. *Anal. Chim. Acta* **232**:77–106.

Stewart, A.J., and R.G. Wetzel. 1980. Fluorescence:absorbance ratios: A molecular-weight tracer of dissolved organic matter. *Limnol. Oceanogr.* **25**:559–564.

Stoertz, G.E., W.R. Hemphill, and D.A. Markle. 1969. Airborne fluorometer applicable to marine and estuarine studies. *Mar. Technol. Soc. J.* **3(6)**:11–26.

Vodacek, A. 1992. An explanation of the spectral variation in freshwater CDOM fluorescence. *Limnol. Oceanogr.* **37**:1808–1813.

Vodacek, A., S.A. Green, and N.V. Blough. 1994. An experimental model of the solar-stimulated fluorescence of chromophoric dissolved organic matter. *Limnol. Oceanogr.* **39**:1–11.

Zepp, R.G., and P.F. Schlotzhauer. 1981. Comparison of the photochemical behavior of various humic substances in water. III. Spectroscopic properties of humic substances. *Chemosphere* **10**:479–486.

4

Comparative Biodegradation Kinetics of Simple and Complex Dissolved Organic Carbon in Aquatic Ecosystems

R.E. HODSON and M.A. MORAN
Department of Marine Sciences, University of Georgia, Athens,
Georgia 30602–2206, U.S.A.

ABSTRACT

Many aspects of the flux of dissolved organic carbon (DOC) through bacteria is poorly understood for both marine and freshwater environments. Recent experimental work suggests that chemically complex pools of DOC are composed of compounds with a range of biodegradability and that some fractions, traditionally viewed as highly refractory, may be taken up by bacteria on biologically relevant time scales. Rates and kinetic patterns of uptake of compounds from these complex pools are predictably diverse. Bacterial uptake of simple compounds from bulk DOC is likewise kinetically diverse. Experimental evidence suggests that bacteria utilize multiple uptake strategies that maximize potential carbon and energy flow as substrate concentrations vary both spatially and temporally.

INTRODUCTION

Dissolved organic carbon (DOC) in the world's oceans is the second largest reservoir of nonliving organic matter (NLOM) on the planet. Concentrations of seawater DOC are determined by the balance between production at the expense of both autochthonous processes (e.g., excretion by phytoplankton, inefficient grazing of phytoplankton by protists and zooplankton) and terrestrial inputs, and turnover (primarily via uptake by bacterioplankton).

Role of Nonliving Organic Matter in the Earth's Carbon Cycle
Edited by R.G. Zepp and Ch. Sonntag © 1995 John Wiley & Sons Ltd.

Although total concentrations average one to several milligrams per liter, most studies to date suggest that much of the DOC is highly refractory to biodegradation and assimilation by bacteria. Estimates based on short-term laboratory incubations place the "utilizable" components of DOC at 20% or less of the total. Although still mostly uncharacterized, the utilizable components are assumed to be comprised of low molecular weight organics such as mono- and polysaccharides and amino acids (Thurman 1985; Benner et al. 1992). Such compounds have turnover times on the order of minutes to hours in surface waters (Azam and Hodson 1977). Much less qualitative information is available for the slow-cycling components, except that a significant fraction can be categorized functionally as humic substances or hydrophilic acids, categories defined primarily by bulk chemical properties (Thurman 1985). Turnover times for the more refractory components are also poorly understood; however, current estimates indicate average radiocarbon ages of at least several thousand years for marine DOC (Williams and Druffel 1987; Bauer et al. 1992).

During the past two decades, aquatic ecologists have come to acknowledge the quantitative importance of carbon and energy flow through DOC to aquatic food webs via bacterial secondary production. This flow, termed the microbial loop (Azam et al. 1983), may account for upwards of 50% of total primary production in some aquatic systems (Cole et al. 1988). Considering this fact as well as the size of the DOC reservoir globally, understanding the processes controlling turnover rates of DOC is important from both trophodynamic and geochemical perspectives.

A variety of models have been applied to describe the kinetics of turnover of slow-cycling and labile components of DOC. For slow-cycling compounds, the simplest and most often applied models have been based on first-order decay (simple exponential) equations. For labile components, various modifications of Michaelis–Menten, hyperbolic kinetics have been most often used (Wright and Hobbie 1966). Recent studies, however, indicate that such models often have serious deficiencies when applied to natural DOC utilized by mixed microbial populations and can result in large (one to several orders of magnitude) over- or underestimations of turnover rates, depending upon the complexity of the DOC mixture and the concentrations of individual components.

Here we discuss how the "utilizable" components of DOC, i.e., those turning over at biologically relevant rates, may be more difficult to identify and model than current wisdom suggests.

BACTERIAL TURNOVER OF COMPLEX DISSOLVED NLOM

Views of the biological role of slow-cycling compounds in aquatic environments have changed substantially in recent years from that of biologically inert compounds of little trophodynamic significance to that of potentially important carbon and energy sources for bacteria, and ultimately, for microbial food webs (Meyer et al. 1987; Hessen et al. 1990). The earlier notion that compounds falling within the higher molecular weight classes of DOC were biologically unimportant derived from their complex structure, their chemical heterogeneity, and their likely origin as refractory remains or condensates from the decomposition of plant and animal biomass. Yet recent studies employing experimental approaches have demonstrated that some fraction of this material is indeed turning over on biologically relevant time scales and that the turnover is driven by bacterial activity (Moran and Hodson 1990a; Tranvik 1990; Tulonen et al. 1992; Leff and Meyer 1991). This new perspective suggests that further study of the turnover of compounds falling within slow-cycling categories of DOC, those with long *average* turnover times, is warranted. Below, we look at two examples of slow-cycling dissolved NLOM that are quantitatively important in freshwater and coastal marine environments.

Lignocellulose-derived DOC

As vascular plant detritus degrades in terrestrial environments and coastal marshes, a significant fraction of the detritus is not directly mineralized but rather converted to dissolved decomposition products. These compounds are released into soil pore water, rivers, and estuaries, and some portion is subsequently transported to coastal marine environments (Moran et al. 1991; Moran and Hodson 1994). The fate of lignocellulose-derived DOC, i.e., whether it enters aquatic food webs via assimilation by bacterioplankton or whether it is transported in relatively unaltered form to the open ocean, depends in large part upon the rates and kinetics of its utilization. The presence of lignocellulose-derived NLOM in open ocean water (Meyers-Schulte and Hedges 1986), at great distance from its probable point of origin, provides evidence that some fraction of this pool is highly resistant to microbial utilization. However, more labile components may also be formed during lignocellulose degradation that, because of a higher turnover rate, would be more difficult to detect.

The pool of dissolved NLOM formed during the decomposition of lignocellulose is quantitatively significant, for example accounting for up to 30% of the total decomposition products of *Spartina alterniflora* lignocellulose degrading in salt marshes of the southeastern U.S. (Moran and Hodson 1989). We have used radiolabeled preparations of *S. alterniflora* lignocellulose as a model substrate for studies of the bacterial utilization of lignocellulose-derived DOC in salt marsh microcosms with natural bacterial assemblages (Moran and Hodson 1989, 1990b). Kinetics of utilization of lignocellulose-derived DOC in these studies suggest a heterogeneous pool of compounds with both fast- and slow-cycling components. For example, about 12% of lignocellulose-derived DOC from *Spartina* is mineralized by bacteria within the first 16 hours, while only an additional 18% is mineralized over the following 30 days. Rates of bacterial utilization are nearly identical regardless of the age of the particulate lignocellulose from which the pools of dissolved NLOM are formed: even older plant detritus that has been degrading for several months produces DOC which is rapidly mineralized (Figure 4.1).

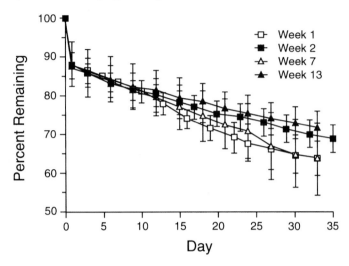

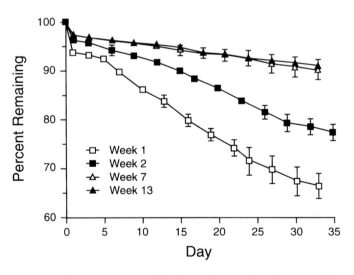

Figure 4.1 Bacterial utilization of DOC derived from *Spartina alterniflora* lignocellulose for DOC formed (top) and accumulated (bottom) at 1, 2, 7, and 13 weeks after decomposition began (from Moran and Hodson 1989).

One can also look at the pool of DOC accumulating during decomposition of particulate lignocellulose, i.e., the reservoir of dissolved compounds that remains unassimilated. Utilization kinetics of pools collected at various intervals during decomposition of *Spartina* lignocellulose indicate that the quality of the dissolved NLOM decreases significantly with stage of decomposition. For example, 6% of the DOC accumulated after only 1 week was used in a 16 hr incubation, while only 2.5% of DOC accumulated after two months was used in the same time interval. Likewise, the slower-cycling fraction of this pool shows greater recalcitrance with age, with percent mineralization over the next 30 days decreasing with age of the accumulating pool from 28% to 5% (Figure 4.1). The turnover rate of the accumulating lignocellulose-derived DOC thus decreases with time.

Because particulate lignocellulose is generally considered a highly refractory carbon source, it might have been assumed that all dissolved intermediates produced during lignocellulose degradation would also be refractory. The rate-limiting step in decomposition of lignocellulose may be cleavage of the lignin polymer, however, which permits access of microbial exoenzymes to structural polysaccharides. Results of chemical analyses of lignocellulose-derived dissolved NLOM indicate that carbohydrate content is high in newly formed pools (Moran and Hodson 1989) and therefore that soluble polysaccharide derivatives may constitute a significant proportion of the more labile components. A likely contributor to the pool of more refractory compounds is lignin-derived phenolics.

The complex kinetics observed during microbial utilization of lignocellulose-derived dissolved NLOM may reflect a polysaccharide-rich fast-cycling component combined with a lignin-rich slow-cycling component. Analysis of DOC formed from degrading *Spartina* lignocellulose over periods of up to six months indicates that lignin is the source of as much as 30% of the accumulated pool. Since lignin makes up only 7% of the weight of intact *Spartina* lignocellulose (structural polysaccharides account for 93%), lignin is highly overrepresented in the lignocellulose-derived DOC (Moran and Hodson 1990b). Evidence indicates that this overrepresentation is not the result of a higher rate of release of lignin-derived compounds from the particulate lignocellulose, but instead reflects a slower rate of bacterial turnover of these compounds relative to the bulk DOC pool. Slow rates of turnover of lignin-derived DOC are consistent with hypotheses that lignin moieties are important precursors in the formation of humic substances. The pathway for humic substance formation may involve transformation of lignin to humic and fulvic acids through degradative processes or degradation of lignin to low molecular weight phenolics, which later condense into humic and fulvic acids (Hatcher and Spiker 1988; Hedges 1988).

Kinetics of microbial utilization of lignocellulose-derived DOC have a further element of complexity in that there is also evidence for preferential bacterial utilization of some compounds relative to others within the lignin-derived subcomponent. Controlled chemical oxidation of lignin produces a suite of characteristic compounds falling into three categories: vanillyl phenols (*V*), syringyl phenols (*S*), and cinnamyl phenols (*C*) (Ertel et al. 1984). While undegraded *Spartina* lignocellulose has a phenol signature typical of nonwoody angiosperms (*S:V* ratio of 0.78 and *C:V* ratio of 0.33), dissolved lignocellulose-derived compounds have an average *S:V* ratio of 0.13 and *C:V* ratio of 0.06 (Moran and Hodson 1990b). The divergence of lignin phenol signatures of dissolved degradation products from those of the undegraded source plant is apparently due to differential bacterial assimilation of the classes of phenols from the pool of dissolved lignin-derived compounds, leading to a relatively shorter turnover time for some (*S* and *C* phenols) and the overrepresentation of others (*V* phenols). Discrepancies between these ratios in source particulate material and resulting DOC greatly complicates using such analyses for determining the types of plants from which the DOC is derived. Similarly low *S:V* and *C:V* ratios relative to those in presumed source plants have also been found for Amazon River DOC (Ertel et al. 1986).

Thus the picture emerging from our studies of lignocellulose-derived DOC is that of a highly heterogeneous pool with complex bacterial utilization kinetics. At least two main fractions can be identified: one polysaccharide-rich component with turnover times that rival those measured for very simple compounds (on the order of hours) and one lignin-rich component that cycles more slowly (on the order of months to years). Within these pools, additional subcomponents can be identified, each having their own characteristic rates of bacterial turnover. These studies suggest that even within a well-defined pool of slow-cycling NLOM, preferential bacterial degradation and assimilation of certain components leads to complex uptake kinetics and modified composition of the remaining pool. Understanding the biogeochemical fate of lignocellulose-derived DOC in natural environments is an important goal not just in terrestrial systems, where particulate NLOM pools are dominated by vascular plants, but also in coastal marine systems to which the dissolved products from vascular plant decomposition are transported. In salt marsh drainages in the southeastern U.S., for example, we estimate that as much as 40% of the DOC is derived from lignocellulose (Moran and Hodson 1990b), while in nearshore regions in close proximity to major sources of lignocellulose, such as salt marshes or high DOC rivers, contributions from vascular plants may account for 20% to 30% of the bulk DOC (Moran et al. 1991).

Dissolved Humic Substances

Evidence that dissolved humic substances may be turning over on time scales relevant to biological processes, not only to long-term geochemical processes, is found in recent studies of freshwater systems, which demonstrate measurable effects of high concentrations of humic substances on natural bacterial communities. For example, Hessen (1985) and Tranvik (1988) found positive correlations between bacterioplankton abundance and humic substance concentrations for a series of lakes. Similarly, the ratios of bacterial biomass to phytoplankton biomass (Jones 1990) and of annual bacterial production to phytoplankton production (Hessen et al. 1990) were also found to be higher in aquatic systems with abundant dissolved humic substances, suggesting that bacteria are more numerous or active in these systems than would be expected based on phytoplankton activity alone.

A more direct approach to assessing the biological role of humic substances involves experiments in which humic substances from bulk DOC are physically isolated by passage of water samples through columns of standard XAD-8 resin (Aiken 1985). Humic substance turnover rates can then be determined following inoculation with natural bacterial populations by tracking changes in DOC concentrations or, more sensitively, changes in biomass of natural bacteria growing at the expense of the DOC. In such studies of a highly humic blackwater marsh in the southeastern U.S., where humics typically account for 80% of the bulk DOC, humic substances were found to support 50% of the bacterial growth occurring in batch cultures. In a southeastern lake, a system having lower and more typical humic substances concentrations (approximately 50% of bulk DOC), humics supported about 20% of the bacterioplankton growth (Moran and Hodson 1990a).

There is evidence that humic substances from various aquatic ecosystems may be turning over at significantly different rates. In the above study of two freshwater systems, bacterioplankton growth calculated per weight of humic carbon was 15-fold greater on humic substances isolated from the lake environment than from the blackwater marsh (Table 4.1). In marine systems (a

Table 4.1 Bacterial turnover of dissolved humic substances from freshwater and marine environments.

Environment	Bacterial Growth [mg bacterial C (mg humic C)$^{-1}$]	Humic C Utilization (%)[a]
Freshwater Environments:		
Lake[b]	26.6	8.9
Blackwater Marsh[b]	1.8	0.6
Marine Environments:		
Mangrove Swamp[c]	11.0	3.7
Nearshore Planktonic[c]	33.0	11.0
Gulf Stream[c]	21.3	7.1

[a]Calculated assuming 30% bacterial growth efficiency (Meyer et al. 1987; Moran and Hodson 1989).
[b]From Moran and Hodson (1990a).
[c]Moran and Hodson, unpublished data.

Bahamian mangrove swamp, a coastal site adjacent to the mangrove swamp, and the Gulf Stream between Miami, FL, and the Bahamas), up to 3-fold variation in yield of bacterial biomass from dissolved humic substances has been observed (Table 4.1). The concentration of lignin-derived compounds may be an important influence on the relative biological availability of humics: the two environments in which primary production is dominated by vascular plants (the blackwater marsh and mangrove swamp sites) supported the least bacterial growth per weight of humic carbon (Table 4.1).

The biologically mediated turnover of dissolved humic substances may also be studied by preparing radiolabeled humics of known age and source. ^{14}C-labeled dissolved humic substances formed from degrading salt marsh grass (S. alterniflora) were found to be readily assimilated by marine bacterio-plankton (24% of the pool was used by bacteria during a 7-week incubation; Moran and Hodson 1994). These recently formed (average age, 3 weeks) vascular plant-derived humics are more biologically labile than humic substances collected from natural environments (only 1–11% assimilated; Table 4.1). These studies point out the significant heterogeneity of those compounds included within the humic substances definition. While some humics accumulating in natural environments are highly resistant to biological attack, others are relatively labile compounds with ecologically relevant turnover times.

Kinetics of bacterial utilization of dissolved humic substances may be influenced by properties other than chemical structure. For example, recent studies have shown that photooxidation converts humics into biologically available compounds, either through the formation of nonhumic, small molecular weight molecules from the humic substances (Kieber et al. 1989, 1990) or perhaps through chemical modifications of core humic materials (Geller 1986; Mopper et al. 1991). Photochemical processes may therefore gradually modify the chemical structure of the humic substances, continually converting some of the humic carbon pool to more biologically available compounds.

Results of studies with dissolved humic substances point out two considerations important for modeling the fate of slow-cycling dissolved compounds in aquatic environments: First, within a complex NLOM pool, some fractions may cycle rapidly relative to the bulk pool and others more slowly. For dissolved humic substances in aquatic environments, the rapidly cycling fraction appears to account for 1% to 15% of the total, although newly labile compounds may be continually produced by biological and photochemical processes. Second, complex NLOM pools from various environments, although defined and isolated by identical procedures, can have distinctly different biological lability and therefore different turnover times.

BACTERIAL TURNOVER OF SIMPLE DISSOLVED NLOM

Until recently, microbial uptake of simple dissolved organic compounds from natural waters was generally modeled assuming that kinetics followed saturable, Michaelis–Menten kinetics. When uptake of compounds, such as simple sugars and amino acids, was examined over narrow ranges of concentration, this Michaelis–Menten model closely approximated uptake by natural aquatic microbial populations (Wright and Hobbie 1965, 1966; Crawford et al. 1973). When plotted according to Wright and Hobbie's modified Lineweaver–Burk linearization, uptake rates supported by natural marine or freshwater microbial populations increased with increasing substrate concentration at low substrate levels and then appeared to saturate. Using this approach, the concentration of added substrate [A] is plotted on the X axis against t/f (incubation time, t, divided by f, the fraction of added substrate used in time t) on the Y axis (Figure 4.2, top). Generally, a least squares linear regression then was used to force a straight line through the data. The Wright–Hobbie plot eliminates the need to determine the natural substrate concentration (S_n) and the actual uptake rate (v) and graphically yields the microbial maximum uptake rate, V_{max} (reciprocal of the slope), and $[K_t + S_n]$, a value equal to the sum of the transport constant and the natural substrate concentration (intercept on the X axis), as well as the turnover time, T_t, of the compound (intercept on the Y axis). A linear relationship between t/f and [A] was taken to indicate that a single rectangular hyperbola described the relationship between v and [S]. Generally, a linear model fits these data quite well over narrow substrate concentration ranges, and the occasional occurrence of nonlinearity in Wright–Hobbie plots was attributed to experimental problems or to the existence of appreciable diffusion-mediated uptake of the substrate (i.e., not via saturable uptake systems). Later, this approach also was adapted to model rates of removal of toxic and xenobiotic substances from natural waters (Pfaender and Bartholomew 1982; Schmidt et al. 1985).

In retrospect, there appear not to have been obvious compelling reasons why an entire population of bacteria should take up a substrate by a single kinetic system, or multiple systems all possessing identical kinetic parameters. In fact, uptake of simple compounds, such as sugars and amino acids, and even xenobiotic compounds, such as methyl parathion (Lewis et al. 1985), by individual strains often follows complex kinetic patterns as a result of contributions from multiple systems with distinctly different kinetic parameters. Again, however, it should be remembered that, prior to that time, the water column environment had been considered to be highly homogeneous with regard to concentrations of dissolved substrates (i.e., uniformly very low). Thus, it could be reasonably argued that bacteria adapted for long periods to such a uniform environment might be under selective pressure to evolve very similar (and very low) K_t values (Wright and Burnison 1979). During the late 1960s and into the

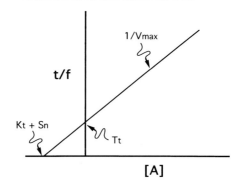

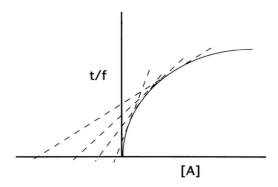

Figure 4.2 Wright–Hobbie plots for bacterial uptake of labile DOC showing the theoretical linear relationship expected for a single uptake system (top) and the curvilinear relationship indicative of multiphasic kinetics (bottom).

1970s, this approach was widely used, and it was common to see papers in the literature wherein a single V_{max} calculated from a Wright–Hobbie plot was offered as a measure of the "heterotrophic potential" of a given water sample (Wright and Hobbie 1965, 1966).

In the 1970s, the aquatic ecologists began to appreciate the true heterogeneous nature of substrate distribution in the water column environment. Zones of transiently high substrate concentrations were viewed as being interspersed among regions of low, bulk substrate concentration. This new model of NLOM distribution cast doubts on the reasonableness of assuming uniform K_t and V_{max} values across all members of a bacterial population. At this same time, availability of new radiolabeled substrates, such as high specific activity tritiated D-glucose, made possible uptake kinetic studies in which substrate concentration could be varied by many orders of magnitude (beginning with sub-nanomolar concentrations nearly as low as those *in situ*). Armed with these substrates, Azam and Hodson (1981) found that Wright–Hobbie plots for uptake of simple organic substrates by marine bacterioplankton were themselves curvilinear, with decreasing slope as initial substrate concentration was increased. This work suggested that the concept of the kinetically homogeneous microbial population in natural aquatic ecosystems needed to be reevaluated. As shown in the hypothetical plot of Figure 4.2 (bottom), the apparent V_{max} and K_t values over individual substrate concentration ranges can be determined from the slopes of the straight lines tangent to the nonlinear line of best fit to these data. Apparent V_{max} and K_t values can vary over many orders of magnitude with the concentration of added substrate. Within the same population of marine microorganisms, and for the same sampling time, K_t values were found to vary from a few nanomolar to many millimolar (Azam and Hodson 1981). As no single V_{max} value exists over the entire range of potential substrate concentrations, the once popular concept of "heterotrophic potential" appears not to be applicable to the vast majority of natural aquatic microbial assemblages studied to date. Rather, as substrate concentration increases, high-capacity, low-affinity uptake mechanisms within the population increasingly dominate uptake, and rate of uptake tends not to saturate. Monophasic uptake kinetics (i.e., uptake by a population exhibiting single apparent V_{max} and K_t values) have been observed only on rare occasions, and then mostly for the uptake of unusual compounds (those other than simple sugars or amino acids) (Fuhrman and Ferguson 1986) or by very low-diversity microbial communities (Azam and Hodson 1981). Uptake of individual compounds by individual species of marine bacteria can be monophasic (Hodson and Azam 1979) and follow saturable Michaelis–Menten kinetics in species where uptake is mediated by single transport systems. In other marine bacterial species, uptake follows more complex, multiphasic kinetic patterns owing to the simultaneous action of multiple transport systems, each having distinctly different K_m and V_{max} values (Nissen et al. 1984). In a recent report, it was demonstrated theoretically that nonsaturable Fickian diffusion across bacterial cellular membranes (as opposed to either active transport or facilitated transport) could yield uptake kinetic patterns that would also be multiphasic (Logan and Fleury 1993). However, results of numerous studies of bacterial membrane transport using mutant bacteria lacking specific transport systems have demonstrated that such strains generally cannot support measurable growth on water soluble substrates at very high external concentrations when supplied purely by diffusion (Saier and Mocydlowski 1978). Considering the exceedingly low concentrations of most available organic substrates in natural waters, it thus appears unlikely that strains acquiring carbon and energy sources by simple diffusion could effectively compete with bacteria possessing highly efficient transport systems.

Such "kinetic diversity" within marine microbial assemblages appears to fit the new model of the pelagic environment as temporally and spatially heterogeneous in its chemical composition (e.g., Azam and Cho 1987). These multiphasic kinetic patterns are attributable to the presence within a single population of a range of uptake systems with widely varying kinetic parameters. Low K_t, low V_{max} transport systems tend to dominate under ultra-oligotrophic, bulk seawater substrate concentrations, while higher K_t, higher V_{max} systems contribute most of the total uptake at high substrate concentrations typical of enriched microzones (near surfaces of degrading organic detrital particles or phytoplankton undergoing grazing by zooplankton).

In general, the values of both apparent population K_t and V_{max} increase with increasing substrate concentration. However, for most compounds, and in most natural microbial assemblages examined, the two parameters do not always change to an equal extent. A plot of the ratio V_{max}/K_t versus initial substrate concentration reveals that the ratio changes very little if at all with changes in substrate concentration, [S], at high substrate concentrations (> one to a few micromoles). For example, Lewis et al. (1988) show the relationship between V_{max}/K_t and [S] for the utilization of a variety of toxic pollutants by the natural microbial populations in freshwater Lago Lake (Athens, GA) or in seawater from Chub Cay, Bahamas (Figure 4.3). V_{max}/K_t values are highest at lowest [S] values and decrease with increasing [S], reaching essentially constant (and many fold lower) values at concentrations higher than a few micromoles. The extent of change in the value of V_{max}/K_t, and the [S] above which the ratio becomes constant, vary among compounds and between microbial populations. As Wright–Hobbie plots reveal that both V_{max} and K_t are, in fact, increasing with additional [S] over this concentration range, both are increasing proportionately. However, at lower substrate concentrations, V_{max}/K_t increases sharply with decreasing [S] (Figure 4.3). As the actual uptake rate, v, is proportional to V_{max}/K_t, substrate uptake by the microbial population is effectively enhanced at low [S]. This relationship has been observed for uptake of a variety of naturally occurring and synthetic organic compounds by the microbial assemblages of a range of marine ecosystems (Azam and Hodson 1981), the highly acidic freshwaters of the Okefenokee Swamp (Murray and Hodson 1984), as well as the microbial population of a groundwater aquifer (Hwang et al. 1989). These complex uptake kinetic patterns are the norm for uptake by both free-living and particle-attached bacteria in the water column and also seem to be common in the heterotrophic components of biofilms attached to submerged surfaces (Lewis et al. 1988).

This uptake kinetic diversity by natural microbial populations has important implications for predicting the flow of carbon from dissolved NLOM to the microbial loop in aquatic food webs. Using the Michaelis–Menten model to fit a simple rectangular hyperbola to the uptake of a dissolved substance will result in gross errors if rates of uptake are extrapolated to concentrations

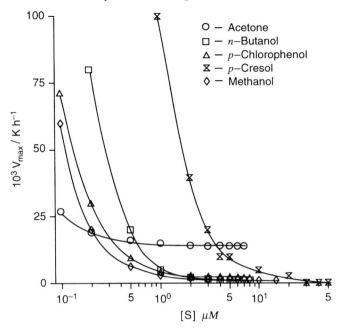

Figure 4.3 Kinetic parameters (V_{max}/K) for low concentrations of various organic substrates added to replicate water samples from marine and freshwater environments (from Lewis et al. 1988).

outside the range used in the experimental study. In most such studies, the concentrations of added substrate are kept as low as possible (dictated by the highest available specific activity of the radiolabeled form of the chosen compound) in order to produce the least possible disturbance to the natural system. Thus, with tritiated compounds, for example, a range of concentrations in the sub-nanomolar to low nanomolar range might be seen as ideal. However, if multiphasic kinetics dominate the system, then the actual rate of uptake at much higher (e.g., 100 to 1000 nmol l^{-1}) substrate concentrations could exceed the calculated V_{max} value by as much as a factor of 10^4. This discrepancy is the result of the fact that kinetic diversity greatly extends the range of substrate concentrations over which uptake rates do not saturate (Azam and Hodson 1981). Likewise, if one measures uptake kinetics using substrate concentrations that fall at the high end of natural concentrations, calculates simple Michaelis–Menten parameters K_t and V_{max} from the data, and uses these parameters to estimate uptake rate, v, at much lower substrate concentrations, then the resulting values will be gross underestimates of actual uptake rate (again, by as much as 10^4; Lewis et al. 1988).

The enhancement of uptake rates at very low substrate concentrations can be directly visualized using plots of ln [S] versus time. Using this graphical representation, the slope of the plot is equal to the specific rate of substrate utilization, k. In this experimental design, a series of individual experiments are run in which initial $[S]$ is varied and the concentration of S remaining after time intervals is followed. At relatively high initial substrate concentrations ranges over which V_{max}/K_t is constant, specific utilization is nearly constant, or changes little with decreasing $[S]$. However, below a certain value of initial $[S]$, the specific rate of uptake increases markedly (i.e., the slope of the plot becomes increasingly negative). In Figure 4.4 (top), from Lewis et al. (1988), the average specific rate of removal of 2,4–DME by the marine microbial populations in Bahamian seawater was followed at five different initial $[S]$ values. Specific removal rates increased by approximately 100-fold as initial $[S]$ was decreased from $2.6\,\mu$mol l^{-1} to $60\,$nmol l^{-1}. Similarly, Figure 4.4 (bottom) illustrates how the specific rate of removal of D-glucose by the natural microbial populations in a freshwater lake (Lago Lake, Athens, GA) increased by nearly 4-fold during a 24-h experiment (Lewis et al. 1988). Although the definitive studies have yet to be conducted, one can assume that the extent of rate enhancement with decreasing $[S]$ might increase until the point at which uptake was mediated solely by the component substrate uptake system with the highest ratio of V_{max} to K_t within the microbial population.

CONCLUSION

Results of recent studies show that the uptake and degradation of both the simple and complex pools of dissolved NLOM from natural waters follow involved kinetic patterns. Neither simple first-order nor Michaelis–Menten models faithfully describe the microbially mediated flow of carbon from highly heterogeneous pools. The pool of humic substances, for example, is comprised of compounds representing a wide range of biodegradability, and, hence, with widely differing turnover times in aquatic environments. Although the humic substances have generally been considered to be practically inert to short-term microbial utilization, recent studies suggest that they may support measurable percentages of the total bacterial secondary production in certain aquatic ecosystems.

In the utilization of low molecular weight, nominally labile, dissolved NLOM, natural aquatic bacterial populations appear to utilize multiple uptake strategies that maximize potential carbon and energy flow as substrate concentrations vary both temporally and spatially. Rather than being adapted to either bulk, ultra-oligotrophic substrate conditions, or to the high substrate concentrations characteristic of enriched microzones, the populations typically show great diversity with respect both to their affinity (K_t) and overall capacity (V_{max}) to acquire organic nutrients. Any attempts to model the flow of carbon and energy from dissolved NLOM pools (either simple or complex) into the

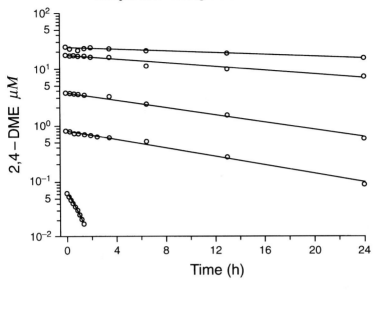

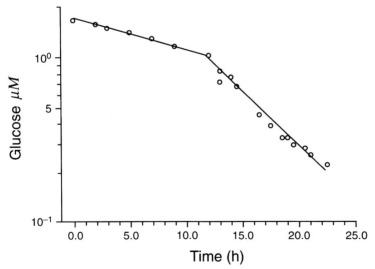

Figure 4.4 Enhancement of removal of 2,4–DME (top) and D-glucose (bottom) from replicate water samples from freshwater environments (from Lewis et al. 1988).

microbial loop in marine and freshwater environments would benefit from consideration of both the high degree of chemical and functional heterogeneity of the available organic substrates and the kinetic diversity with which the substrates are assimilated by the resident populations.

Several important questions for future research would benefit from an increased knowledge of the kinetics of DOC utilization by natural bacterial populations. First, to further our understanding of the rates and mechanisms of carbon flux through bacterial pools, we need to identify which components of the dissolved NLOM pool are available as bacterial carbon sources. Previous research has focused on simple, chemically identifiable compounds (such as carbohydrates and amino acids) as models compounds for characterizing rates and kinetics of bacterial uptake. Yet approaches that encompass the complex, chemically heterogenous DOC in considerations of bacterial turnover of dissolved NLOM are needed as well. Second, we need to reconcile short-term data sets on the biological turnover of DOC, such as bacterial uptake and kinetics studies, with long-term data sets on the geochemical fate of DOC in aquatic environments, such as ^{14}C dating. Finally, we now suspect that, even in the ocean, there exists significant heterogeneity in the concentrations of utilizable substrates. We need to recognize, therefore, that kinetics are concentration dependent and that models based on simple Michaelis–Menten approaches are inappropriate unless used over a very narrow range of substrate concentrations.

REFERENCES

Aiken, G. 1985. Isolation and concentration techniques for aquatic humic substances. In: Humic Substances in Soil, Sediment, and Water, ed. G.R. Aiken, D.M. McKnight, R.L. Wershaw, and P. MacCarthy, pp. 363–385. Chichester: Wiley.

Azam, F., and B.C. Cho. 1987. Bacterial utilization of organic matter in the sea. In: Ecology of Microbial Communities, ed. M. Fletcher, T.R.G. Gray, and J.G. Jones, pp. 261–281. Cambridge Univ. Press.

Azam, F., T. Fenchel, J.G. Field, J.S. Gray, L.-A. Meyer-Reil, and F. Thingstad 1983. The ecological role of water-column microbes in the sea. *Mar. Ecol. Prog. Ser.* **10**:257–263.

Azam, F., and R.E. Hodson. 1977. Dissolved ATP in the sea and its utilization by marine bacteria. *Nature* **267**:696–698.

Azam, F., and R.E. Hodson. 1981. Multiphasic kinetics for D-glucose uptake by assemblages of natural marine bacteria. *Mar. Ecol. Prog. Ser.* **6**:213–222.

Bauer, J.E., P.M. Williams, and E.R.M. Druffel. 1992. ^{14}C activity of dissolved organic carbon fractions in the north-central Pacific and Sargasso Sea. *Nature* **357**:667–670.

Benner, R., J.D. Pakulski, M. McCarthy, and J.I. Hedges. 1992. Bulk chemical characteristics of dissolved organic matter in the ocean. *Science* **255**:1561–1564.

Cole, J.J., S. Findlay, and M.L. Pace. 1988. Bacterial production in fresh and saltwater ecosystems: A cross-system overview. *Mar. Ecol. Prog. Ser.* **43**:1–10.

Crawford, C.C., J.E. Hobbie, and K.L. Webb. 1973. Utilization of dissolved organic compounds by microorganisms in an estuary. In: Estuarine Microbial Ecology, ed.

L.H. Stevenson and R.R. Colwell, pp.169–177. Columbia, SC: Univ. of South Carolina Press.

Ertel, J.R., J.I. Hedges, A.H. Devol, J.E. Richey, and M.N.G. Riberio. 1986. Dissolved humic substances of the Amazon River system. *Limnol. Oceanogr.* **31**:739–754.

Ertel, J.R., J.I. Hedges, and E.M. Purdue 1984. Lignin signature of aquatic humic substances. *Science* **223**:485–487.

Fuhrman, J.A., and R.L. Ferguson. 1986. Nanomolar concentrations and rapid turnover of dissolved free amino acids in seawater: Agreement between chemical and microbiological measurements. *Mar. Ecol. Prog. Ser.* **33**:237–242.

Geller, A. 1986. Comparison of mechanisms enhancing biodegradability of refractory lake water constituents. *Limnol. Oceanogr* **31**:755–764.

Hatcher, P.G., and E.C. Spiker. 1988. Selective degradation of plant biomolecules. In: Humic Substances and Their Role in the Environment, ed. F.H. Frimmel and R.F. Christman, pp. 59–74. Dahlem Workshop Report LS 41. Chichester: Wiley.

Hedges, J.I. 1988. Polymerization of humic substances in natural environments. In: Humic Substances and Their Role in the Environment, ed. F.H. Frimmel and R.F. Christman, pp. 45–58. Dahlem Workshop Report LS 41. Chichester: Wiley.

Hessen, D.O. 1985. The relation between bacterial carbon and dissolved humic compounds in oligotrophic lakes. *FEMS Microbiol. Ecol.* **31**:215–223.

Hessen, D.O., T. Anderson, and A. Lyche. 1990. Carbon metabolism in a humic lake: Pool sizes and cycling through zooplankton. *Limnol. Oceanogr* **35**:84–99.

Hodson, R.E., and F. Azam. 1979. Occurrence and characterization of phosphoenol-pyruvate: Glucose phosphotransferase system in a marine bacterium, *Serratia marinorubra*. *Appl. Environ. Microbiol.* **38**:1087–1091.

Hwang, H.M., R.E. Hodson, and D. Lewis. 1989. Microbial degradation kinetics of toxic organic chemicals over a wide range of concentrations in natural aquatic systems. *Environ. Toxicol. Chem.* **8**:65–74.

Jones, R.I. 1990. Phosphorus transformations in the epilimnion of humic lakes: Biological uptake of phosphate. *Freshwat. Biol.* **23**:323–337.

Kieber, D.J., J. McDaniel, and K. Mopper. 1989. Photochemical source of biological substrates in sea water: Implications for carbon cycling. *Nature* **341**:637–639.

Kieber, R.J., X. Zhou, and K. Mopper. 1990. Formation of carbonyl compounds from UV-induced photodegradation of humic substances in natural waters: Fate of riverine carbon in the sea. *Limnol. Oceanogr.* **35**:1503–1515.

Leff, L.G., and J.L. Meyer. 1991. Biological availability of dissolved organic carbon along the Ogeechee River continuum. *Limnol. Oceanogr.* **36**:315–323.

Lewis, D., R.E. Hodson, and L.F. Freeman III. 1985. Multiphasic kinetics for transformation of methyl parathion by *Flavobacterium* species. *Appl. Environ. Microbiol.* **50**:553–557.

Lewis, D.L., R.E. Hodson, and H.-M. Hwang. 1988. Kinetics of mixed microbial assemblages enhance removal of highly dilute organic substrates. *Appl. Environ. Microbiol.* **54**:2054–2057.

Logan, B.E., and R.C. Fleury. 1993. Multiphasic kinetics can be an artifact of the assumption of saturable kinetics for microorganisms. *Mar. Ecol. Prog. Ser.* **16**:155–160.

Meyer, J.L., R.T. Edwards, and R. Risley. 1987. Bacterial growth on dissolved organic carbon from a blackwater river. *Microb. Ecol.* **13**:13–29.

Meyers-Schulte, K.J., and J.I. Hedges. 1986. Molecular evidence for a terrestrial component of organic matter dissolved in ocean water. *Nature* **321**:61–63.

Mopper, K., X. Zhou, R.J. Kieber, D.J., Kieber, R.J., Sikorski, and R.D. Jones. 1991. Photochemical degradation of dissolved organic carbon and its impact on the oceanic carbon cycle. *Nature* **353**:60–62.

Moran, M.A., and R.E. Hodson. 1990a. Bacterial production on humic and nonhumic components of dissolved organic carbon. *Limnol. Oceanogr.* **35**:1744–1756.

Moran, M.A., and R.E. Hodson. 1990b. Contributions of degrading *Spartina alterniflora* lignocellulose to the dissolved organic carbon pool of a salt marsh. *Mar. Ecol. Prog. Ser.* **62**:161–168.

Moran, M.A., and R.E. Hodson. 1994. Dissolved humic substances of vascular plant origin in a coastal marine environment. *Limnol. Oceanogr.* **39**:762–771.

Moran, M.A., L.R. Pomeroy, E.S. Sheppard, L.P. Atkinson, and R.E. Hodson. 1991. Distribution of terrestrially-derived dissolved organic matter on the southeastern U.S. continental shelf. *Limnol. Oceanogr.* **36**:1134–1149.

Murray, R.E., and R.E. Hodson. 1984. Microbial biomass and utilization of dissolved organic matter in the Okefenokee Swamp ecosystem. *Appl. Environ. Microbiol.* **47**:685–692.

Nissen, H., P. Nissen, and F. Azam. 1984. Multiphasic uptake of D-glucose by an oligotrophic marine bacterium. *Mar. Ecol. Prog. Ser.* **16**:155–160.

Pfaender, F.K., and G.W. Bartholomew. 1982. Measurement of aquatic biodegradation rates by determining heterotrophic uptake of radiolabeled pollutants. *Appl. Environ. Microbiol.* **44**:159–164.

Saier, M.H., and E.G. Mocydlowski. 1978. The regulation of carbohydrate transport in *Escherichia coli* and *Salmonella typhimurium.* In: Bacterial Transport, ed. B.P. Rosen, pp. 103–125. New York: Marcel Dekker.

Schmidt, S.K., S. Simpkins, and M. Alexander. 1985. Models for the kinetics of biodegradation of organic compounds not supporting growth. *Appl. Environ. Microbiol.* **50**:323–331.

Thurman, E.M. 1985. Organic geochemistry of natural waters. Boston: Martinus Nijhoff/ Dr. W. Junk.

Tranvik, L.J. 1988. Availability of dissolved organic carbon for planktonic bacteria in oligotrophic lakes of differing humic content. *Microb. Ecol.* **16**:311–322.

Tranvik, L.J. 1990. Bacterioplankton growth on fractions of dissolved organic carbon of different molecular weights from humic and clear waters. *Appl. Environ. Microbiol.* **56**:1672–1677.

Tulonen, T., K. Salonen, and L. Arvola. 1992. Effects of different molecular weight fractions of dissolved organic matter on the growth of bacteria, algae and protozoa from a highly humic lake. *Hydrobiol.* **229**:239–252.

Williams, P.M., and E.R.M. Druffel. 1987. Radiocarbon in dissolved organic matter in the central North Pacific Ocean. *Nature* **330**:246–248.

Wright, R.T., and B.K. Burnison. 1979. Heterotrophic activity measured with radiolabeled organic substrates. In: Native Aquatic Bacteria: Enumeration, Activity, and Ecology, ed. J.W. Costerton and R.R. Colwell, pp. 140–155. STP 695. Philadelphia: Am. Soc. Test. Mat.

Wright. R.T., and J.E. Hobbie. 1965. The uptake of organic solutes in lake water. *Limnol. Oceanogr.* **10**:22–28.

Wright, R.T., and J.E. Hobbie. 1966. Use of glucose and acetate by bacteria and algae in aquatic ecosystems. *Ecology* **47**:447–464.

Standing, left to right:
Jérome Balesdent, Hans-Hermann Richnow, Neil Blough, Ellen Druffel, Georg Guggenberger, Ron Benner, Fritz Frimmel, Dan Repeta
Seated, left to right:
Russ Christman, Marc Alperin, Roger Swift

5

Group Report: How Can We Best Characterize and Quantify Pools and Fluxes of Nonliving Organic Matter?

M.J. Alperin, Rapporteur
J. Balesdent, R.H. Benner, N.V. Blough, R.F. Christman,
E.R.M. Druffel, F.H. Frimmel, G. Guggenberger,
D.J. Repeta, H.H. Richnow, R.S. Swift

INTRODUCTION

Nonliving organic matter (NLOM) represents one of the largest pools of carbon on Earth (Hedges 1992). Because of its immense size, small changes in the pool can lead to significant perturbations in the global carbon cycle. Compared to the living biomass and inorganic carbon reservoirs, NLOM has received relatively little attention. There are significant uncertainties regarding the pool sizes, reactivity, sources, and transformations of NLOM. Quantitative information regarding cycling of NLOM is required to parameterize models designed to predict the effect of climate change on the Earth's carbon cycle. In addition, observations of temporal changes in NLOM pools and fluxes may provide a means of detecting and confirming perturbations to the natural carbon cycle.

There are fundamental differences between NLOM cycling in terrestrial and marine systems. Thus it is useful to classify NLOM pools according to environment. Pools of NLOM in the terrestrial sphere include litter and soil organic matter (SOM), dissolved organic matter (DOM), and particulate organic matter (POM) in fresh and ground water, and the organic matter contained within

Role of Nonliving Organic Matter in the Earth's Carbon Cycle
Edited by R.G. Zepp and Ch. Sonntag © 1995 John Wiley & Sons Ltd.

sedimentary rocks (kerogen). In the marine environment, NLOM pools include DOM and POM in the water column and organic matter contained in ocean sediments. The atmosphere also contains NLOM in the form of airborne particles, DOM, and organic gases.

In this report we focus on those NLOM pools that have the potential to perturb the global carbon cycle on the time scale of decades. This includes pools with large fluxes that are sensitive to changes in temperature, precipitation, primary production, microbial heterotrophy, and land-use practices. The pools that meet these criteria are (a) litter and soil organic matter, (b) marine water column organic matter, and (c) marine sedimentary organic matter. We also consider the freshwater organic matter pool because it can be strongly impacted by anthropogenic pollution.

To understand the dynamics of global-scale NLOM cycling, three parameters must be characterized. First, the masses of organic matter in the NLOM pools must be evaluated. Second, the fluxes of material to and from the pools must be estimated. Third, the sources and fates of the reactive fraction of NLOM pools must be identified. In this report, we discuss a variety of methods that have been shown to be useful for estimating each of the above parameters. The discussion also describes a number of novel techniques that are currently undergoing development which have the potential to be useful in the near future. We also consider artifacts associated with the various characterization methods and discuss the precautions that must be applied when interpreting results from these techniques. Finally, we present a series of recommendations regarding future research directions.

WHAT METHODS ARE MOST USEFUL FOR MAKING GLOBAL ESTIMATES OF NLOM POOL SIZES?

Litter and Soil Organic Matter

Soil organic matter (SOM) is defined as the nonliving component of organic matter in soil. Although litter is not usually considered to be part of the SOM pool, organic matter that constitutes the litter fraction plays a major role in the terrestrial carbon cycle. For example, litter decomposition is a major contributor to respiration on the forest floor (Goudriaan and Ketner 1984; Emanuel et al. 1984) and the main source of carbon monoxide emitted from savannas (Sanhueza et al. 1994; Miller et al. 1993).

The soil profile is typically divided into several horizons. The upper layers usually contain the highest concentration of organic matter and hence have the greatest biological activity. The lower horizons contain less organic matter, but can receive decomposition products that percolate from above. Despite having lower organic matter concentrations, the deeper layers constitute a large NLOM pool owing to the thickness of the soil column.

The conventional approach in estimating the global pool size of the soil organic matter reservoir is to classify the Earth's surface according to soil types and evaluate the organic carbon content characteristic of each soil. A variety of vegetation- or climate-based divisions have been used as classification schemes (e.g., Schlesinger 1977; Post et al. 1982). Estimates of global soil organic matter from these different classification systems are in reasonable agreement and indicate a pool size of 1400–1500 Pg C. These estimates generally consider only the upper meter of the soil column. The precision of the estimate of the soil NLOM pool is mainly limited by integration uncertainties due to the high degree of horizontal heterogeneity characteristic of soil systems. In addition, the deep horizons have been poorly studied and the carbon content is not well constrained. The accuracy of the soil NLOM pool size would be improved by additional soil data from large geographic regions that have been poorly studied.

Remote sensing of soil organic matter levels is an attractive possibility but will be difficult to achieve for a number of reasons. First, vegetative cover frequently interferes with the remotely sensed signal from the soil organic matter. Second, even in the absence of vegetative cover, the remotely sensed signal is only indicative of the organic matter level in the surface layer of the soil. Newer remote sensing techniques involving multiple correlation of signals obtained from different parts of the electromagnetic spectrum together with effective "ground-truthing" offer possibilities of overcoming these problems.

Marine Water Column Organic Matter

Dissolved organic matter is by far the largest reservoir of NLOM in the ocean water column. Seawater DOM distributions are less heterogeneous than organic matter in soil systems. This is because marine DOM is subject to mixing that tends to reduce spatial variability. Hence global estimates of the size of the water column DOM pool are less sensitive to integration uncertainties than for the soil pool. Uncertainty in the DOM pool size is largely due to analytical difficulties caused by dilute concentrations and the high salt matrix. The large disparity between DOM methods reported in Hedges et al. (1993) has been for the most part resolved (Sharp et al. 1993). Recent estimates indicate that the DOM pool contains about 700 Pg C (E. Peltzer, pers. comm. as cited in Siegenthaler and Sarmiento 1993).

The quantity of particulate organic matter in seawater is typically 0.1% (deep water) to 5% (surface water) of the DOM pool. Both sinking and suspended particles are included in the definition of POM. POM concentrations are highly variable both spatially and temporally. This variability makes it difficult to estimate the size of the global ocean water column POM pool accurately. The ratio of living to nonliving organic matter in the POM pool is also highly variable. Recent estimates of biomass that include bacteria suggest that half the POM pool in meso- and oligotrophic regions could be composed of living organic matter (Cho and Azam 1990).

Remote sensing techniques offer a possible means of determining variations in the surface aquatic NLOM pool over broad spatial and temporal scales, primarily through passive and active airborne and satellite ocean color measurements (Carder et al. 1991; Hoge et al. 1993; Blough and Green, this volume). In particular, active lidar measurements represent a very promising remote technique for determining the distribution and nature of aquatic NLOM. While providing broad coverage, these techniques suffer from several disadvantages. First, remote sensing techniques only view the very surface of the oceans (through the first optical depth). Second, an estimate of total DOM is only possible if the chromophoric compounds represent a constant fraction of the DOM or vary in a well-defined way with the total. Further research is needed to determine whether satellite measurements can ultimately provide accurate estimates of total DOM.

Marine Sedimentary Organic Matter

Measurements of sediment organic matter concentrations are available for many regions of the world's oceans (Romankevich 1984). Based on these data, the pool of sedimentary NLOM is estimated to be 150 Pg C (Emerson and Hedges 1988). This estimate includes only the mixed surface layer (taken to be the upper 20 cm in continental margin sediments and the upper 10 cm in open ocean sediments).

Freshwater Organic Matter

Although the contribution of freshwater organic matter to the global carbon budget is small, there is evidence that pollution can have a major impact on NLOM pools in rivers, lakes, and ground water. Some freshwater systems receive large quantities of anthropogenic organic pollutants in addition to the NLOM derived from natural sources. Many questions remain regarding turnover rates of anthropogenic compounds, conversion of xenobiotics into NLOM, and fates of the biochemical reaction products.

WHAT METHODS ARE MOST USEFUL FOR MAKING GLOBAL ESTIMATES OF FLUXES TO AND FROM NLOM POOLS?

The potential of an NLOM pool to perturb the global carbon cycle depends much more upon the flux of material passing through the pool than on the pool size. For pools that have a large flux, even slight perturbations in turnover rate could have a major impact on the global carbon cycle. It is possible that changes in flux to or from key NLOM pools could provide a more sensitive indicator of perturbations to the global carbon cycle than changes in pool size.

Direct flux measurements are one approach for estimating global fluxes associated with NLOM pools. A variety of techniques exist for measuring NLOM fluxes in local environments (e.g., chamber techniques, sediment traps, radiochronometric tracers). Global fluxes can be estimated by extrapolating discrete measurements to global scales. Although direct measurements provide process-level information regarding the factors that control fluxes, extrapolation leads to highly uncertain global fluxes due to spatial and temporal heterogeneity.

Large-scale models provide an alternative technique for estimating global-scale fluxes through NLOM pools. In this approach, direct flux measurements made at specific sites are used to parameterize those processes that control NLOM fluxes. The models are then used to make the large-scale extrapolations required for global flux estimates.

Time series measurements of atmospheric composition provide useful constraints on changes in fluxes from NLOM pools. Temporal changes in atmospheric O_2 (Keeling and Shertz 1992), CO_2 and CH_4 (Boden et al. 1990), and other trace gases (e.g., Najjar et al., this volume) serve as large-scale integrators of changes in net transfer rates between NLOM, biomass, and atmospheric carbon pools, and provide checks on fluxes inferred from measurements and models.

Satellite and airborne platforms can provide high spatial and temporal resolution measurements of many ocean properties such as sea-surface temperature, surface wind speeds, phytoplankton biomass, and potentially primary productivity. This information can be extremely useful in determining global-scale fluxes of NLOM within the oceans.

WHAT METHODS ARE MOST USEFUL FOR CHARACTERIZING THE REACTIVITY OF NLOM?

The flux of carbon to and from NLOM pools is a direct function of the reactivity of the organic matter in that pool. The reactivity of NLOM depends upon (a) the intrinsic reactivity of the molecules that comprise the NLOM pool and (b) external factors such as matrix effects, temperature, redox potential, pH, and the presence or absence of macrobenthic organisms. As described by Zepp and Sonntag (this volume), NLOM that has a chemical structure resulting in low intrinsic reactivity is considered to be resistant whereas NLOM that is protected from decomposition by external factors is defined as persistent. For a discussion on the effect of chemical structure, biological processes, and environmental factors on NLOM reactivity, see Davidson et al. (this volume). This section focuses instead on the influence of matrix effects on NLOM reactivity.

Classical techniques for categorizing NLOM reactivity have largely been based on distinguishing between humic and nonhumic substances (Hayes et al. 1989). Humic compounds are thought to have low reactivity owing to their complex chemical structures. Substances in the nonhumic fraction have been further categorized on the basis of compound classes such as polysaccharides, peptides, lipids, and lignins. Although these classification and fractionation techniques have been very useful—leading to a better understanding of the nature and composition of NLOM—they have been less successful in identifying and predicting the amounts of biodegradable soil organic matter or the rates at which degradation is likely to occur (Frimmel and Christman 1988; Hayes et al. 1989).

Traditionally, it has been assumed that chemical composition and structure are largely responsible for the low reactivity that leads to accumulation and preservation of NLOM. More recently, it has become clear that physical protection is an equally important factor in controlling NLOM reactivity. Components of NLOM can be protected from rapid decomposition through adsorption, occlusion, clustering, and mineral aggregate formation. An organic compound that might otherwise be readily biodegradable can persist for long periods of time due to physical protection.

It is difficult to separate the influence of physical protection and chemical structure on biodegradability. To understand better those processes that influence the decomposition and turnover of NLOM, it is necessary to combine the concepts of chemical stability and physical protection into a unified classification system. An example of such an approach for soil NLOM is given below. The classification system assumes that there are two categories of nonliving soil organic matter: labile (quickly degrading) and resistant (slowly degrading). Furthermore, it assumes that there are three categories of microenvironments: free-particulate (unprotected), occluded (physically entrapped in an abiotic microenvironment), and adsorbed (attached to a mineral or other surface by physical or chemical mechanisms). This classification gives rise to six distinct categories of NLOM:

1. free-particulate: labile,
2. free-particulate: resistant,
3. occluded-particulate: potentially labile but currently protected due to inaccessibility,
4. occluded-particulate: resistant and inaccessible,
5. adsorbed: potentially labile but currently protected by adsorption,
6. adsorbed: resistant and further protected by adsorption.

The significance of this classification scheme is that it allows for the protection of NLOM that would otherwise be readily biodegraded and for the enhanced protection of materials which are already stabilized due to their chemical composition and structure.

It is important to point out that the effect of adsorption on NLOM reactivity is complex and not well understood. Adsorption onto clay minerals can stabilize organic matter, thereby allowing reactive compounds to persist (Oades 1988). Alternatively, high clay contents can promote NLOM decomposition (Gregorich et al. 1991; Schäfer 1992) owing to the high microbial activity on clay surfaces (Amato and Ladd 1992).

By using a variety of physical and chemical separation techniques, it is possible to isolate NLOM fractions containing organic matter of differing reactivities. In some cases, the reactivity and turnover rates for an NLOM fraction can be evaluated by heterotrophic uptake experiments. In other cases, bomb-derived radiocarbon can provide a constraint on NLOM turnover (see Trumbore and Druffel, this volume). Bomb-derived ^{14}C was produced by thermonuclear weapons testing during the late 1950s and early 1960s, causing ^{14}C levels in atmospheric CO_2 to nearly double by 1963. As negligible bomb ^{14}C has been produced since then, this isotope of carbon provides an excellent spike tracer for examining short-term exchange processes within the carbon cycle (see Post et al., this volume). For example, the presence of bomb-derived ^{14}C in an NLOM pool indicates turnover of the carbon contained within this pool during the past 30 to 40 years.

In addition, methods are now becoming available for quantifying turnover rates of extremely fast reactions. Recently, the sunlight-induced production of carbon monoxide (CO) and carbonyl sulfide (COS) from NLOM in seawater and wetlands has been reported (see Najjar et al., this volume). Due to the large global area covered by aquatic environments, the photochemical transformation of NLOM into CO and COS has a significant effect on biosphere–atmosphere fluxes of these trace gases. It is evident that the formation of rapidly reacting intermediates plays a major role in photochemical reactions. Steady-state concentrations of many reactive oxygen species and other photochemically produced intermediates have been estimated for the photic zones of natural waters (Cooper 1989; Hoigné et al. 1989; Blough and Zepp 1995). Two main techniques have been used to investigate the production rates, quantum yields, and action spectra for the reactive transients involved in NLOM photoreactions: (a) laser flash spectroscopy with time resolution down to the nanosecond range (Frimmel et al. 1987) and (b) continuous irradiations with selective chemical probes added to the system to monitor transient formation and decay.

M.J. Alperin et al.

WHAT METHODS ARE MOST USEFUL FOR CHARACTERIZING SOURCES AND FATES OF NLOM?

The chemical composition of NLOM can be characterized at different levels of structural detail. Each level of detail provides source information with varying degrees of specificity. Techniques that measure bulk (unfragmented) properties of NLOM are usually less source specific than techniques that operate at the molecular level. However, molecular-level analyses require extensive alteration of the sample (see discussion below) and are quantitatively restricted to a fraction (usually small) of the total pool. The relationship between molecular-level properties and the entire NLOM pool is often unclear.

Methods used to characterize sources and fates of NLOM differ widely in their sampling requirements and sensitivity. In general, the potential for introducing artifacts and contaminants increases with the specificity of the technique. Characterization techniques also differ greatly in the amount of time and expense needed to process a sample, hence some molecular-level techniques are currently not suitable for large-scale surveys of NLOM. Examples of analyses providing different levels of structural information regarding NLOM sources are given in Table 5.1.

A variety of bulk characterization techniques are useful indicators of the geographic, taxonomic, and biochemical sources of NLOM. Visual inspection and microscopic analysis can be used to identify plant tissue fragments, pollen, soot, and other macromolecular components of known source. Measurements

Table 5.1 Methods used to characterize sources of NLOM.

A. Bulk Properties
　　Visual inspection of NLOM
　　Elemental analysis
　　Absorption/fluorescence spectroscopy
　　Isotopic composition

B. Intermediate Level
　　Functional group analysis
　　Compound class analysis
　　Molecular weight and size distribution
　　Charge analysis
　　^{13}C- and ^{1}H-NMR

C. Molecular Level
　　Biomarkers
　　Compound specific isotope analysis
　　Analytical pyrolysis
　　Chemical degradation GC/MS

---

of bulk organic carbon and nitrogen content of NLOM and the molar ratio of these elements (C:N) are useful source indicators in transition zones between terrestrial and marine environments. Additional information about the biochemical source of NLOM can be obtained from determinations of organic hydrogen and oxygen and the molar ratios of O:C and H:C.

Spectroscopic techniques provide a broad view of chemical structure. Terrestrial and marine materials have characteristic absorption and fluorescence spectra that can be indicative of NLOM source. Optical techniques can be developed into field-deployable measurements that provide real-time *in situ* information regarding sources and transport of NLOM.

Stable isotopes, particularly ^{13}C, are very useful indicators of NLOM source (see Trumbore and Druffel, this volume). Marine- and terrestrial-derived NLOM can be distinguished on the basis of stable carbon isotope compositions due to the large difference (8‰) in $^{13}C:^{12}C$ ratios of atmospheric CO_2 and marine dissolved inorganic carbon. Terrestrial vegetation derived from C_3 (Calvin cycle) and C_4 (Hatch-Slack) photosynthetic pathways can often be distinguished and the contributions of these two vegetation types to NLOM thereby determined. Other isotopes, such as ^{15}N and ^{34}S, are also useful indicators of organic matter source and in combination with ^{13}C can provide a multidimensional picture of NLOM sources. The contribution of microbial biomass to an NLOM pool can be inferred from the stable carbon and nitrogen isotopic composition, given knowledge of the isotopic shift associated with heterotrophic metabolism.

Recent advances in liquid chromatographic systems (e.g., gel-permeation and reversed-phase chromatography) have been developed that allow on-line detection of DOM (Frimmel and Huber 1992). By combining size-exclusion chromatography with high-sensitivity DOM analysis and simultaneous UV and fluorescence detection, the molecular size and polarity distribution of aquatic NLOM can be fingerprinted.

Measurement of macromolecular charge is an example of a technique that provides intermediate-level information on organic matter structure, which may be useful in elucidating sources of NLOM. A common property of natural macromolecules is the existence of a net electrical charge. The presence of charge greatly affects the properties and behavior of these macromolecules. The charge may arise from the ionization of acidic functional groups, protonation of nitrogen groups, and/or coordination with hydrated metals. Thus, macromolecular charge can be positive or negative and can change depending upon the chemical environment. The measurement of net charge, particularly in combination with macromolecular size and polarity, may constitute a useful characterization tool for organic matter in both marine and terrestrial NLOM pools. With the increased sensitivity of newer DOM detection methods and the recent availability of commercial capillary-zone electrophoresis units capable of separation efficiencies of several hundred thousand theoretical plates, it may be possible to measure charge densities on small fractions of liquid chromatographic effluents from gel permeation columns.

^{13}C-NMR spectroscopy is another powerful tool providing intermediate-level detail on the structure of NLOM. A major advantage of NMR is that it allows for direct investigation of the sample without chemical or physical pretreatment. By conducting uptake experiments with ^{13}C-labeled substrates, ^{13}C-NMR spectroscopy allows us to monitor changes in the chemical structure of organic matter associated with incorporation into microbial biomass (e.g., Baldock et al. 1989).

Molecular-level biomarker analyses provide the most specific source information. Concurrent radiocarbon or stable isotope analysis may yield an even higher level of source information. Molecular-level analyses suffer from two major prerequisites: (a) chemical degradation precedes the analysis whereby the sample is severely altered, and (b) molecular-level analyses usually operate on only a quantitatively small subfraction of the total NLOM. Two areas in need of urgent development are (a) techniques to fractionate NLOM into analytically useful subfractions with minimum loss of sample integrity, and (b) a better understanding of the relationship between molecular-level observations and the bulk NLOM pool.

Analytical pyrolysis represents another powerful tool for obtaining molecular-level information regarding NLOM sources. In this method, high molecular weight organic matter is fractured into smaller subunits that are representative of the building blocks of the parent macromolecule. Unfortunately, the pyrolytic degradation is relatively nonspecific to distinct bonds (e.g., ester, ether, sulfur, and alkyl) and therefore no specific information can be obtained about the arrangement of the structural units in the macromolecular matrix.

Chemical degradation techniques combined with GC/MS have been applied to release structural units from macromolecular organic substances (for a review, see Rullkötter and Michaelis 1990). This approach is designed to degrade specific bonds or linkages in the macromolecular matrix and therefore provides a more precise picture of the abundance and position of different types of linkages in macromolecular substances.

In summary, most of the information on NLOM sources has come from a combination of isotopic, spectroscopic, and molecular-level biomarker analyses (Richnow et al. 1992; Kohnen et al. 1992). Elucidating the source of carbon to the seawater DOM pool provides an example of the power of combining multiple characterization techniques. The low or undetectable amounts of lignin oxidation products found in humic extracts in seawater indicate that terrestrial-derived carbon accounts for less than 10% of the total DOM pool (Meyers-Shulte and Hedges 1986; J. Ertel, unpublished data). This implies that most of the DOM originated as a result of production by marine organisms, rather than transport of organic carbon produced on land. In addition, specific organic compounds identified within the DOM pool are indicative of marine organism sources (Lee and Wakeham 1988), and humic extracts studied using ^{13}C-NMR are quite different from riverine-derived humic extracts (Hedges et al. 1992). However, the existing data do not preclude the possibility that a significant portion of the DOM enters the sea via diffusion from sediments or emission from hydrothermal vents/seeps, as these sources have not been adequately quantified.

HOW DO METHODS THAT CHARACTERIZE NLOM AFFECT THE RESULTS?

The most important step in any analysis involving NLOM is the sampling procedure. It is critical that the sampling strategy take into account the temporal and spatial variability inherent in natural systems. In addition, artifacts arising from the sampling procedure itself must be carefully controlled. This is particularly important if the data are to be extrapolated to provide global-scale estimates.

In addition to problems associated with sampling, many of the organic matter characterization methods described in this report require that the sample be subjected to some form of extraction, isolation, or concentration procedure as part of the analysis. It is important to keep in mind that the chemical structure of the sample could be severely altered by the analytical procedure. Analytical pyrolysis techniques are particularly prone to artifacts. Products can undergo primary or secondary thermal rearrangement during or following the pyrolytic degradation. Furthermore a heavy tar residue is typically formed during pyrolysis, which makes it difficult to evaluate quantitative aspects of this degradation technique.

One approach to evaluate whether the characterization procedure alters the structure is to monitor the chemical structure before and after sample treatment. This requires a method that functions over a large concentration range. Unfortunately, there are a limited number of characterization techniques that can be applied to both bulk and treated samples. The techniques that do exist (e.g., UV absorption, fluorescence, and elemental ratios) tend to provide relatively nonspecific structural information.

An alternative approach to quantifying the magnitude of artifacts caused by sample preparation is to use internal standards. Unfortunately, internal standards for most components of NLOM pools are not yet available. In some cases it is possible to produce natural internal standards, as in the case of ^{13}C-labeled algae. However, the problem that the matrix may affect the natural sample in a way that is different from the internal standard still remains.

Soil carbohydrate analyses provide a specific example of the effect that matrix can have on analytical results. There is evidence that acid hydrolysis methods and ^{13}C-NMR spectroscopy provide reasonable estimates of the

M.J. Alperin et al.

carbohydrate content of organic matter from the forest floor (Kögel et al. 1988). In mineral soil, however, the carbohydrate content of NLOM determined by acid hydrolysis and ^{13}C-NMR spectra differ widely (Oades et al. 1988; Guggenberger, pers. comm.). It appears that chemical or physical association with the soils mineral component increases the chemical stability of NLOM resulting in poor analytical yields.

AREAS OF FUTURE RESEARCH

1. Development of analytical methods which allow *in situ* quantification or characterization of unaltered NLOM.
2. Development of methods to characterize the size, shape, and charge of compounds that constitute the dissolved NLOM pools.
3. Measurements of isotopic composition (Δ^{14}C, δ^{13}C, δ^{15}N) of individual organic compounds that represent key indicators of sources and transformation reactions within NLOM pools.
4. Development of fractionation methods for NLOM pools that minimize contamination, degradation, or alteration of molecular structure.
5. Improvement of methods used to fractionate NLOM into subpools that are useful in models of global carbon cycling.
6. Development of methods to evaluate the effect of the sample matrix on wet-chemical and analytical pyrolysis techniques.
7. Improvements in our understanding of the relationship between molecular-level and bulk characterizations of NLOM.
8. Direct measurements of NLOM reactivity in the oceans in conjunction with measurements of chemical composition and structure.
9. Increased use of remote sensing techniques to obtain global estimates of the oceanic NLOM pools and processes.
10. Continued development of methods to investigate molecular-level changes and reactions in NLOM pools that have characteristic turnover times of less than one second.

REFERENCES

Amato, M., and J.N. Ladd. 1992. Decomposition of ^{14}C-labeled glucose and legume material in soils: Properties influencing the accumulation of organic residue C and microbial biomass C. *Soil Biol. Biochem.* 24:455–464.

Baldock, J.A., J.M. Oades, A.M. Vassallo, and M.A. Wilson. 1989. Incorporation of uniformly labelled ^{13}C-glucose into the organic fraction of the soil carbon balance and CP/MAS ^{13}C-N.M.R. measurements. *Austr. J. Soil Res.* 27:725–746.

Blough, N.V., and R.G. Zepp. 1995. Reactive oxygen species (ROS) in natural waters. In: Reactive Oxygen Species in Chemistry, ed. C.S. Foote and J.S. Valentine. New York: Chapman and Hall, in press.

Boden, T.A., P.A. Kanciruk, and M.P. Farrell. 1990. Trends 90: A Compendium of Data on Global Change. Oak Ridge, TN: Oak Ridge Natl. Laboratory.

Carder, K.L., S.K. Hawes, K.A. Baker, R.C. Smith, R.G. Steward, and B.G. Mitchell. 1991. Reflectance model for quantifying chlorophyll a in the presence of productivity degradation products. *J. Geophys. Res.* **96**:20,599–20,611.

Cho, B.C., and F. Azam. 1990. Biogeochemical significance of bacterial biomass in the oceans euphotic zone. *Mar. Ecol. Prog. Ser.* **63**:253–259.

Cooper, W.J. 1989. Sunlight-induced photochemistry of humic substances. In: Aquatic Humic Substances: Influence on Fate and Treatment of Pollutants, ed. I.H. Suffet and P. McCarthy, pp. 333–361. Washington, D.C.: Am. Chemical Soc.

Emanuel, R.F., G.G. Killough, W.M. Post, and H.H. Shugart. 1984. Modeling terrestrial ecosystems in the global carbon cycle with shift in carbon storage capacity by land use changes. *Ecology* **65**:970–983.

Emerson, S., and J.I. Hedges. 1988. Processes controlling the organic carbon content of open ocean sediments. *Paleoceanography* **3**:621–634.

Frimmel, F.H., H. Bauer, and J. Putzien. 1987. Laser flash photolysis of dissolved humic material and the sensitized production of singlet oxygen. *Environ. Sci. Technol.* **21**:541–545.

Frimmel, F.H., and R.F. Christman, eds. 1988. Humic Substances and Their Role in the Environment. Dahlem Workshop Report LS 41. Chichester: Wiley.

Frimmel, F.H., and S. Huber. 1992. A liquid chromatographic system with multidetection for the direct analysis of hydrophilic organic compounds in natural waters. *Fresenius J. Anal. Chem.* **342**:198–200.

Goudriaan, J., and P. Ketner. 1984. A simulation study for the global carbon cycle including man's impact on the biosphere. *Clim. Change* **6**:167–192.

Gregorich, E.G., R.P. Voroney, and R.G. Kachanoski. 1991. Turnover of carbon through the microbial biomass in soils with different textures. *Soil Biol. Biochem.* **23**: 799–805.

Hayes, M.H.B., P. MacCarthy, R.L. Malcolm, and R.S. Swift, eds. 1989. Humic Substances II: In Search of Structure. Chichester: Wiley.

Hedges, J.I. 1992. Global biogeochemical cycles: Progress and problems. *Mar. Chem.* **39**:67–93.

Hedges, J.I., B.A. Bergamaschi, and R. Benner. 1993. Comparative analyses of DOC and DON in natural waters. *Mar. Chem.* **41**:121–134.

Hedges, J.I., P.C. Hatcher, J.R. Ertel, and K.J. Meyers-Shulte. 1992. A comparison of dissolved humic substances from seawater with Amazon River counterparts by ^{13}C-NMR spectroscopy. *Geochim. Cosmochim. Acta* **5b**:1753–1757.

Hoge, F.E., A. Vodacek, and N.V. Blough. 1993. Inherent optical properties of the ocean: Retrieval of the absorption coefficient of chromophoric dissolved organic matter from fluorescence measurements. *Limnol. Oceanogr.* **38**:1394–1402.

Hoigné, J., B.C. Faust, W.R. Haag, F.E. Scully, Jr., and R.E. Zepp. 1989. Aquatic humic substances as sources and sinks of photochemically produced transient reactants. In: Aquatic Humic Substances: Influence on Fate and Treatment of Pollutants, ed. I.H. Suffet and P. McCarthy, pp. 363–381. Washington, D.C.: Am. Chemical Soc.

Keeling, R.F., and S.R. Shertz. 1992. Seasonal and interannual variations in atmospheric oxygen and implications for the global carbon cycle. *Nature* **358**:723–727.

Kögel, I., R. Hemplfing, W. Zech, P.G. Hatcher, and H.-R. Schulten. 1988. Chemical composition of the organic matter in forest soils: 1. Forest litter. *Soil Sci.* **146**:124–136.

Kohnen, M.E.L., S. Schouten, J.S. Sinninghe Damste, J.W. de Leeuw, D.A. Merrit, and J.M. Hayes. 1992. Recognition of palaeobiochemicals by a combined molecular sulfur and isotope geochemical approach. *Science* **256**:358–362.

Lee, C., and S.G. Wakeham. 1988. Organic matter in seawater: Biogeochemical processes. In: Chemical Oceanography, ed. J.P. Riley, vol. 9, pp. 1–51. London: Academic.

Meyers-Schulte, K.J., and J.I. Hedges. 1986. Molecular evidence for a terrestrial component of organic matter dissolved in ocean water. *Nature* **321**:61–63.

Miller, W.L., R.G. Zepp, and M.A. Tarr. 1993. Carbon monoxide dynamics in an African savanna: Potential for tropospheric gas exchange. *EOS Trans. AGU* **74**:136.

Oades, J.M. 1988. The retention of organic matter in soils. *Biogeochemistry* **5**:35–70.

Post, W.M., W.R. Emanuel, P.J. Zinke, and A.G. Tangenberger. 1982. Soil carbon pools and world life zones. *Nature* **298**:156–159.

Richnow, H.H., A. Jenisch, and W. Michaelis. 1992. Structural investigations of sulfur-rich macromolecular oil fractions and a kerogen by sequential chemical degradation. In: Advances in Organic Geochemistry 1991, ed. C. Eckard, J. Maxwell, and D.A.C. Manning, vol. 19, pp. 351–370. Oxford: Pergamon.

Rullkötter, J., and W. Michaelis. 1990. The structure of kerogen and related materials. A review of recent progress and future work. In: Advances in Organic Geochemistry 1989, ed. B. Durand and F. Behar, vol. 16, pp. 829–852. Oxford: Pergamon.

Romankevich, E.A. 1984. Geochemistry of Organic Matter in the Ocean. Heidelberg: Springer.

Sanhueza, E., L. Donoso, D. Scharffe, and P.J. Crutzen. 1994. Carbon monoxide fluxes from natural, managed, or savannah grasslands. *J. Geophys. Res.* **99**:16,421–16,434.

Schäfer, D. 1992. Einfluss der mineralischen Bodensubstanz auf den Abbau von Buchenstreu und den Aufbau organo-mineralischer Komplexe. Master's thesis, University of Bayreuth, Germany.

Schlesinger, W.H. 1977. Carbon balance in terrestrial detritus. *Ann. Rev. Ecol. Syst.* **8**:51–81.

Sharp, J.H., R. Benner, L. Bennett, C.A. Carlson, R. Dow, and S.E. Fitzwater. 1993. A reevaluation of high temperature combustion and chemical oxidation measurements of dissolved organic carbon in seawater. *Limnol. Oceanogr.* **38**:1774–1782.

Siegenthaler, U., and J.L. Sarmiento. 1993. Atmospheric carbon dioxide and the ocean. *Nature* **365**:119–125.

6

The Role of Nonliving Organic Matter in Soils

D.W. ANDERSON
Department of Soil Science, University of Saskatchewan, Saskatoon,
Saskatchewan, S7N 0W0 Canada

ABSTRACT

The nonliving organic matter (NLOM) or humic substances contained in soils is an important part of the Earth's carbon (C) cycle, estimated at 1,500 to perhaps 2,000 Pg, almost three times that present in the biota. The main role of NLOM is to link the biota with the mineral part of the Earth, adding stability to ecosystems. NLOM, produced by living organisms, in turn makes the soil a better environment for plants and other organisms. NLOM functions as a reserve store for nutrients, enhances the structural integrity of soils, improves the intake and storage of soil moisture, and is able to complex potentially biotoxic elements.

The amount of NLOM stored in a soil will be a balance of C produced by the biota, in equilibrium with decomposition as driven by the energy and nutrient requirements of the biota, and influenced by environmental factors, particularly temperature and moisture. The maximum power principle (MPP) as postulated by Lotka (1922), suggests that soil ecosystems will prevail that maximize stores and flows of useful energy or C. A corollary of this principle is that the NLOM in a particular soil will be as biologically resistant to microbial decomposition as required for persistence in that particular environment, but not so resistant to limit severely the nutrient supply from NLOM. Comparisons of otherwise similar soils along a moisture gradient indicate increased proportions of structurally complex humic acids in the more moist soils with higher organic C contents. Despite greater production and nutrient cycling in moist soils, nutrients such as N or S occur in more biologically resistant NLOM components in the moist as compared to drier soils.

Role of Nonliving Organic Matter in the Earth's Carbon Cycle
Edited by R.G. Zepp and Ch. Sonntag © 1995 John Wiley & Sons Ltd.

INTRODUCTION

In this chapter, I focus on the nonliving organic matter (NLOM) contained in soil and provide a general estimate of the size of the world's soil carbon (C) pool, factors influencing C storage and cycling in soils, and the role of NLOM in the functioning of soil ecosystems, with particular attention to nutrient cycling.

Soil scientists generally designate NLOM in soils as humus or humic substances, differentiating humus from the living components or microbial and microfaunal biomass, and from partly decomposed organic fragments that still retain the character of the plant or animal source. Ironically, as pointed out by Hillel (1991), the word *humus* in its initial context referred to "the stuff of life in the soil". This indicates an obvious connection between the dark material in soil and fertility that was obvious even to ancient people, whereas our modern convention is to use humus for NLOM and terms such as the *active fraction* or *microbial biomass* for the living, more dynamic components. Humic substances are mostly large, highly random biopolymers with complex aromatic and aliphatic structures. NLOM also includes simple, more soluble compounds (about 2% to 5% of the organic C in most soils), identifiable compounds such as carbohydrates, amino acids and lignin, and root or microbial exudates.

Humus, however, is the bridge that connects the biota, mainly vegetation, with the mineral soil that supports the biota. In a paper describing the soils on the glacially derived deposits of western Canada, Ellis (1938) wrote: "For the maintenance of life through the ages since glaciation nature evolved her own conservation policy. The living forms were maintained in equilibrium with their environment, and the soil, developed by life, continued as the supporter of life". Perry et al. (1989) discussed the important connections between plants and soils that sustain them both. Plants hold soil in place and provide the time for soil formation. Root and microbial exudates, such as siderophores and the organic acids produced in microbial decomposition, enhance the release of nutrients from minerals, leading to the formation of clay minerals from primary minerals such as feldspars or micas. The addition of organic matter makes soil a better medium for roots, enhances water entry into the soil and its storage there, contributes to the porosity that provides air to roots and the microflora, and imparts structural integrity. The buildup of NLOM enlarges pools of moderately available nutrients and provides negatively charged sites for the retention of nutrient cations against downward or leaching fluxes.

Humus, as a soil science term, includes the alkali-soluble but acid-insoluble humic acids, the alkali- and acid-soluble fulvic acids, and humin. Humin includes that wide variety of humic materials which are insoluble in alkali extracts or associated so intimately with mineral soil colloids that they cannot be removed with cold alkali extractant. Conventionally, studies of humic substances required their extraction and fractionation, processes that may alter the nature of the material isolated or able to remove only a small and, perhaps, nonrepresentative fraction of the humus. A modern approach, however, is to study organic matter without extracting it from the soil. Methods such as cross polarization, magic angle spinning nuclear magnetic resonance (CP/MAS ^{13}C-NMR) spectroscopy (Baldock et al. 1992) or pyrolosis-field ionization mass spectrometry (Schulten et al. 1992) can be used. The techniques permit the determination of the chemical structure of humic components *in situ*, in whole soils, or in the less dense or finer size fractions where most of the NLOM occurs.

THE SIZE OF THE SOIL CARBON POOL

Recent estimates of the global pool of sequestered C in soil have been mainly in the range of 1200 to 1500 Gt. The global estimates still require refinement. Eswaran et al. (1993), for example, estimate 390 Pg of C in Histosol or peat soils or about one-quarter of the world's C store of an estimated 1555 Pg. By comparison, Bolin and Fung (1992) estimate the global soil C pool at 1200 to 1500 Pg, with 150 Pg in peatlands.

Estimates of global pools have relied primarily on deriving reasonable estimates of soil organic C content for various taxonomic groups of soils, usually orders and suborders, from published data (Eswaran et al. 1993). Many more data are available for soil samples from the intensively managed lands of North America, Western Europe, and similar regions. The data are much fewer for certain critical areas, such as the high-latitude soils with large C stores (Townsend et al. 1992) and for desert, subtropical, and tropical regions. In many cases, soils data are limited by the absence of estimates of the bulk density of the soil horizons sampled, an important factor for calculating the mass of C on an areal basis from organic C concentration data. Studies to estimate organic C losses with cultivation have shown that the mass of C present in a unit depth of surface horizon may increase, despite decreases in C content, due to an increase in the density of the layer.

The well-known relationships between biomass production and decomposition rates of organic matter, as related to soil environment (particularly moisture and temperature), indicate that estimates of global C pools based on soil orders and suborder do have relevance. It should be pointed out, however, that these higher categorical levels are, to a considerable degree, mental constructs used in taxonomy, not necessarily real bodies of soil. Areas shown on small-scale, generalized maps as one soil order generally contain many other kinds of soils, on those parts of the land where regionally representative climate and vegetation characteristics are not present.

Areas shown on soil maps as dominantly one soil order or great group have a high degree of variability at the landscape level. For example, in the

Canadian prairie region with semiarid Brown or Boroll soils, C storage ranges from about 4 kg m^{-2} on dry uplands to three or four times that amount in moist depressional areas. Cultivation tends to increase landscape variability, as erosion and tillage redistribute soil and the C that it contains from convex uplands to concave, depositional areas. In many instances, agricultural research has concentrated on evaluating the impact of cultivation on humus contents of cultivated horizons from a perspective of soil fertility and soil tilth. Only recently has the emphasis shifted from those concerns to regional estimates of C stores within soil landscapes and regions, and evaluating the effects of land use. Soil scientists, however, can contribute substantively to improved estimates of stores, with the soil survey at various levels of detail used to extrapolate from measurement points and toposequences to more valid regional estimates (Anderson 1991). Soil survey information generally provides a hierarchical grouping of soils, with more general levels most useful where the map delineations are based not only on dominant kind of soil, but related physiographic properties such as parent material, climate, landscape vegetation, and land use. The data and structure provided by soil survey, when combined with land-use and vegetation data from remote sensing using a geographical information, will lead to better regional and global estimates of soil C stores. Soil C data, usually for typical or modal soils, can be extrapolated more reliably when related physiographic and land-use properties are documented as well.

CARBON OR ENERGY STORES IN SOIL

The amount of NLOM stored in a soil will reflect a balance between the amount of C produced by the biota (mainly plants), in equilibrium with decomposition as driven by the energy and nutrient requirements of the biota, and influenced by environmental factors, particularly temperature, moisture, and the clay content of the soil.

In a theoretical discussion of the energetics of evolution, Lotka (1922, p. 148) considered that "natural selection will so operate as to increase the total mass of the organic system, to increase the rate of circulation of matter through the system, and to increase the total energy flux through the system, so long as there is presented an un-utilized residue of matter and available energy". This idea came to be known as the maximum power principle (MPP) and has been discussed further by Odum (1983), who postulated that systems which maximize the flow of useful energy will prevail or survive. Veizer (1988) suggests that natural systems have a tendency to self-organize in order to maximize useful power, i.e., store more energy that can be fed back or depreciated to catalyze the inflow of additional energy. The MPP is, I believe, applicable to soil ecosystems and, more particularly, to organic matter cycling. It does not

imply that ever-increasing amounts of organic C will accumulate in soil ecosystems with time. In virtually all cases, C storage is limited by factors such as
nutrient availability and moisture supplies to the plant community; however,
the tendency will be for systems to develop and prevail that maximize stores of
energy (and nutrients) in the NLOM. This idea is consistent with Lotka's
original premise that natural selection tends to make the energy flux through
the system at a maximum, so far as compatible with the constraints to which
the system is subject. The latter point was stressed by Månsson and McGlade
(1993) in their critical evaluation of Odum's use of the MPP in ecology. Indeed,
the C balance in soils will reflect the storage of energy in organic matter
countered by the requirements for energy and nutrients of the biota of succeed-
ing generations. Soil C stores, therefore, will reflect a balance between organic
inputs mainly from plant residues, and decomposition driven by the energy and
nutrient demands of the microbial population, and plant requirements for
nutrients. In addition, it must be remembered that Lotka, Odum, and others
were referring to organic systems, not just the soil.

Many ecosystems, for example, the grasslands of semiarid to subhumid
regions, store much more C in the soil than in the vegetation. Carbon or
NLOM stored below ground, particularly when in complex association with
mineral soil, is a secure C pool and a major store for nutrients. Tropical forest
ecosystems, in comparison, have a much higher proportion of their energy and
nutrient capital in the phytomass and in rapidly decomposing organic matter
on the forest floor than in the soil. The stability of the tropical forest systems is
maintained by rapid and efficient cycling of nutrients and C; however, the
systems lack resilience because reserve stores in the soil are low. Odum
(1983) compared a near-climax tropical forest in Puerto Rico with an adjacent
exotic yield plantation. The energy captured by photosynthesis in the plantation goes into yield with little structure or diversity maintained. The wild forest
with many more species puts its energy into structure and diversity, thus
capturing more of the solar energy, recycling better, and generating higher
gross production over the long term.

The soils of tundra regions accumulate large stores of C, mostly in the
organic or peat layers on the soil surface, because decomposition rates in
cold and often wet environments are reduced to a greater degree than production. The buildup of large stores of partly decomposed humic materials has
been decried as the tying up of an inordinate proportion of nutrients in the
peaty materials, thereby limiting production. However, if one considers that
the decomposition rate for the material is constant and limited by temperature,
the larger the pool of nutrients in the NLOM the more nutrients that will be
released. Warmer temperatures can be expected to increase decomposition, but
at the same time increase nutrient supply and, in turn, production. The result
may well be organic cycles with increased circulation of nutrients and energy
(described by Lotka [1922] as a wheel that turns faster), not smaller stores of C

within the system as some have suggested. In an experiment on a black spruce (*Picea mariana*) forest in Alaska, Van Cleve et al. (1990) found that warming the forest floor 8° to 10°C above ambient temperature increased decomposition. The increased decomposition resulted in greater mineralization of N and P and elevated N, P, and K concentrations in spruce needles. This indicates the potential for increased production. Similarly, as discussed by Melillo et al. (this volume), soil warming in a mixed, deciduous forest in Massachusetts resulted in increased CO_2 emission (decomposition) and nearly doubled nitrogen mineralization. Melillo et al. postulated that soil warming may lead to increased C storage in the ecosystem.

A corollary of the idea of maximum storage of energy within a system is that the organic fraction of a particular soil will be tough (recalcitrant) enough to exist or endure in that particular edaphic environment, but not so tough that decomposition and the release of nutrients is limited severely. A detailed comparison of the amount and structural characteristics of the organic matter in a sequence of soils from the semiarid Brown (Aridic Boroll) to humid Black (Udic Boroll) soil zones across Saskatchewan supports the contention that organic matter will be as tough as required to maintain maximum storage in a soil (Anderson et al. 1974). This is a good region for such comparisons in that all factors except temperature and available moisture (soil parent material, fertility, time, etc.) can be kept constant. A gradient of temperature and moisture, as they interact to influence inputs to the NLOM and decomposition, can be studied. Organic matter in Brown soils has relatively more of the partially decomposed residues, simple and weakly humified humic materials with a high proportion of hydrolyzable or labile components, and more easily mineralized nutrients. This composition points to relatively weak decomposition pressure in the comparatively dry and cool environment. Black soils of more humid environments have higher humus contents, more strongly decomposed or humified materials, a strong presence of condensed alkyl-aromatic structures, and a larger proportion of nonhydrolyzable or chemically resistant components, with less easily mineralized nutrients. This is the kind of material that is expected where decomposition pressure is more intense. These differences are consistent with the idea that humus will be resistant to decomposition as required to exist at a particular equilibrium level in a soil, but not so resistant to decomposition that nutrients supply to the biota will be limited severely.

ROLE OF NLOM IN SOILS

The benefits or role of soil organic matter to crop productivity have been recognized for centuries and were summarized by Stevenson (1982):

1. It serves as a slow-release source of N, P, and S for plant nutrition and microbial growth.
2. It possesses considerable water-holding capacity and thereby helps to maintain the water regime of the soil.
3. It acts as a buffer against changes in pH of the soil.
4. Its dark color contributes to absorption of energy from the sun and heating of the soil.
5. It acts as a "cement" for holding clay and silt particles together, thus contributing to the crumb structure of the soil and to resistance against soil erosion.
6. It binds nutrient metal ions in the soil that otherwise might be leached out of surface soils.
7. Organic constituents in the humic substances may act as plant-growth stimulants.

This discussion will provide more depth on some of the points listed above as well as to consider some additional role of NLOM in soils and make reference to natural systems.

Certainly NLOM is an important, long-term store for nutrients such as N and S, which do not exist in mineral form in well-drained, oxidized soils. Phosphorus, which is derived almost wholly from the soil parent material and decreases gradually with time of soil formation (Smeck 1985), has an important, moderately available but reasonably secure storage in NLOM. Nutrients that are an integral, structural part of humus or that are associated by adsorption to more inert humus component or mineral colloids vary considerably in their availability. Drawing upon work from the Canadian prairie region and on published material (Jenkinson and Rayner 1977, for example), I have proposed three organic material nutrient pools of different lability (Anderson 1979). One, the recent products of microbial metabolism, root exudates, and the like, constitutes a nutrient- and energy-rich substrate important to nutrient supply in the short term. The second, described as a medium-term store, includes those potentially labile materials that are stabilized by adsorption to clay or to other humic materials of more complex structure. The medium-term pool, accounting for up to one-half of the total NLOM, provides a consistent and reliable supply of nutrients over the decades and contributes to the resilience of soil ecosystems. The third pool, postulated to be mainly structurally complex alkyl-aromatic structures with nutrient elements an integral part of the structures (Schulten et al. 1992), exists in the soil for hundreds to thousands of years. Nutrient supply from this pool is, therefore, much reduced proportionally to other organic material pools. The amount of nutrients mineralized can be significant where the strongly humified, resistant pool is large.

The basic ideas behind this proposition (Anderson 1979 and others) have been included in the CENTURY model for soil organic matter (Parton et al. 1987). This model has been used in a variety of agricultural, natural grassland, and forest scenarios to predict impacts of global warming on NLOM stores in soil.

The role of NLOM in nutrient supply varies with kind of soil, the edaphic environment, and nature of the ecosystem. Soils with a large store of NLOM within the soil in intimate association with clays are considered to be fertile soils. The grassland soils of North America, for example, were able to supply adequate supplies of nutrients for most crops upon first cultivation and to sustain supplies for several decades. Actually, supplies of N mineralized from NLOM were often in excess of crop needs, resulting in nitrate leaching beyond the root zone. The NLOM fractions reduced over the course of 85 years of cultivation were first the partially decomposed residues and microbial residues, then organic material associated with fine clay fractions (Tiessen and Stewart 1983). NLOM was reduced by 50%, but the NLOM remaining still provides most of the nutrients required by crops. In addition, on the young nutrient-rich parent materials that are characteristic of the region, supplies of nutrients such as K and P are replenished by weathering of primary minerals.

Tropical ecosystems, particularly those on old, strongly weathered parent materials where mineral supplies of nutrients have long since washed away, are strongly dependent upon rapid and efficient organic cycling of nutrients, with a proportionately large store in the forest vegetation and forest floor (Andriesse and Schelhaas 1987). Singh et al. (1989) have discussed the nutrient-conserving mechanisms that appear operative in nutrient-poor tropical soils. They postulate that the microbial biomass acts as a sink and source for nutrients, accumulating and conserving nutrients in a biologically active form during the dry period when plants have low nutrient demands. During the monsoon period, the microbial biomass becomes active and the nutrients are released rapidly to supply plant growth.

These reports indicate a reduced role for NLOM in nutrient and energy cycles of tropical ecosystems, at least in comparison to temperate soils. A comparison of semiarid grassland (Boroll) of western Canada, Oxisol soils of Venezuela and the Amazon region of Brazil (Tiessen et al. 1994) indicates the importance of NLOM to the fertility and resilience of the grassland soil, with reduced significance in the Oxisols. The more active pools (mainly biomass and roots) of the Oxisol soils are used up within a few years of clearing the land for farming, with about 75% of the organic material, the NLOM, remaining and of little or no potential for nutrient supply. Land may be abandoned and the forest allowed to grow again. Small cleared areas with less disturbance appear to recover their productivity under forest. Larger, more severely disturbed areas often do not recover as well, because nutrient cycles via the biota have been disrupted completely and reserves in the soil are minimal. Where land-use pressure is intense and the period of bush fallow becomes shortened, land is cleared and used again, with ever-decreasing levels of fertility (and potential for C storage) over time.

Returning to the list of Stevenson (1982), point 5, or the role of NLOM as a cement for holding clay and silt particles together to form aggregates, warrants in-depth consideration. Certainly, the formation of clay-humus complexes is an area where soil scientists must acquire more knowledge, but there can be little doubt of the importance of the forming of complexes. Jacks (1963) may have overstated the case with his statement, "The union of mineral and organic matter to form the organo-mineral complex is a synthesis as vital to the continuance of life, and less understood, than photosynthesis"; however, organo-mineral complexes are an important part of the Earth's C cycle. The importance to nutrient supply has been alluded to earlier, as has the role of organo-mineral complexes and aggregates to improving the rooting environment for plants. Soils with more clay generally contain more organic material, other factors being equal, than less clayey soils. Higher organic material contents may be partly a consequence of higher plant productivity due to more fertile soils, but the protection of organic material by complexing with clay is definitely a factor. The CENTURY model includes a clay-protection factor, which has resulted in simulations more similar to experimental findings (Parton et al. 1987). Biogeochemically, clay-humus complexing can be regarded as important, by retaining nutrients in moderately labile pools and protecting against the formation of the most strongly humified, highly refractory alkylaromatic structures of limited biological activity.

One important role for NLOM in soils concerns the complexing by humic substances of cations, especially metals and trace elements (Stevenson and Fitch 1986). The complexing of cations by organic ligands is important to the weathering of rocks and minerals, to enhancing the availability of metal ions to plants, and by reducing toxicity effects due to free cations. The complexing of free cations by soluble humic materials extends in significance from the soil to aquatic environments. Most research has focused on metals, e.g., copper, cadmium, lead, with particular attention to the application of sewage sludges to land. Humus complexing is considered a means of tying up of metals and reducing toxicity. It must be recognized, however, that complexing with NLOM may limit the availability of certain nutrient elements, particularly trace elements such as copper.

In this discussion, I would like to consider the role of humic substances in controlling soil solution levels of Al^{3+} cations in acid soils and adjacent waters. Podzolization is the soil-forming process by which Al released by the weathering of aluminosilicate minerals in the upper part of acid soils, is complexed by humic substances and thereby transported downward to accumulate in the subsoil or podzolic B horizon. Substantial amounts of humus complexed with Al (and Fe, and possibly other metals) accumulate in podzolic B horizons and may be sequestered there in an innocuous form for hundreds to thousands of years. Al cations moving to the aquatic environment or accumulating in strongly acid soils are biotoxic. Experience has indicated that the mineral acids contained in acid precipitation can remobilize Al^{3+}, moving it to streams and lakes with decidedly negative impacts on aquatic life. Podzolization, therefore, may be more than an interesting pedogenetic process. By sequestering a toxic or potentially toxic element such as Al, podzolization is a means of maintaining ecosystem productivity in the long term. Peterson (1989) found that 66 microequivalents of aluminum were tied up by each milliequivalent of organic acid contained in leachate from the organic layer of a maple forest. At the same time, NLOM complexed in B or Bh horizons of Podzol or Spodosol is removed from active cycling and represents a significant portion of the soil C pool. Estimates based on the evaluation of 244 Spodosol profiles or pedons in Florida indicated a total C store of 809 Tg, with 431 Tg in the Bh horizons of the Spodosol soils of that state (Stone et al. 1993).

CONCLUDING COMMENTS

Humic substances or NLOM in the soil are important to the functioning of soils as parts of ecosystems and are therefore important to the Earth's carbon cycle. As noted by MacCarthy et al. (1990), some of the key questions remaining relate to the structure of NLOM in relation to its function and persistence in soil. In this discussion I have mentioned that the biological resistance of NLOM varies with the intensity of decomposition in different soil environments and is related to factors such as the nature of initial organic residues, the complex chemical structure of humic substances, its association with clay or cations in stable complexes, and the inaccessibility of organic material in aggregates or within deeper soil horizons. At the Chapman Conference on the Gaia Hypothesis, MacCarthy and Rice proposed that the resistance of humic substances to microbial decay may be explained by the very complex or random structure of the materials (see MacCarthy et al. 1990). The disordered structure of humic structures in relation to more ordered biopolymers would not serve as a receptive substrate for enzymatic decomposition. A highly complex mixture of irregular molecules makes sense ecologically, because it results in material that contributes to essential properties of the soil, and thereby to plant growth and ecosystem stability, and is also persistent in the environment. Persistence or biological resistance is required so that the organic molecules that confer desirable properties have the opportunity to be effective, thereby moving ecosystems as far as possible towards maximum storage of useful energy or stores of NLOM.

REFERENCES

Anderson, D.W. 1979. Processes of humus formation and transformation in soils of the Canadian Great Plains. *J. Soil Sci.* **30**:77–84.

Anderson, D.W. 1991. Long-term ecological research: A pedological perspective. In: Long-Term Ecological Research, ed. Paul G. Risser, pp. 115–134. SCOPE 47.

Anderson, D.W., D.B. Russell, R.J. St. Arnaud, and E.A. Paul. 1974. A comparison of humic fractions of Chernozemic and Luvisolic soils by elemental analyses, UV and ESR spectroscopy. *Can. J. Soil Sci.* **54**:447–456.

Andriesse, J.P., and R.M. Schelhaas. 1987. A monitoring study of nutrient cycles in soils used for shifting cultivation under various climatic conditions in tropical Asia. II. Nutrient stores in biomass and soil. *Agr. Eco. Environ.* **19**:285–310.

Baldock, J.A., J.M. Oades, A.G. Waters, X. Peng, A.M. Vasallo, and M.A. Wilson. 1992. Aspects of the chemical structure of soil organic materials as revealed by solid-state ^{13}C-NMR spectroscopy. *Biogeochemistry* **16**:1–42.

Bolin, B., and I. Fung. 1992. The carbon cycle revisited. In: Modeling the Earth System, ed. D. Ojima, pp. 151–164. Boulder, CO: UCAR.

Ellis, J.H. 1938. The Soils of Manitoba. Winnipeg: Dept. of Agriculture and Immigration, Province of Manitoba.

Eswaran, H., E. Van Den Berg, and P.F. Reich. 1993. Organic carbon in soils of the world. *Soil Sci. Soc. Am. J.* **57**:192–195.

Hillel, D. 1991. Out of the Earth: Civilization and the Life of the Soil. New York: The Free Press (a division of MacMillan Inc.).

Jacks, G.V. 1963. The biological nature of soil productivity. *Soils Fert.* **26**:147–150.

Jenkinson, D.S., and J.H. Rayner. 1977. The turnover of soil organic matter in some of the classical Rothamsted experiments. *Soil Sci.* **123**:298–305.

Lotka, A.J. 1922. Contributions to the energies of evolution. *Proc. Natl. Acad. Sci.* **8**:151–154.

MacCarthy, P., P.R. Bloom, C.E. Clapp, and R.L. Malcolm. 1990. Humic substances in soil and crop sciences: An overview. In: Humic Substances in Soil and Crop Sciences, Selected Readings, ed. P. MacCarthy, C.E. Clapp, R.L. Malcolm, and P.R. Bloom, pp. 261–272. Madison, WI: Am. Soc. of Agronomy.

Månsson, B.Å, and J.M. McGlade. 1993. Ecology, thermodynamics and H.T. Odum's conjectures. *Oceologia* **93**:582–596.

Odum, H.T. 1983. Systems Ecology. New York: Wiley.

Parton, W.J., D.S. Schimel, C.V. Cole, and D.S. Ojima. 1987. Analysis of factors controlling soil organic matter levels in Great Plans Grasslands. *Soil Sci. Soc. Am. J.* **51**:1173–1179.

Perry, D.A., M.P. Amaranthus, J.G. Borchers, S.L. Borchers, and R.E. Brainerd. 1989. Bootstrapping in ecosystems. *BioScience* **39**:230–236.

Peterson, R.C., Jr. 1989. The contradictory biological behavior of humic substances in the aquatic environment. In: Humic Substances in the Aquatic and Terrestrial Environment, ed. B. Allard, H. Borén, and A. Grimvall, pp. 369–390. Berlin: Springer.

Schulten, H.-R., P. Leinweber, and G. Reuter. 1992. Initial formation of soil organic matter from grass residues in a long-term experiment. *Biol. Fert. Soils* **14**:237–245.

Singh, J.S., A.S. Raghubanshi, R.S. Singh, and S.C. Srivastava. 1989. Microbial biomass acts as a source of plant nutrients in dry tropical forest and savanna. *Nature* **338**:499–500.

Smeck, N.E. 1985. Phosphorus dynamics in soils and landscapes. *Geoderma* **36**:185–199.

Stevenson, F.J. 1982. Humus Chemistry: Genesis, Composition, Reactions. New York: Wiley.

Stevenson, F.J., and A. Fitch. 1986. Chemistry of complexation of metal ions with soil solution organics. In: Interaction of Soil Minerals with Natural Organics and Microbes, ed. P.M. Huang and M. Schnitzer, pp. 29–58. SSSA Spec. Publ. 17. Madison, WI: Am. Soc. of Agronomy.

Stone, E.L., W.G. Harris, R.B. Brown, and R.J. Kuehl. 1993. Carbon storage in Florida Spodosols. *Soil Sci. Soc. Am. J.* **57**:179–182.

Tiessen, H., E. Cuevas, and P. Chacon. 1994. The role of soil organic matter in sustaining soil fertility. *Nature* **371**:783–785.

Tiessen, H., and J.W.B. Stewart. 1983. Particle-size fractions and their use in studies of soil organic matter. II. Cultivation effects on organic matter composition in size fractions. *Soil Sci. Soc. Am. J.* **47**:509–514.

Townsend, A., S. Frolking, and E. Holland. 1992. Carbon cycling in high-latitude ecosystems. In: Modelling the Earth System, ed. D. Ojima, pp. 315–326. Boulder, CO: UCAR.

Van Cleave, K., W.C. Oechel, and J.L. Horn. 1990. Response of black spruce (*Picea mariana*) ecosystems to soil temperature modification in interior Alaska. *Can. J. For. Res.* **20**:1530–1535.

Veizer, J. 1988. The evolving exogenic cycle. In: Chemical Cycles in the Evolution of the Earth, ed. C.B. Gregor, R.M. Garrels, F.T. Mackenzie, and J.B. Maynard, pp. 175–220. New York: Wiley.

7

Late Quaternary Accumulation of Nonliving Organic Matter in Low-latitude Oceans: Implications for Climate Change

T.F. Pedersen and S.E. Calvert
Department of Oceanography, University of British Columbia,
Vancouver, B.C., Canada V6T 1Z4

ABSTRACT

Increased burial of sedimentary nonliving organic matter (NLOM) in low-latitude oceans during the last glacial represents enhanced primary production, which may have contributed to the atmospheric CO_2 drawdown. Higher productivity in such areas has been explained by strengthened trade-wind-driven upwelling, but this is unlikely to be the only causal mechanism because excess nutrients are often present in modern surface waters of these regions. Although enhanced upwelling probably played a role, increased delivery of iron to the equatorial regions during the arid, windy Last Glacial Maximum (LGM) may instead have been fundamental in generating the higher production indicated by the sedimentary record. While some continental margin settings show similarly increased organic carbon burial during the LGM, others, such as the continental slope off western North America, do not. Unlike the equatorial regions, it is not clear whether the net burial on margins was higher or lower during the LGM. Hence, the role of continental margin areas in climate change remains unresolved.

Recent applications of the relationship between $\delta^{13}C_{plankton}$ and the molecular CO_2 concentration of surface waters have shown that the eastern equatorial Pacific and the low-latitude Atlantic continued to act as CO_2 sources rather than sinks throughout the last glacial. In the Pacific case, this efflux is thought to represent the dominance of an upwelling-induced increase in degassing of CO_2 over the export of carbon (as particulate matter) attributable to the stimulation of primary production by the input to surface waters of iron-releasing dust. Future work on low-latitude sediments using the $\delta^{15}N$ of marine organic matter may allow discrimination between these two processes. In the

Role of Nonliving Organic Matter in the Earth's Carbon Cycle
Edited by R.G. Zepp and Ch. Sonntag © 1995 John Wiley & Sons Ltd.

absence of published $\delta^{15}N_{org}$ data for the equatorial regions, we predict that in the eastern equatorial Pacific, the $\delta^{15}N$ of sedimentary marine organic matter will decrease in sediments deposited during the LGM, reflecting an increased supply of nitrate relative to its rate of utilization.

The results from the equatorial Pacific and low-latitude Atlantic imply that enhanced sequestration of CO_2 by phytoplankton and increased export (relative to the total production) and burial of particulate organic matter (i.e., a stronger "biological pump") were not directly responsible for the atmospheric CO_2 drawdown observed in the ice core data. Instead, increased export and consequent degradation of organic matter in the deep sea may have contributed to the glacial atmospheric CO_2 deficit indirectly via the alkalinity feedback loop described by Boyle (1988), which operates only in the deep ocean. By implication, this diminishes the importance of continental margins as frontline players in the game of climate change.

INTRODUCTION

Carbon is one of a suite of elements that is particularly important in regulating the climate of the evolving Earth; nowhere is this more obvious than in the growing concern about the influence on the planetary thermostat of the CO_2 released from fossil-fuel emissions and deforestation. The spectacular results derived from the analysis of ice cores recovered by recent drilling on the Greenland and Antarctic ice caps have focused much attention on the intimate coupling between climate change and atmospheric pCO_2. As the global thermometer has risen in the recent geological past, so in near-lockstep has the CO_2 concentration (Lorius et al. 1990). Perhaps equally importantly, recent direct measurements of the $\delta^{18}O$ of air trapped in the ice have shown that the CO_2 and temperature increases precede the melting of continental ice sheets during deglaciations by as much as several thousand years (Sowers et al. 1991). The observed CO_2 variations must be shepherded by the ocean, which hosts about sixty times as much CO_2 as the atmosphere. But how have changes in the cycling of carbon in the sea caused the pCO_2 variations observed in the ice core data? One of the leading ideas is that of Boyle (1988), who suggested that a change in the vertical distributions of nutrients and CO_2 in the oceans could have mediated the atmospheric pCO_2 contrast between glacial and interglacial stages. According to this hypothesis, by shifting downward the depth at which the degradation of particulate organic matter (POM) occurs, more respired CO_2 (and nutrients) is added to the deep waters of the ocean. This fosters enhanced dissolution of $CaCO_3$ because the CO_2 added to deep waters reacts with CO_3^{2-} ions to form HCO_3^-, rendering the bottom waters more corrosive with respect to solid carbonate phases. In such a scenario, the result is a buildup of alkalinity in the ocean, because the riverine input of alkalinity (principally carbonate and bicarbonate ions) is no longer balanced by an equivalent output (i.e., deposition of calcium carbonate). After a geologically brief lag, a new ocean equilibrium is established in which the resulting higher mean carbonate alkalinity sequesters a larger proportion of the atmospheric CO_2 inventory via the equilibrium reaction:

$$CO_3^{2-} + CO_2 + H_2O = 2HCO_3^-. \tag{7.1}$$

Although Boyle was not able to specify with any certainty how such a transfer of nutrients and CO_2 to deeper waters could be accomplished, he provided benthic foraminiferal carbon isotope and Cd:Ca ratio data as evidence that such a nutrient reorganization had occurred. One of several possibilities suggested to account for the shift was that increased productivity during glacial periods increased the POM flux to deep waters, which would have augmented the nutrient and CO_2 inventories at depth. This suggestion is consistent with two principal aspects of the dynamics of POM cycling in the ocean: higher production is accompanied either by (a) inefficient grazing by zooplankton ("sloppy feeding," Eppley and Peterson 1979), which reduces recycling within the euphotic zone and increases the export of labile organic matter, relative to production, to the deep sea, or (b) rapid settling during blooms which leads to underexploitation by grazers ("mismatch theory," Sieracki et al. 1993). Direct observations of green phytodetritus on the sediment–water interface during or shortly after plankton blooms in the equatorial Pacific (McManus 1993) and elsewhere (Lampitt 1985) provide graphic evidence that high productivity delivers enhanced amounts of POM, apparently episodically, into the deep sea. These observations suggest that the normal tight coupling between phytoplankton, herbivores, and saprovores seen in the open ocean can frequently break down.

Berger and Wefer (1991) recently reviewed a wide range of scenarios, in addition to the Boyle (1988) model, that have been proposed to account for glacial CO_2 drawdown. Variable oceanic productivity plays a fundamental role in almost all such hypotheses, and it is this subject that will form the nucleus of this article. Our overall objective is to examine the possibility that changes in primary production fundamentally contributed to the observed pCO_2 fluctuations. We will draw on recent biological and chemical information from the modern ocean and the sedimentary record in an effort to weigh critically the postulated linkages between oceanic primary production and atmospheric CO_2, particularly during the Late Quaternary interval represented by the last full glacial–interglacial cycle.

INFLUENCES ON PRODUCTION IN THE MODERN OCEAN

Marine organic matter is derived overwhelmingly from photosynthesis in the surface mixed layer of the ocean where both nutrients and light are replete.

Because stratification severely limits delivery of nutrients to the surface waters in central gyres, production in these regions is low. In contrast, productivity is high in most coastal regions, where nutrients are furnished to the surface waters by upwelling or by rivers, and along much of the equator, where upwelling marks the divergence induced by trade-wind-driven Ekman transport in the mixed layer. Seasonal light limitation at high latitudes restricts production to the illuminated half of the year in both the fertile Southern Ocean and the North Atlantic and Pacific, where deep convective mixing supplies an overabundance of nutrients to the euphotic zone.

Despite their limited aggregate area, production in coastal and equatorial regions dominates both carbon fixation and the export flux of organic matter from the photic zone. Berger et al. (1989) estimate that 20% to 50% of the global export flux ("new" production) occurs in shelf seas, while Chavez and Barber (1987) calculated that as much as half of the new production in the world's oceans occurs in the equatorial region of the Pacific. Much of the remainder must be attributed to the extensive seasonal fixation of carbon in the nutrient-replete high latitudes.

The presence of rather high concentrations of unutilized NO_3^- and PO_4^{3-} in surface waters in the equatorial Pacific and in the high-latitude regions under light-replete conditions has posed a conundrum for biological oceanographers: what limits productivity in these regions? Although food-web structure, particularly the amount of grazing of primary producers by zooplankton, may be an important factor in limiting the extent of nutrient utilization in these areas, the old idea that the concentration of micronutrients, especially dissolved iron, might prove limiting has recently received much attention, thanks to the insightful but controversial suggestions by John Martin and colleagues (Martin and Fitzwater 1988; Martin and Gordon 1988; see also the review by Berger and Wefer 1991). The rationale is as follows: large phytoplankton are unable to assimilate the available nitrate because insufficient Fe is available in the mixed layer of the open ocean to support the synthesis by cells of the required complement of the Fe-bearing enzyme nitrate reductase. Under such circumstances, the lack of Fe availability selects a flora dominated by small cells that assimilate ammonium. This allows dominance of herbivory by micro-grazers, which efficiently recycle the nitrogen as NH_3 (Miller et al. 1991). Such recycling constrains the utilization of nitrate, which then persists in the mixed layer. Any injection of dissolved Fe into the mixed layer, however, for example via the deposition of Fe-bearing aerosols onto the sea surface and release of exchangeable Fe, could promote enhanced nitrate (and also phosphate) uptake by stimulating the growth of the nitrate-utilizing species, resulting in higher primary productivity. In turn, higher production may be associated with enhanced export as a result of faster settling of larger cells, streamlining of the trophic structure (fewer trophic transfers) and "sloppy feeding." Increased settling of POM during the spring bloom has recently been observed in the

North Atlantic, in response to changes in the planktonic community structure, and has been attributed to such phenomena (P. Boyd, pers. comm.). Empirical and experimental support for the ideas promoted by Martin's group have come both from observations that episodic dust deposition in the North Pacific promotes plankton blooms (e.g., Young et al. 1991) and from recent growth experiments in the equatorial Pacific, where the addition of Fe to surface seawater samples doubled the phytoplankton growth rate (Price et al. 1991). Although the biological jury is still out (see Banse 1991 and Sunda et al. 1991), there appears to be significant merit to the iron hypothesis and, as discussed later, this has important, but speculative, implications for productivity history and marine carbon cycling in the Quaternary.

THE ACCUMULATION AND PRESERVATION OF NONLIVING ORGANIC MATTER IN SEDIMENTS

Of the organic matter synthesized in and exported from the euphotic zone, only a small fraction is actually buried in the underlying sediments. This reflects the combined effects of oxidation during particle settling, utilization by benthic epifauna and infauna at or near the sediment surface, and continuing (largely microbial) degradation in the deposits (Pedersen and Calvert 1990). These processes have been recently reviewed by Middelburg et al. (1993), who point out that the fraction of export production that reaches the sediment surface in deep pelagic areas of the ocean is very small on average due to remineralization during the transit to the sediments in such areas. Middelburg et al. (1993) note in contrast that the bulk of organic carbon burial in marine sediments occurs in coastal areas, where settling organic material reaches the shallower seafloor much more quickly. Furthermore, the settling flux of organic material to the sediment–water interface in such regions, as a proportion of the export production, is enhanced by the less-efficient consumption by pelagic grazers that characterizes marginal upwelling regions. Consequently, the oligotrophic gyre regions play a small role in the carbon cycle in the oceans.

Degradation of organic matter that reaches the sediments proceeds via bacterially mediated oxidation reactions in which electrons are transferred from reduced carbon to a number of oxidants. These are used and sequentially depleted in the order O_2, NO_3^-, MnO_2 (s), $FeOOH$ (s), and SO_4^{2-}. Should the organic supply be sufficient to promote depletion of all of these electron acceptors, disproportionation to CO_2 and CH_4 (methanogenesis) becomes the dominant mechanism by which degradation proceeds. However, the high-settling flux of organic matter to coastal and hemipelagic sediments typically fosters depletion of dissolved oxygen at very shallow subbottom depths; consequently SO_4^{2-}, which is the most abundant oxidant in seawater, is the principal electron acceptor in most coastal sediment settings. Furthermore, recent information suggests that the amount of degradation by methanogenesis may have been underestimated in the past (Middelburg et al. 1993) and that reevaluation of the importance of this process to the total anaerobic mineralization budget is necessary. Despite these observations, it is widely believed that degradation proceeds more efficiently under oxygenated conditions than in anaerobic settings and that, as a result, variations in the organic carbon accumulation rate with depth in marine sediments could reflect changes in the concentration of oxygen in the bottom waters with time. We have not been able to find compelling evidence that this is so, except in infrequent and specific circumstances. Indeed a growing body of data (summarized in Calvert and Pedersen 1992) suggests that the presence or absence of oxygen in bottom waters makes little difference to the accumulation of organic matter in the underlying sediments. Instead, the settling flux of organic carbon, which is primarily a reflection of the amount of primary productivity, the structure of the planktonic community, and water depth provide a first-order control on the amount of organic matter that becomes buried in marine deposits. Thus, temporal increases in the burial flux of organic carbon fundamentally reflect increased production in the overlying surface waters. We will adopt this stance in our discussion of the sedimentary record of organic carbon burial.

THE SEDIMENTARY RECORD

Abundant evidence now demonstrates that the accumulation rate of nonliving organic matter (NLOM) in marine sediments was significantly higher in many areas of the ocean during the Last Glacial Maximum (LGM) (Figure 7.1). Using this information and the relationship between the production rate, the sedimentary carbon accumulation rate and the water depth at a large number of sites in the modern ocean, Sarnthein et al. (1988) calculated that new production increased globally by roughly 40% during the LGM. In their view, this increase resulted from higher meridional trade-wind speeds, which would have been induced by the steeper equator-to-pole temperature gradient during the last glacial period, and the corresponding enhancement of the Hadley Cell circulation. Higher wind stress at low and middle latitudes would increase upwelling rates and, in theory, strengthen the "biological pump" by enhancing the export flux of organic matter from the photic zone relative to the total primary production. However, there are two mechanistic problems with this model as it relates to CO_2 fluxes.

First, because nutrients are present in excess in surface waters of some key low-latitude areas in the modern ocean (e.g., the eastern equatorial Pacific), increasing the rate of nutrient delivery to the surface waters of such areas during the glacials would not necessarily enhance new production, unless

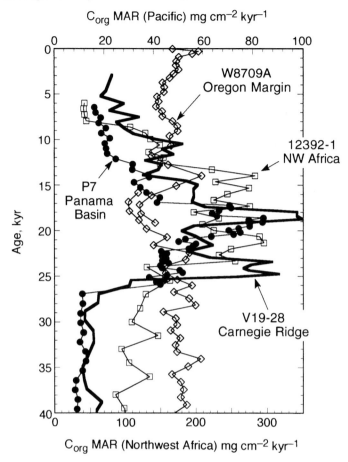

Figure 7.1 Sedimentary organic carbon mass accumulation rate (MAR) profiles from several areas of the oceans. Oregon Margin data are represented by a stacked profile (trigger and piston cores) from Lyle, Zahn et al. (1992), plotted on a time scale derived from ^{14}C and δ^{18}O data; Panama Basin data are from Pedersen et al. (1991) and are plotted on a ^{14}C time scale adjusted to conform with an absolute U/Th age; the NW African Margin profile is estimated from %C_{org}, age-depth and porosity information in Muller et al. (1983) and is plotted on a time scale derived from δ^{18}O data; and the Carnegie Ridge (eastern equatorial Pacific) data are from Lyle et al. (1988), plotted on a sidereal time scale. Note that the accumulation rate in the Northwest African Margin core is higher than the others and is plotted relative to the lower scale.

some factor other than upwelling controls production in these areas. Increased export of carbon relative to total production would be needed to strengthen the "biological pump" and draw down CO_2. Second, increased upwelling would also increase the CO_2 efflux from ocean to atmosphere because cold, newly upwelled and CO_2-rich water warms at the surface and degasses. It is for this reason that the productive equatorial Pacific today is actually a major supplier of CO_2 to the atmosphere rather than a sink. Did low-latitude upwelling belts become sinks for atmospheric CO_2 during the LGM? This important question can be addressed by using a recently refined stable isotopic tracer, which promises to permit calculation of the PCO_2 of ocean surface waters.

Photosynthesis via the common C_3 pathway discriminates against ^{13}C so that under fully CO_2-replete conditions, marine algae have a $\delta^{13}C_{org}$ value depleted by as much as 30‰ relative to the dissolved inorganic carbon substrate taken up by the cell, where $\delta^{13}C$ is defined as

$$1000 \times \{[(^{13}C{:}^{12}C)_{sample} / (^{13}C{:}^{12}C)_{standard}] - 1\}. \qquad (7.2)$$

However, as the dissolved molecular CO_2 concentration decreases (for example, in warm subtropical or tropical waters as compared to cold polar waters), the $\delta^{13}C_{org}$ increases (becomes heavier) in an approximately linear fashion. For example, phytoplankton in the cold waters of the Weddell Sea typically have $\delta^{13}C_{org}$ values of –28‰ (relative to the Pee Dee Belemnite standard) while equatorial plankton are nearer –19‰ (Rau 1994). The contrast appears to be related directly to the difference in concentration of molecular CO_2, which is roughly 23 μmol l^{-1} in the Weddell Sea and 10–12 μmol l^{-1} in the low-latitude regions. These observations have opened up the intriguing possibility that the CO_2 concentration in the surface mixed layer in the past can be estimated by measuring the $\delta^{13}C$ of marine organic matter in sediment cores. Rau (1994) recently supplied a rigorous discussion of this approach to PCO_2 hindcasting and confirmed that the method has considerable potential, providing that consideration is given to diagenetic effects (which appear to be minor in most cases but could alter the $\delta^{13}C_{org}$ of the residual organic matter after burial), contamination by terrestrial organic matter (which has an isotopically lighter mean $\delta^{13}C_{org}$ value), and past temperature changes (which influence the molecular CO_2 content).

Applications of this approach in low-latitude upwelling regions have been useful for assessing the direction of the ocean–atmosphere CO_2 flux during the Late Quaternary. For example, Pedersen et al. (1991) and Jasper et al. (1994) found that the efflux of CO_2 to the atmosphere in the equatorial Pacific has persisted throughout the last 50,000 years despite an apparent major increase in primary productivity and organic carbon export from the photic zone during the LGM. Indeed, the ocean–atmosphere pCO_2 difference (ΔpCO_2) was higher by about 80 ppmv during the LGM, suggesting that the efflux was even larger than today. Muller et al. (1994) similarly found that the productive waters of the Angola Current in the subequatorial Atlantic, where upwelling is well-developed, have continuously evaded CO_2 to the atmosphere during the last 160,000 years. However, in contrast to the equatorial Pacific results of

Pedersen et al. (1991), these authors noted that the ΔpCO_2 during glacial periods was diminished, suggesting that this area proved a weaker but still positive CO_2 source during the LGM. They attribute this reduction to the lower sea-surface temperatures and higher productivity evinced by paleotemperature estimates and observed increases in the organic carbon accumulation rate during glacial periods.

The equatorial Pacific case is particularly intriguing in light of potential feedbacks between the ocean and atmosphere. Earlier we noted that a substantial pool of unutilized nutrients characteristically exists in the eastern equatorial Pacific. Given that upwelling rates were indeed higher during the LGM (see Lyle, Prahl et al. 1992), which would have augmented the nutrient inventory, and given that the accumulation rate of sedimentary organic carbon was substantially higher (Figure 7.1), there would have been increased utilization of the nutrient pool during the glacial. Although direct evidence is lacking, Pedersen et al. (1991) suggested that increased delivery of dust to surface waters in this area could have accounted for the enhancement of production in an area that already has a surfeit of nutrients. The $\delta^{13}C_{org}$ data and the derived higher $\Delta pCO_{2(ocean-atm)}$ imply, however, that the negative influence of the increased export of carbon on the CO_2 efflux during the LGM must have been more than balanced by increased wind-driven vertical advection and the warming and degassing of the CO_2-rich upwelled waters. More simply stated, in the context of CO_2 fluxes, upwelling dominated productivity during the LGM just as it does today. Presumably, not enough aerosol iron reached the eastern equatorial Pacific during the LGM to turn it from a source to a CO_2 sink.

Can we assess whether the enhanced CO_2 efflux in the equatorial Pacific during the LGM was a result of dominance by increased upwelling or by a relative paucity of iron? Stable nitrogen isotopes promise to provide at least a provisional answer to this question. Francois and Altabet (1992) have recently shown that because plankton preferentially take up $^{14}NO_3^-$ during photosynthesis in nutrient-replete areas, settling particulate matter is initially isotopically lighter than the nutrient substrate. However, $\delta^{15}N_{plankton}$ increases with progressive nitrate depletion in surface waters because isotope discrimination decreases as surface nutrients are completely utilized. Thus the $\delta^{15}N$ of exported organic matter shows promise as a nutrient utilization index and, encouragingly, the preliminary results of Francois and Altabet (1992) suggest that sedimentary $\delta^{15}N_{org}$ records this signal reasonably faithfully. There are as yet no published down-core $\delta^{15}N_{org}$ data for the equatorial Pacific; however, based on the PCO_2 results of Pedersen et al. (1991) and in consideration of Redfield systematics (which couple nitrate and CO_2 concentrations in the upwelled waters), we forecast that the dominance of nutrient delivery by upwelling during the LGM over the utilization of the nutrient pool resulting from increased iron input will be marked by lower sedimentary $\delta^{15}N_{org}$ values in LGM sediments compared with the Holocene, particularly in the eastern equatorial Pacific where the modern excess nutrient inventory is highest.

IMPLICATIONS FOR THE ROLE OF UPWELLING

The limited data available to date quite strongly imply that increased production during the LGM did not draw down the atmospheric CO_2 content via a direct strengthening of the biological pump. That is, enhanced CO_2 sequestration in the photic zone and export and burial of particulate carbon did not apparently produce CO_2 deficits in the surface waters that then fostered an enhanced invasion of CO_2 into the ocean. However, there is little doubt that increased primary production was a widespread phenomenon during the LGM, and there is no doubt that export of particulate carbon increases disproportionately as productivity rises. Thus there must have been a substantially higher transfer of POM to the deep sea during the last glacial than there is today. The degradation of this material in the *deep* glacial ocean would have contributed to an indirect atmospheric CO_2 drawdown via the alkalinity feedback loop espoused by Boyle (1988).

Productivity variations on continental margins do not have the same effect for two reasons. First, the CO_2 produced by degrading the settled organic matter may be insufficient to cause undersaturation with respect to calcite of the normally supersaturated shallow- to intermediate-depth bottom waters. Thus, the alkalinity effect would not operate like it does where POM is injected into those regions of the deep sea where calcite is more prone to dissolution, should addition of metabolic CO_2 occur. The data of Boyle (1988) indeed show that intermediate-depth waters (which are the waters that bathe the upper continental margins) contained less respired CO_2 during the LGM than today. Second, and perhaps more importantly, intermediate-depth waters are in contact with only a circum-oceanic ribbon of sediments on the continental margins, unlike the deep sea where the deep waters overlie a continuous sedimentary carpet. Thus, respiration processes at the sediment–water interface are likely to have more of an effect on adjacent waters in the deep ocean rather than at intermediate depths (see Jahnke and Jackson 1992).

The implication of this view is that on continental margins it is only the direct burial of organic matter that can temporally influence the atmospheric CO_2 reservoir by removing carbon from the ocean–atmosphere cycle. However, the Late Quaternary record of organic carbon accumulation on continental margins is equivocal: increases are recorded in some areas during the LGM, such as off Northwest Africa, while decreases are seen elsewhere, such as off Oregon (Figure 7.1). At this stage it is not possible to estimate the mean effect for the last glacial, i.e., whether more or less organic matter was buried in margin sediments globally relative to today. In the absence of such an estimation, we suggest that the contribution of carbon cycling on the margins to the atmospheric CO_2 concentration during the Late Quaternary was probably minor in comparison to the role played by variations in production in the surface waters of the deep sea. This view emphasizes the fact that the proportion of organic carbon that is actually buried in marine sediments is a mere one-thousandth of that produced in the photic zone. Where the other nine hundred and ninety-nine thousandths are mineralized is critical, because the rheostat for the alkalinity feedback loop exists only in the deep ocean.

SUMMARY

Primary production in the oceans is concentrated primarily in coastal and equatorial regions and in the high latitudes of both hemispheres. High-standing stocks of nutrients in high-latitude regions and in the equatorial Pacific may reflect iron limitation on productivity.

Late Quaternary variations in the accumulation rate of organic matter have been profound and cyclical in many of these areas, reflecting climatically driven changes in production. The observed variations are thought to be independent of the bottom water oxygen content. Ice core data have shown that atmospheric pCO_2 during the LGM was reduced by about 30% ($\sim$80 ppmv) and it is tempting to ascribe this decline to the widespread increase in organic carbon burial that occurred in the equatorial Pacific and Atlantic and along the Northwest African Margin at the same time. This increase has been attributed to enhanced trade-wind-driven upwelling during the glacial. However, this suggestion is inadequate because excess nutrients are found in surface waters in some of these areas today. Increased delivery of iron to the equatorial regions during the arid, windy LGM, in addition to enhanced upwelling, may have been responsible for the higher production recorded by the sedimentary record of organic carbon burial. Higher burial rates may also reflect a greater export of organic carbon relative to total production, thereby having the potential to affect atmospheric CO_2 levels. However, not all margin settings show increased organic carbon burial during the LGM: compared to the Holocene, carbon accumulation was reduced during this period on the continental slope off western North America. Thus, unlike the equatorial regions, it is not clear whether the net burial on continental margins was higher or lower during the LGM.

It has been recently suggested that the stable carbon isotopic composition of marine organic matter can be used to hindcast the PCO_2 of oceanic surface waters, because the fractionation of carbon isotopes by phytoplankton appears to be dependent upon the molecular CO_2 concentration in surface waters. Using this relationship, it has been shown that the eastern equatorial Pacific and the low-latitude Atlantic continued to act as CO_2 sources rather than sinks

throughout the last glacial. In the Pacific case, this efflux is thought to represent the dominance of upwelling and consequent degassing over iron-induced
productivity increases.

A second tracer, the $\delta^{15}N$ of marine organic matter, promises to provide
proxy information on the NO_3^- concentration or the extent of NO_3^- utilization in the mixed layer in the past, because plankton discriminate against ^{15}N
during photosynthesis when nitrate is replete but more faithfully record the
heavier $\delta^{15}N$ of residual nitrate as nutrients are consumed. In theory, this
new tracer should allow discrimination between increased drawdown of
nitrate (if, for example, iron is added to the surface waters) and increased
upwelling (which will renew the stock of isotopically light NO_3^-). As yet
there are no published $\delta^{15}N_{org}$ data available for the equatorial regions,
but we predict that in the eastern equatorial Pacific, the $\delta^{15}N$ of sedimentary
marine organic matter will decrease in sediments deposited during the LGM.
This prediction is consistent with indications from the hindcast PCO_2 data
that CO_2 fluxes in the mixed layer were dominated by upwelling and not
export productivity.

The recent results from the equatorial Pacific and low-latitude Atlantic
imply that enhanced sequestration of CO_2 by phytoplankton and increased
export of POM (a strengthening of the "biological pump") were not directly
responsible for the atmospheric CO_2 drawdown observed in the ice core data.
Instead, we suggest that increased export productivity contributed (possibly
substantially) to the glacial atmospheric CO_2 deficit via the alkalinity feedback
loop described by Boyle (1988).

Finally, we urge the joint use of $\delta^{13}C_{org}$, $\delta^{15}N_{org}$, and marine organic carbon
accumulation-rate measurements in future studies of the sedimentary record.
This approach should prove highly useful in assessing the relationships
between past organic carbon export and nutrient and CO_2 dynamics in all
major areas of the surface ocean.

ACKNOWLEDGEMENTS

We are grateful to Philip Boyd for stimulating discussion, to John Farrell
for carefully commenting on the draft manuscript, to Larry Mayer for his
helpful comments, and to Mitchell Lyle for providing age and organic
carbon accumulation rate data for cores from the equatorial Pacific and
the Oregon Margin. This paper is respectfully dedicated to the memory of
the late Dr. John Martin, whose insight was and will continue to be of
benefit to us all.

REFERENCES

Banse, K. 1991. Iron availability, nitrate uptake, and exportable new production in the subarctic Pacific. *J. Geophys. Res.* **96**:741–748.

Berger, W.H., V.S. Smetacek, and G. Wefer. 1989. Ocean productivity and paleoproductivity: An overview. In: Productivity of the Ocean: Present and Past, ed. W.H. Berger, V.S. Smetacek, and G. Wefer, pp. 1–34. Dahlem Workshop Report LS 44. Chichester: Wiley.

Berger, W.H., and G. Wefer. 1991. Productivity of the glacial ocean: Discussion of the iron hypothesis. *Limnol. Oceanogr.* **36**:1899–1918.

Boyle, E.A. 1988. The role of vertical chemical fractionation in controlling Late Quaternary atmospheric carbon dioxide. *J. Geophys. Res.* **93**:15,701–15,714.

Calvert, S.E., and T.F. Pedersen. 1992. Organic carbon accumulation and preservation in marine sediments: How important is anoxia? In: Productivity, Accumulation and Preservation of Organic Matter in Recent and Ancient Sediments, ed. J.K. Whelan and J.W. Farrington, pp. 231–263. New York: Columbia Univ. Press.

Chavez, F.P., and R.T. Barber. 1987. An estimate of new production in the equatorial Pacific. *Deep-Sea Res.* **34**:1229–1243.

Eppley, R.W., and B.J. Peterson. 1979. Particulate organic matter flux and planktonic new production in the deep ocean. *Nature* **282**:677–680.

Francois, R., and M. Altabet. 1992. Glacial to interglacial changes in surface nitrate utilization in the Indian sector of the Southern Ocean as recorded by sediment $\delta^{15}N$. *Paleoceanography* **7**:589–603.

Jahnke, R.A., and G.A. Jackson. 1992. The spatial distribution of sea floor oxygen consumption in the Atlantic and Pacific oceans. In: Deep-Sea Food Chains and the Global Carbon Cycle, ed. G.T. Rowe and V. Pariente, pp. 295–307. Dordrecht: Kluwer.

Jasper, J.P., F.G. Prahl, A.C. Mix, and J.M. Hayes. 1994. Photosynthetic ^{13}C fractionation and estimated CO_2 levels in the central Equatorial Pacific over the last 255,000 years. *Paleoceanography* **9**, in press.

Lampitt, R.S. 1985. Evidence for the seasonal deposition of detritus to the deep-sea floor and its subsequent resuspension. *Deep-Sea Res.* **32**:885–897.

Lorius, C., J. Jouzel, D. Raynaud, J. Hansen, and H.L. Treut. 1990. The ice core record: Climate sensitivity and future greenhouse warming. *Nature* **347**:139–145.

Lyle, M., D.W. Murray, B.P. Finney, J. Dymond, J.M. Robbins, and K. Brookforce. 1988. The record of late Pleistocene biogenic sedimentation in the eastern tropical Pacific Ocean. *Paleoceanography* **3**:39–60.

Lyle, M.W., F.G. Prahl, and M.A. Sparrow. 1992. Upwelling and productivity changes inferred from a temperature record in the central equatorial Pacific. *Nature* **355**:812–815.

Lyle, M., R. Zahn, F. Prahl, J. Dymond, R. Collier, N. Pisias, and E. Suess. 1992. Paleoproductivity and carbon burial across the California Current: The Multitracers transect, 42N. *Paleoceanography* **7**:251–272.

Martin, J.H., and S.E. Fitzwater. 1988. Iron deficiency limits phytoplankton growth in the northeast Pacific subarctic. *Nature* **331**:341–343.

Martin, J.H., and R.M. Gordon. 1988. Northeast Pacific iron distributions in relation to phytoplankton productivity. *Deep-Sea Res.* **35**:177–196.

McManus, J. 1993. EqPac benthic Leg completed. *EOS* **74**:180–181.

Middelburg, J.J., T. Vlug, and F.J.W.A. van der Nat. 1993. Organic matter decomposition in the marine environment. *Glob. Planet. Change* **8**:47–58.

Miller, C.B., B.W. Frost, P.A. Wheeler, M.R. Landry, N. Welschmeyer, and T.M. Powell. 1991. Ecological dynamics in the subarctic Pacific, a possible iron-limited system. *Limnol. Oceanogr.* **36**:1600–1615.

Muller, P.J., H. Erlenkeuser, and R. von Grafenstein. 1983. Glacial–interglacial cycles in oceanic productivity inferred from organic carbon contents in eastern north Atlantic sediment cores. In: Coastal Upwelling, Part B, ed. J. Thiede and E. Suess, pp. 365–398. New York: Plenum.

Muller, P.J., R. Schneider, and G. Rutland. 1994. Late Quaternary pCO_2 variations in the Angola Current: Evidence from organic carbon and $\delta^{13}C$ alkenone temperatures. In: Carbon Cycling in the Glacial Ocean: Constraints on the Ocean's Role in Global Change, ed. R. Zahn, T.F. Pedersen, M. Kaminski, and L.D. Labeyrie, pp. 343–366. Berlin: Springer.

Pedersen, T.F., and S.E. Calvert. 1990. Anoxia vs. productivity: What controls the formation of organic-carbon-rich sediments and sedimentary rocks? *Am. Assn. Petrol. Geol. Bull.* **74**:454–466.

Pedersen, T.F., B. Nielsen, and M. Pickering. 1991. The timing of Late Quaternary productivity pulses in the Panama Basin and implications for atmospheric CO_2. *Paleoceanography* **6**:657–677.

Price, N.M., L.F. Andersen, and F.M.M. Morel. 1991. Iron and nitrogen nutrition of equatorial Pacific plankton. *Deep-Sea Res.* **38**:1361–1378.

Rau, G. 1994. Variations in sedimentary organic $\delta^{13}C$ as a proxy for past changes in ocean and atmospheric CO_2 concentrations. In: Carbon Cycling in the Glacial Ocean: Constraints on the Ocean's Role in Global Change, ed. R. Zahn, T.F. Pedersen, M. Kaminski, and L.D. Labeyrie, pp. 307–321. Berlin: Springer.

Sarnthein, M., K. Winn, J.-C. Duplessy, and M. Fontugne. 1988. Global variations of surface ocean productivity in low and mid latitudes: Influence on CO_2 reservoirs of the deep ocean and atmosphere during the last 21,000 years. *Paleoceanography* **3**:361–399.

Sieracki, M.E., P.G. Verity, and D.K. Stoecker. 1993. Plankton community response to sequential silicate and nitrate depletion during the 1989 North Atlantic spring bloom. *Deep-Sea Res.* **40**:213–225.

Sowers, T., M. Bender, D. Raynaud, Y.S. Korotkevich, and J. Orchardo. 1991. The $\delta^{18}O$ of atmospheric O_2 from air inclusions in the Vostok Ice Core: Timing of CO_2 and ice volume changes during the penultimate deglaciation. *Paleoceanography* **6**:679–696.

Sunda, W.G., D.G. Swift, and S.A. Huntsman. 1991. Low iron requirement for growth in oceanic phytoplankton. *Nature* **351**:55–57.

Young, R.W., et al. 1991. Atmospheric iron inputs and primary productivity: phytoplankton responses in the North Pacific. *Glob. Biogeochem. Cyc.* **5**:119–134.

8

Modeling the Air–Sea Fluxes of Gases Formed from the Decomposition of Dissolved Organic Matter: Carbonyl Sulfide and Carbon Monoxide

R.G. Najjar[1,2], D.J. Erickson III[2], and S. Madronich[2]

[1]Department of Meteorology, 503 Walker Building, Pennsylvania State University, University Park, Pennsylvania 16802–5013, U.S.A.
[2]National Center for Atmospheric Research, P.O. Box 3000, Boulder, Colorado 80307–3000, U.S.A.

ABSTRACT

An overview of field and laboratory studies of carbonyl sulfide (COS) and carbon monoxide (CO) in surface ocean waters is given with the aim of developing models of the air–sea fluxes of these gases. A simple one-dimensional model of COS in the surface ocean mixed layer is then constructed and used to explore the sensitivity of COS to various forcing parameters. The model predicts that surface COS concentrations are very sensitive to vertical mixing within the mixed layer and are relatively insensitive to air–sea gas exchange. Due to the similarities between COS and CO, we believe the same conclusion applies to CO. This contrasts a previous study, which maintained that air–sea gas transfer exerts a greater influence on surface CO concentrations than does vertical mixing. In the model used here, wind speed and mixed layer depth greatly affect the intensity of vertical mixing within the mixed layer and thus are important parameters in governing surface concentrations of COS and CO. The model also predicts that COS concentrations are extremely sensitive to temperature, due to the temperature dependence of COS hydrolysis. The possible effects of increased atmospheric CO_2 and decreased stratospheric ozone on surface ocean COS are also explored. In the absence of other changes, our mixed layer model predicts that increased atmospheric CO_2, through its effect on the basic hydrolysis of COS, will cause surface ocean COS to increase significantly. Using a model of atmospheric UV radiation, we estimate that the

Role of Nonliving Organic Matter in the Earth's Carbon Cycle
Edited by R.G. Zepp and Ch. Sonntag © 1995 John Wiley & Sons Ltd.

change in ozone from 1979 to 1992 has caused a significant increase in marine COS production. The percent increase is between 2% and 5% per decade in most regions poleward of 30° to as much as 15% per decade under the ozone hole.

INTRODUCTION

An often overlooked role of marine dissolved organic matter (DOM) in global biogeochemical cycles is its role as a major reservoir from which trace gases are formed and released to the atmosphere. Of concern here are two such gases: carbonyl sulfide (COS) and carbon monoxide (CO). COS is the most abundant sulfur gas in the atmosphere and its photolysis is believed to be the main source of the background stratospheric aerosol layer. This aerosol layer, particularly when perturbed by volcanic eruptions, significantly alters the radiative budget of the atmosphere and enhances the destruction of stratospheric ozone. The reaction of CO with OH in the troposphere is of utmost importance in atmospheric chemistry; it is the main sink of OH, which is the primary oxidant of many trace gases. Methane is one such gas, and there is some concern that increased levels of atmospheric CO may decrease OH and amplify the buildup of methane. Reaction with CO can also affect levels of tropospheric ozone, being either a sink or source of O_3, depending upon the levels of NO_x. In very low NO_x environments, such as the marine atmosphere, O_3 is consumed by reaction with CO.

The ocean is a source of COS and CO to the atmosphere. The most recent estimates place the oceanic source of COS[1] between 8 and 20 Gmol yr^{-1} (Andreae and Ferek 1992; Mihalopoulos et al. 1992; Erickson and Eaton 1993), making the oceans perhaps the single largest source of COS to the atmosphere. The total of the nonmarine sources is estimated to be between 11 and 114 Gmol yr^{-1}, with a best guess of 44 Gmol yr^{-1} (Khalil and Rasmussen 1984). Less emphasis has been given to the oceanic source of CO to the atmosphere, since it is relatively small compared to total CO sources. Seiler and Conrad (1987) estimate total CO sources[2] at 120±60 Tmol yr^{-1}, natural CO sources at 60±30 Tmol yr^{-1}, and the marine CO source at 3.6±3.2 Tmol yr^{-1}. Even though the latter is relatively small, modeling studies suggest it may have an appreciable effect on atmospheric chemistry in the marine boundary layer far from land (Erickson and Taylor 1992). This is because the lifetime of CO is only about two months globally and can be much shorter locally, depending upon the ambient OH concentration (Cicerone 1988).

Errors in the estimates of the global marine sources of COS and CO are one reason for developing models of the air–sea fluxes of these gases. Models can

---

[1] Gmol COS = 0.032 TgS (COS)
[2] Tmol CO = 28 Tg CO

be used to make theoretically sound extrapolations of existing data to the many undersampled regions of the ocean. A second reason for developing such models is for use in making estimates of past and future air–sea fluxes of COS and CO.

The purpose of this chapter is to demonstrate that plausible models of the air–sea fluxes of COS and CO can currently be made. After reviewing field and laboratory studies of COS and CO, we propose a model of COS in the oceanic mixed layer based on these studies and on theoretical considerations. We then use the model to evaluate the sensitivity of surface ocean COS to various forcing parameters. Finally, we discuss how the air–sea fluxes of COS and CO may change in the future as a result of increased atmospheric CO_2 and decreased stratospheric ozone.

THE BIOGEOCHEMISTRY OF COS AND CO IN SURFACE SEAWATER

Distributions

The most distinctive characteristic of surface ocean COS and CO distributions is their high degree of supersaturation. Surface concentrations of COS and CO are often an order of magnitude greater than the saturation concentration (the concentration in water at equilibrium with the atmosphere). Surface COS and CO distributions are also characterized by a dramatic diurnal cycle and are highly correlated with solar radiation, lagging by 2 to 3 hours (Figures 8.1 and 8.2). Vertical distributions of CO generally show a decrease with depth, even within the so-called mixed layer (Conrad et al. 1982). Unfortunately, there are no vertical profile data of COS outside those in Delaware Bay. These data show a decrease in COS with depth (Ferek and Andreae 1984), but no simultaneous data on the mixed layer depth were given. The diurnal and vertical distributions of COS and CO suggest that the sources of these gases are light dependent.

On time scales longer than a day, COS and CO are correlated with chlorophyll. Figure 8.3 shows that the surface CO concentration, when normalized to light intensity, is highly correlated with chlorophyll. Though similar normalizations have not been done for COS, there does appear to be a positive correlation between chlorophyll and surface COS (Johnson and Harrison 1986; Andreae and Ferek 1992). Coastal waters, which are highly productive, therefore tend to have much higher concentrations of CO and COS than the less-productive waters of the open ocean. In fact, Andreae and Ferek (1992) suggest that coastal waters, despite their relatively small area, dominate the marine source of COS to the atmosphere.

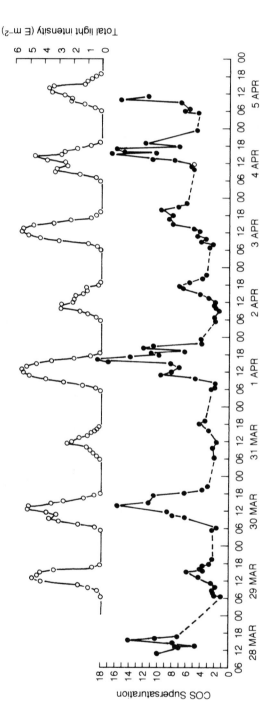

Figure 8.1 COS supersaturation ratio and light intensity measured in surface waters during a cruise in Chesapeake and Delaware Bays during the spring of 1983. The supersaturation ratio is defined as the measured water concentration divided by the water concentration at equilibrium with the atmosphere. Reproduced from Ferek and Andreae (1984).

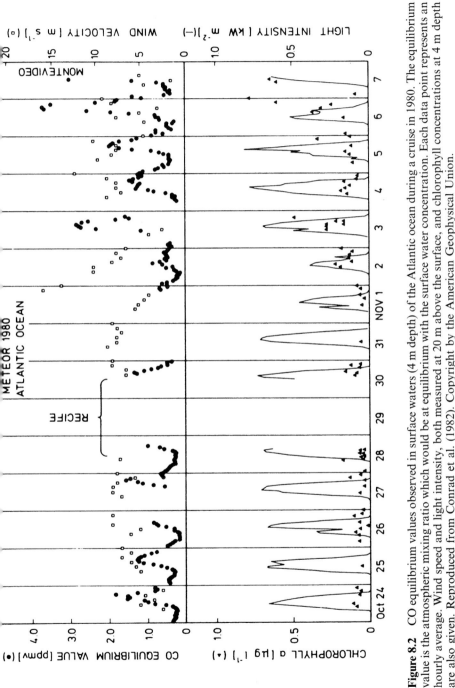

Figure 8.2 CO equilibrium values observed in surface waters (4 m depth) of the Atlantic ocean during a cruise in 1980. The equilibrium value is the atmospheric mixing ratio which would be at equilibrium with the surface water concentration. Each data point represents an hourly average. Wind speed and light intensity, both measured at 20 m above the surface, and chlorophyll concentrations at 4 m depth are also given. Reproduced from Conrad et al. (1982). Copyright by the American Geophysical Union.

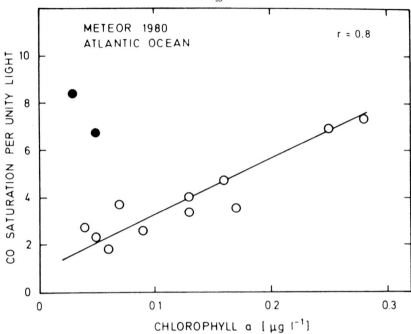

Figure 8.3 Daily mean CO saturation divided by the light intensity plotted as a function of the chlorophyll concentration. Data were collected during a cruise in the Atlantic Ocean in 1980. Solid circles represent periods of low wind speed (2 to 3 m s^{-1}) and open circles represent periods of high wind speed (5 to 9 m s^{-1}). The solid circles were disregarded for the straight line fit. Reproduced from Conrad et al. (1982). Copyright by the American Geophysical Union.

Finally, CO concentrations in surface waters have been found to be inversely correlated with wind speed (Figure 8.4). This has been interpreted as an effect of air–sea gas exchange (Conrad et al. 1982), but later we will show that a more likely explanation involves the role of vertical mixing.

Photochemical Production

Despite the correlation of COS and CO with chlorophyll, these gases are probably not produced directly by phytoplankton in significant quantities. Since irradiation of both filtered and unfiltered seawater produces significant amounts of CO and COS (Conrad et al. 1982; Ferek and Andreae 1984), it is likely that photolytic reactions involving dissolved substances result in the production of CO and COS, and that direct production by organisms is relatively small. Irradiation of dissolved organic compounds known to be found in seawater produces both CO and COS (Wilson et al. 1970; Ferek

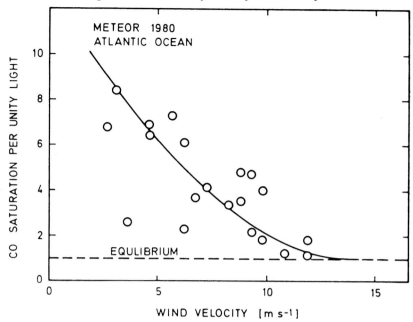

Figure 8.4 Daily average CO saturation normalized to a daily average light intensity of 0.1 kW m^{-2} as a function of the daily average wind speed observed at 20 m height. Data were collected during a cruise in the Atlantic Ocean in 1980. Reproduced from Conrad et al. (1982). Copyright by the American Geophysical Union.

and Andreae 1984). Thus, the correlation of COS and CO with chlorophyll probably indicates that photochemical reactions involving DOM produced directly or indirectly by phytoplankton result in COS and CO production.

Laboratory experiments (Figure 8.5) suggest that the UV portion of the solar spectrum is largely responsible for the photochemical production of COS and CO (Zepp and Andreae 1990; Valentine and Zepp 1993). Valentine and Zepp (1993) have used CO action spectra along with a model of surface UV radiation to show that CO production is at a maximum near 350 nm. Using a solar simulator, Zepp (pers. comm.) has shown that irradiation of water samples from coastal waters of the North Sea and Baltic Sea produces more than an order of magnitude more COS with UV-B radiation (280–320 nm) than without it. Using the COS action spectra in Figure 8.5 and the model of surface UV radiation of Madronich (1992), we have found that COS production peaks at around 315 nm. Presently there are no action spectra for CO production in marine waters. The terrestrial CO spectra in Figure 8.5 may be representative of coastal waters, which may contain significant amounts of terrestrial DOM. The action spectrum in the open ocean for CO, as well as for COS, however, may be quite different

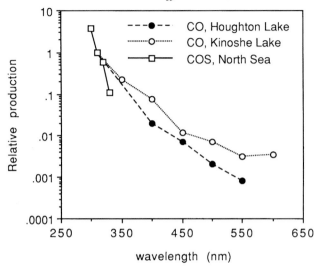

Figure 8.5   Action spectra for photochemical production of COS and CO. The CO spectra were computed from the quantum yields of Valentine and Zepp (1993) and corresponding absorption data from R. Zepp (pers. comm.). The COS spectrum is from North Sea coastal waters and was measured by Zepp and Andreae (1990). The spectra, initially having units of molecules photons-1 m^{-1}, were all normalized to unity at a wavelength of 310 nm.

from those in Figure 8.5, since DOM in the open ocean probably has a marine source.

The actual mechanism of CO and COS production is still under investigation. Zepp and Andreae (1990) suggest that COS is formed from the photosensitized oxidation of organosulfur compounds. This is an indirect photolysis mechanism, whereby a sensitizer molecule is put into an excited state by light absorption. The excited sensitizer molecule then reacts with oxygen to form an excited oxygen molecule, which in turn oxidizes a substrate to a final product (see Zepp 1982). Zepp and Andreae (1990) further suggest that humic substances are important photosensitizers in the production of COS. This would explain the high concentrations of COS in coastal waters, which contain high concentrations of humic substances. A similar mechanism may be responsible for CO formation.

Chemical and Biological Sinks

A strong diurnal cycle in COS and CO suggests not only a diurnal cycle in production, but also a strong sink. Since COS and CO are both highly supersaturated in surface waters, one might guess that gas exchange is a significant sink. This is most likely not the case, however. Under typical marine conditions, the equilibration time of a gas in the mixed layer with respect to exchange with the atmosphere is about one month. This long equilibration time cannot account for the rapid decrease of COS and CO in the late afternoon, which is indicative of a lifetime of several hours, not weeks.

The fact that there is a decrease in CO concentration with depth in the mixed layer also indicates that there is a significant sink of CO within the mixed layer. Assuming a mixed layer depth of 70 m and a diffusion coefficient of 200 cm^2 s^{-1}, the characteristic time scale of mixing in the mixed layer is about 3 days. Therefore, a vertical gradient of CO in the mixed layer can be maintained only if the lifetime of CO in the mixed layer is on the order of a few days or less.

The major sink of COS in seawater is probably neutral and basic hydrolysis:

$$COS + H_2O \rightarrow H_2S + CO_2 \tag{8.1}$$

$$COS + OH^- \rightarrow SH^- + CO_2. \tag{8.2}$$

Recently, Elliot et al. (1987) verified earlier measurements of the rates of the above reactions and found that the first-order rate constant for COS hydrolysis can be described by

$$k_h = k_1 + k_2[OH^-], \tag{8.3}$$

where k_1 and k_2 depend upon temperature only. One can compute [OH$^-$] from the pCO$_2$, total alkalinity, temperature, and salinity. The sensitivity of [OH$^-$] to total alkalinity and salinity is quite small, so that the COS hydrolysis rate constant can be determined primarily as a function of temperature and pCO$_2$. For a given pCO$_2$, the hydrolysis lifetime (k_h^{-1}) decreases with increasing temperature, mainly through the direct effect on k_1 and k_2, but also because [OH$^-$] increases with increasing temperature. For a given temperature, an increase in pCO$_2$ causes [OH$^-$] to decrease and hence the hydrolysis lifetime to increase. We used the equilibrium expressions for the carbonate system given in Peng et al. (1987) and the expression for the hydrolysis rate constant given by Elliot et al. (1987) to compute the variation of k_h with temperature and pCO$_2$, assuming average surface values of 35‰ for salinity and 2320 μeq kg^{-1} for total alkalinity. We found that the hydrolysis lifetime varies significantly over the present range of surface ocean temperature and pCO$_2$. At a pCO$_2$ of 350 μatm, close to the present-day surface ocean global mean, the hydrolysis lifetime varies from 1.07 hr at 30°C to 135 hr at 0°C. The sensitivity of the hydrolysis lifetime to pCO$_2$ is much smaller, but still significant; at 20°C, it increases from 3.55 hr at a pCO$_2$ of 200 μatm to 6.46 hr at a pCO$_2$ of 500 μatm. The range of the COS hydrolysis lifetime under current surface ocean conditions (from hours to days) is consistent with the rapid COS decrease during the late afternoon and evening.

   CO loss in surface seawater, in contrast to COS, appears to be the result of microbial consumption. Using different sized filters, Conrad et al. (1982) showed in a series of dark incubations that CO consumption is significant only when particles between 0.2 and 3 μm in diameter are present, suggesting that bacteria are most likely responsible for CO consumption. With respect to microbial consumption in these incubations, the half-life of CO was found to be about 10 hr, a time scale consistent with the diurnal cycle of CO and the vertical profile of CO in the mixed layer seen during the corresponding cruise. Microbial CO consumption rates can also be much slower. Jones (1991) found CO turnover times between 1 and 19 days in the oligotrophic North Atlantic. Consistent with this finding was a general lack of a diurnal cycle in CO in these waters.

A MODEL OF COS IN SURFACE SEAWATER

Model Formulation

The studies summarized above isolate many of the processes governing the distributions of COS and CO in surface waters, and our model of COS in surface waters uses parameterizations based on these studies. Due to the similarities in the cycles of COS and CO, many of our conclusions based on this model should be applicable to CO as well. The model we propose includes photochemical production, loss by hydrolysis, air–sea gas exchange and vertical mixing. COS is modeled as function of time and depth; horizontal motion is ignored because COS cycling occurs on time scales of a few days at most. The time rate of change of the COS concentration is given by

$$\frac{\partial[\text{COS}]}{\partial t} = \frac{\partial}{\partial z}\left(K\frac{\partial[\text{COS}]}{\partial z}\right) - k_h[\text{COS}] + p(z,t), \qquad (8.4)$$

where K is the eddy diffusion coefficient, $p(z,t)$ is the photochemical production rate of COS, and z is the vertical coordinate, positive downwards. The bottom boundary of the model is fixed at $z_b = 200$ m, where we assume that the vertical diffusive flux is zero:

$$\frac{\partial[\text{COS}]}{\partial z} = 0, \; z = z_b. \qquad (8.5)$$

At the surface, the diffusive flux must match the air–sea gas flux:

$$K \frac{\partial[\text{COS}]}{\partial z} = k_w([\text{COS}] - \alpha p_a), \ z = 0, \tag{8.6}$$

where k_w is the gas transfer velocity for COS, p_a is the partial pressure of COS in the atmosphere, and α is the solubility of COS.

Eddy Diffusion

Very little observational data exists on vertical diffusivity in the mixed layer. To estimate vertical mixing within the mixed layer, we use a parameterization developed by Large et al. (1994). According to their model, the vertical eddy diffusion coefficient is given by

$$K(z) = \kappa w_s z \left(1 - \frac{z}{h}\right)^2, \ 0 < z < h, \tag{8.7}$$

where κ is von Karman's constant (0.40), w_s is a velocity scale, and h is the depth of the planetary boundary layer in water. w_s depends upon the friction velocity in water, u_w^*, and the surface buoyancy flux. For simplicity, we assume that the surface buoyancy flux is equal to zero, in which case $w_s = u_w^*$. We also assume that h is the same as the mixed layer depth, z_{ml}, which is not allowed to vary throughout the day. The friction velocity in water is given by

$$u_w^* = u_a^* \sqrt{\frac{\rho_a}{\rho_w}} = u \sqrt{C_D \frac{\rho_a}{\rho_w}}, \tag{8.8}$$

where u_a^* is the friction velocity in air, ρ_a is the air density, ρ_w is the water density, u is the wind speed at 10 m, and C_D is the drag coefficient at 10 m. We use globally averaged surface values of ρ_a and ρ_w of 1.204 kg m^{-3} (computed at 20°C and 1 atm) and 1025 kg m^{-3}, respectively. For the drag coefficient, we use the formulation proposed by Trenberth et al. (1989) and ignore stability effects. Thus C_D is a function of wind speed only.

We can now write Equation (8.7) as

$$K = \kappa u \sqrt{C_D \frac{\rho_a}{\rho_w}} \, z \left(1 - \frac{z}{z_{ml}}\right)^2, \ 0 < z < z_{ml}. \tag{8.9}$$

K is zero at the surface and the base of the mixed layer and reaches a maximum value at $z = z_{ml}/3$. Since $\sqrt{C_D}$ is only weakly dependent upon wind speed, vertical mixing is nearly proportional to wind speed. At a wind speed of

$7 \, \mathrm{m \, s^{-1}}$ and a mixed layer depth of 70 m, the maximum diffusivity is $340 \, \mathrm{cm^2 \, s^{-1}}$ and the average diffusivity is $190 \, \mathrm{cm^2 \, s^{-1}}$. Instead of letting K go to zero at the base of the mixed layer, we do not allow it to be less than $1.7 \, \mathrm{cm^2 \, s^{-1}}$, a value based on tritium distributions throughout the main thermocline (Li et al. 1984). This value is also used from the base of the mixed layer to the bottom boundary of the model.

Gas Exchange

The gas transfer velocity is thought to depend primarily upon the wind speed and, through the molecular viscosity of water and the molecular diffusivity of the gas in water, the temperature. Based on the natural and bomb radiocarbon distributions in the atmosphere and ocean, the global mean gas transfer velocity for CO_2 is about $20 \, \mathrm{cm \, hr^{-1}}$ (Wanninkhof 1992). The diffusivity of COS in seawater has not been measured, thus the gas transfer velocity of COS is not known. Based on theoretical relationships between the diffusivities of different gases (Wilke and Chang 1955; Tyn and Calus 1975) and critical volume data (Reid et al. 1987), we calculate that the diffusivity of COS is 26% higher than that of CO_2. k_w is thought to be proportional to the square root of the diffusivity; based on theory, the gas transfer velocity of COS would thus be 12% higher than that of CO_2. Given this small difference, we assume that the global mean gas transfer velocity for COS is the same as that for CO_2.

We take p_a to be equal to its average value of 0.5 natm (Mihalopoulos et al. 1991) and use the formulation of COS solubility from Johnson and Harrison (1986), which depends only upon temperature.

Photochemical Production

For simplicity, we assume that photochemical production of COS is proportional to the downwelling irradiance at 315 nm. According to Baker and Smith (1982), irradiance decreases exponentially with depth in the water column. Thus we have

$$p(z, t) = p_o(t)e^{-k_{UV}z}, \qquad (8.10)$$

where $p_o(t)$ is the photochemical production rate just below the sea surface and k_{UV} is the diffuse attenuation coefficient at 315 nm. Using the bio-optical model of Baker and Smith (1982), we compute k_{UV}^{-1} to be 7 m at a chlorophyll concentration of $0.14 \, \mathrm{mg \, m^{-3}}$ (the global mean in surface ocean waters; Yoder et al. 1993). The attenuation of light will be greater in the presence of DOM and detrital material, so we choose k_{UV}^{-1} to be equal to 5 m for all of our simulations.

To parameterize how $p_o(t)$ varies throughout the day, we assume clear skies and dependence upon the solar zenith angle only. We use the results of a study by Behr (1992), who measured the UV radiation weighted for erythema (sunburn), integrated between 295 and 330 nm. This is not an unreasonable data set to use for COS photoproduction, since the COS and erythema action spectra both decrease dramatically with wavelength. Behr (1992) found that the ratio of this weighted UV radiation to the total solar flux over large regions of the Atlantic ocean (between 40°N and 40°S) was primarily a function of the solar zenith angle. A fit to his data reveals that the ratio of the weighted UV radiation to the total solar flux is very nearly proportional to $\cos^2\theta$ where θ is the solar zenith angle. This dependency, which is present in Behr's data regardless of variations in cloudiness and ozone column, reflects the fact that UV radiation is absorbed and scattered much more than longer wavelength radiation. The clear sky transmissivity of the atmosphere for the total solar flux is also a function of the zenith angle. Based on field data from the North Atlantic, Lumb (1964) determined the clear sky transmissivity of the atmosphere to be $0.61 + 0.2\cos\theta$. Lumb's formulation agrees very well with a more comprehensive model of the total clear sky radiation, which includes ozone, water vapor, and aerosols (Frouin et al. 1989). Given that the radiation at the top of the atmosphere is proportional to $\cos\theta$, we can parameterize the diurnal variation in the UV flux, and hence the diurnal variation in COS production, as a function of the zenith angle only:

$$p_o(t) = p_o[\theta(t)] = p_{max}(0.75 + 0.25\cos\theta)\cos^3\theta, \qquad (8.11)$$

where p_{max} is the COS production rate just below the surface when the sun is directly overhead. We have normalized Lumb's formula to be unity when the sun is directly overhead.

Model Results

Our time-dependent model was initialized at zero concentration and run for several days. The vertical resolution was set at 2 m and the timestep for the explicit numerical scheme at 30 seconds for most of the runs. In most cases the solution reached equilibrium in three days or less. The parameters of our standard run are shown in Table 8.1. Except for p_{max}, these parameters were chosen to be equal to their global mean values, when known. p_{max} was chosen to yield a daily mean surface COS concentration equal to the global mean value of 32 pmol l^{-1} (Andreae and Ferek 1992).

The model solution for the standard run during day three is shown in Figure 8.6 along with the photochemical production rate of COS just below the surface. A pronounced diurnal cycle is present in the solution at all depths throughout the mixed layer. Surface waters reach a maximum value of 112 pmol l^{-1}, which is a factor of 12 greater than the saturation concentration (9.4 pmol l^{-1}). The hydrolysis lifetime for the standard run is 5.0 hours and is

Table 8.1 Parameter values for standard run of the mixed layer model.

Parameter	Value	Effect
u, wind speed	$7.0\,\mathrm{m\,s^{-1}}$	vertical mixing
z_{ml}, mixed layer depth	$70\,\mathrm{m}$	vertical mixing
Diffusivity below mixed layer	$1.7\,\mathrm{cm^2\,s^{-1}}$	vertical mixing
k_w, gas transfer velocity	$20\,\mathrm{cm\,hr^{-1}}$	gas transfer
p_a, partial pressure of atmospheric COS	$0.5\,\mathrm{natm}$	gas transfer
temperature	$20°\mathrm{C}$	gas transfer (through solubility), hydrolysis
pCO$_2$, partial pressure of surface ocean CO$_2$	$350\,\mathrm{\mu atm}$	hydrolysis (through [OH$^-$])
salinity	$35\,\mathrm{permil}$	hydrolysis (through [OH$^-$])
alkalinity	$2320\,\mathrm{\mu eq\,kg^{-1}}$	hydrolysis (through [OH$^-$])
p_{max}, maximum surface COS production rate	$68\,\mathrm{pmol\,m^{-3}\,s^{-1}}$	photochemical production
k_{UV}, UV diffuse attenuation coefficient	$0.2\,\mathrm{m^{-1}}$	photochemical production
Julian day	80	photochemical production
Latitude	$30°\mathrm{N}$	photochemical production

sufficiently short so that COS is undersaturated at all depths during most of the night. The diurnal cycles at 21 m and deeper are due primarily to downward mixing of COS produced near the surface. Local production can be ruled out at these depths since the length scale of photochemical production is only 5 m. The maximum concentration at all depths lags the maximum in production, and the lag increases with depth.

Some features of the diurnal cycle differ somewhat from observations. For both COS and CO, surface concentration lags solar radiation by 2 to 3 hours (Andreae and Ferek 1992; Conrad et al. 1982), while we have found difficulty in increasing the lag beyond 1.5 hours within a reasonable parameter space. We suspect that this difficulty is due to our omission of any diurnal physics in the model. Since solar heating decreases the turbulent velocity scale, w_s, and makes the planetary boundary layer smaller, a distinct diurnal cycle in vertical mixing can occur (see, for example, Price et al. 1987). A component of the diurnal cycle in COS and CO, therefore, may be physical. If production is constant through-out a 24-hour period and is concentrated near the surface, but mixing is weaker during the day than at night, then surface COS and CO concentrations will be greatest during the day and smallest at night, consistent with observations. The time of maximum surface concentration may therefore be determined partly by physical processes. On the other hand, there is some evidence that the phase lag between solar forcing and surface COS and CO concentration is indeed photo-

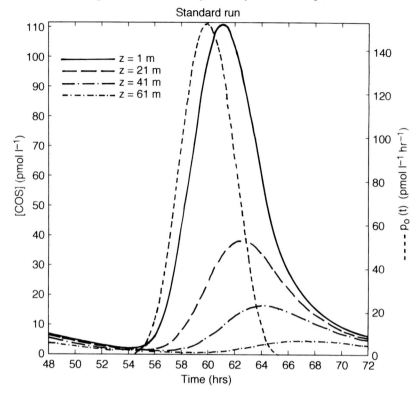

Figure 8.6 Solution of the one-dimensional model of COS in the mixed layer at day three for the standard run (see Table 8.1). The short dashed line is the photochemical production rate of COS just below the surface.

chemical. In laboratory studies Conrad et al. (1982) found that CO production continued even after light had been turned off.

The lag with depth seen in our COS simulation (Figure 8.7) also appears to be inconsistent with some observations. Conrad et al. (1982) noted an absence of a lag between CO concentrations at 4 m and 20 m and concluded that vertical mixing was unimportant. Part of this discrepancy may be due to different scale depths of production for COS and CO. The limited action spectra data discussed earlier suggest that CO will have a greater scale depth of production than COS, since light at 350 nm penetrates deeper than light at 315 nm. The difference is nearly a factor of two for pure water (Baker and Smith 1982). It is also possible that the lag with depth is not adequately resolved in the data presented by Conrad et al. (1982).

R.G. Najjar et al.

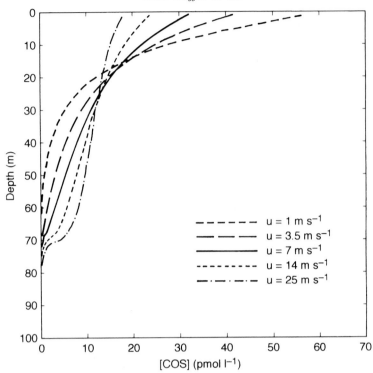

Figure 8.7 Daily mean profiles of COS concentration for the mixed layer model at different wind speeds. All other parameters are the same as for the standard run (Table 8.1).

The daily mean concentrations throughout the mixed layer are also affected by vertical mixing. An increase in wind speed, for example, decreases vertical COS gradients in the mixed layer, making surface concentrations lower and deep concentrations higher. The inverse relationship between surface CO and wind speed (see Figure 8.4) reported by Conrad et al. (1982) was interpreted on the basis of air–sea gas exchange rates. The gas transfer velocity increases with wind speed, so one would expect lower surface concentrations under higher winds. To investigate this, we included in our model the capability of having the gas transfer velocity vary with the square of the wind speed (Wanninkhof 1992). The constant of proportionality between k_w and u^2 was determined by setting k_w at $7 \, \mathrm{m \, s^{-1}}$ (the mean wind speed over the ocean) equal to $20 \, \mathrm{cm \, hr^{-1}}$ (our assumed global mean gas transfer velocity for COS). Figure 8.8 shows the daily mean surface COS concentration as a function of wind speed when (a) mixing is constant (computed at $7 \, \mathrm{m \, s^{-1}}$) and the gas transfer velocity varies

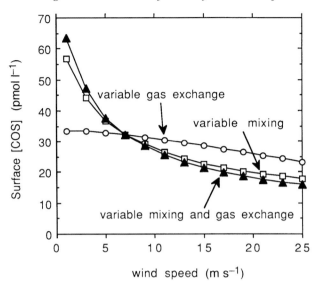

Figure 8.8   Daily mean surface COS concentration for the mixed layer model plotted as a function of wind speed. Solid triangles: mixing and gas exchange vary with wind speed; open squares: gas exchange is constant and mixing varies with wind speed; open circles: mixing is constant and gas exchange varies with wind speed. All other parameters are the same as for the standard run (Table 8.1). Each data point represents an individual model run.

with wind speed, (b) the gas transfer velocity is constant at $20\,\text{cm}\,\text{hr}^{-1}$ and vertical mixing varies with wind speed, and (c) both vertical mixing and the gas transfer velocity vary with wind speed. According to the model, the effect of wind speed on surface COS concentration occurs mainly through the effect on vertical mixing, not air–sea gas exchange. The curves in Figure 8.8 with variable mixing look very much like the relationship between surface CO and wind speed in Figure 8.4. Given the similarities between CO and COS, we suggest that vertical mixing plays a more important role than gas exchange in controlling surface CO concentrations.

Vertical mixing not only varies with wind speed but also with mixed layer depth (Equation 8.9). For reasonable variations in the mixed layer depth, the daily mean vertical profile of COS concentration varies greatly (Figure 8.9). Since mixed layer depth and wind speed are somewhat correlated over the annual cycle (both are higher in winter and lower in summer), we expect that the annual cycle in vertical mixing is large and significantly affects surface COS and CO concentrations. In the absence of other seasonal changes, vertical

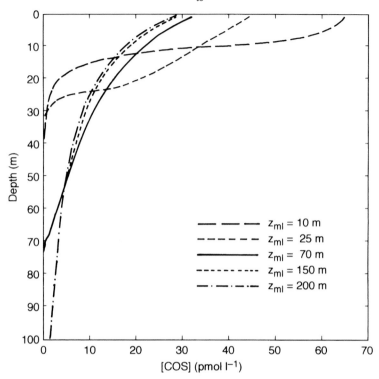

Figure 8.9 Daily mean profiles of COS concentration for the mixed layer model for different mixed layer depths. All other parameters are the same as for the standard run (Table 8.1).

mixing will tend to make surface COS and CO concentrations lower in the winter and higher in the summer.

Variations in temperature, through the hydrolysis rate constant, also greatly affect COS in the mixed layer (Figure 8.10). Varying the temperature in the standard run between $0°$ and $30°C$ causes the daily mean surface COS concentration to vary by a factor of 18. It is surprising that this effect has been little noted in observed data sets. One reason for this may be that solar insolation and temperature, which are correlated meridionally and over the annual time scale, have effects on COS that cancel each other. For example, we found that the surface COS concentration predicted by the model was only 8% lower at $60°$ latitude and $5°C$ than at $0°$ latitude and $27°C$ (with all other parameters having values corresponding to the standard run).

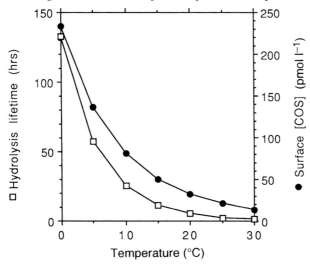

Figure 8.10 Daily mean surface COS concentration and COS hydrolysis lifetime for the mixed layer model plotted as a function of temperature. All other parameters are the same as for the standard run (Table 8.1). Each COS concentration data point represents an individual model run.

POSSIBLE FUTURE CHANGES IN THE AIR–SEA FLUXES OF COS AND CO

Effects of Increased CO_2

Because CO_2 plays the dominant role in the acid-base chemistry of the sea, increases in atmospheric CO_2 can directly affect surface ocean COS concentrations. As atmospheric pCO_2 increases, surface ocean pCO_2 will follow closely and surface ocean OH^- will decrease. This will result in a decrease in the hydrolysis rate constant for COS and an increase in surface ocean COS. A factor of four increase in atmospheric CO_2 from preindustrial times for the standard run results in a 70% decrease in surface ocean $[OH^-]$ and a 60% increase in surface ocean COS (Figure 8.11).

As far as we know, the effects of increased atmospheric CO_2 on marine chemistry have focused solely on calcium carbonate dissolution. This is surprising since the effects of large pH changes could be quite dramatic on the cycling of many elements in the sea. The equilibrium between ammonia and ammonium, for example, is pH dependent, and we expect ammonia and its flux to the atmosphere to decrease as pH decreases. Trace metal cycling is also pH dependent, and changes in pH may therefore affect the biological availability of trace metals and hence biological productivity in the ocean.

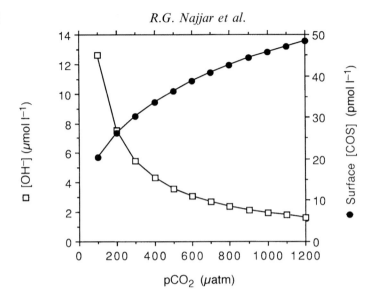

Figure 8.11   Daily mean surface COS and OH⁻ concentrations for the mixed layer model plotted as a function of atmospheric pCO_2. All other parameters are the same as for the standard run (Table 8.1). Each COS concentration data point represents an individual model run.

Increased atmospheric CO_2 may affect surface COS and CO concentrations by indirect means. For example, if surface waters warm as CO_2 increases, then COS hydrolysis will increase and surface COS concentrations will go down. On the other hand, surface warming may increase the stratification of surface waters and decrease mixed layer depths, resulting in decreased vertical mixing and higher surface COS and CO concentrations. Marine biota may also be indirectly affected by increased atmospheric CO_2 through changes in circulation and nutrient supply. Such changes may affect the penetration of UV light in surface waters, as well as the surface cycling of DOM, the substrate for photochemical production of COS and CO.

Effects of Decreased Stratospheric Ozone

Since COS and CO are produced by UV radiation at the surface, ozone depletion probably results in increased production and fluxes of these gases to the atmosphere. We have used a model of atmospheric UV radiation in combination with satellite ozone data to estimate the change in surface ocean COS production between January 1979 and December 1992. We take the COS production rate (molecules $COS \, m^{-3} \, s^{-1}$) to be

$$p(z) = \int_{\lambda_1}^{\lambda_2} E(z, \lambda) A(\lambda) d\lambda = \int_{\lambda_1}^{\lambda_2} E_o(\lambda) e^{-k_{UV}(\lambda)z} A(\lambda) d\lambda, \qquad (8.12)$$

where λ is wavelength (nm), $E(z, \lambda)$ is the downwelling spectral irradiance (photons $m^{-2} s^{-1} nm^{-1}$), $A(\lambda)$ is the COS action (molecules COS photons^{-1} m^{-1}), $E_o(\lambda)$ is the surface downwelling spectral irradiance (photons $m^{-2} s^{-1}$ nm^{-1}), and $k_{UV}(\lambda)$ is now the *spectral* diffuse attenuation coefficient (m^{-1}). The limits of integration (λ_1, λ_2) correspond to the spectral range of the model (280–400 nm). The vertically integrated COS production (molecules COS $m^{-2} s^{-1}$) is then

$$\bar{p} = \int_{\lambda_1}^{\lambda_2} \frac{E_o(\lambda) A(\lambda)}{k_{UV}(\lambda)} d\lambda. \qquad (8.13)$$

$E_o(\lambda)$ is computed from the model of Madronich (1992) as a function of the solar zenith angle and the ozone column. We created an action spectrum for the UV range of the model by making a least squares exponential fit to the data of Zepp and Andreae (1990) (see Figure 8.5). The diffuse attenuation coefficient was computed using the bio-optical model of Baker and Smith (1982), which parameterizes $k_{UV}(\lambda)$ as a function of chlorophyll concentration. We chose a constant value of chlorophyll equal to $0.15 \, mg \, m^{-3}$, which we justify below. No attempt was made to include light attenuation due to dissolved and detrital components. Monthly mean, vertically integrated COS production rates were computed by integration at 15-minute increments over one day during the middle of each the 168 months in 10° latitude bands. Trends in monthly mean COS production rates were computed by linear least squares over the data record, and the slopes are reported as percent per decade trends relative to 1979. In short, the calculation done here is similar to that done by Madronich (1992) to estimate increases in biologically harmful UV radiation at the Earth's surface; the main difference is that we have used the action spectrum for COS divided by the diffuse attenuation coefficient in place of action spectra for biological damage.

Figure 8.12 shows the calculated seasonal and latitudinal variation of the trends in surface ocean COS production during the years 1979–1992. Significant relative increases in COS production range from 2% to 5% per decade at mid-latitudes of both hemispheres to more than 15% under the ozone hole (where and when the zonal mean ice cover is less than 80%). Thus, the model predicts that COS production in the ocean has increased significantly. Of course, our model depends upon the global representativeness of the action spectrum for COS production that has been used. Madronich

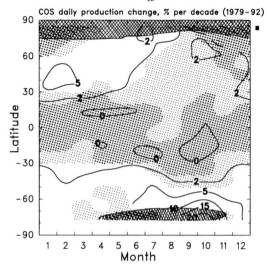

Figure 8.12   Trend in surface COS production rate computed using TOMSv6 ozone column data over the period 1979–1992. White areas indicate regions where the trend differs from zero by more than two standard deviations, light stippling by more than one standard deviation. Dark stippling denotes regions of insignificant trends (in the tropics) and regions of no light (winter poles). Cross-hatching indicates that the zonal average ice coverage is greater than 80%.

(1992) demonstrated the sensitivity of the relative increases in biological UV damage on the chosen action spectra; this should alert us to the importance of obtaining COS action spectra from various regions of the oceans during the various seasons. This also tells us that we cannot simply extend these results to CO, since the action spectra of CO differ significantly from the COS action spectrum used here (see Figure 8.5).

Because the amount of DOM available for COS production is not known, we have no way of computing absolute increases in COS production. In computing relative increases, it is important only that we know the spectral *shapes* of the COS action spectrum and the diffuse attenuation coefficient, not their absolute values. We have found that the spectral shape of the diffuse attenuation coefficient of Baker and Smith (1982) does not depend upon the chlorophyll concentration. Thus our choice of a particular value of chlorophyll in the calculation is not important.

There are a multitude of possible feedbacks between atmospheric chemistry and air–sea exchange of COS and CO. A positive feedback could be set in motion due to stratospheric ozone depletion (Erickson 1989). Decreased stratospheric ozone may lead to increased production of COS and CO in surface waters and hence increased fluxes of these gases to the atmosphere. Increased atmospheric COS could lead to an increase in the stratospheric aerosol layer and a decrease in stratospheric ozone. Increased CO in the marine boundary layer can also lead to tropospheric ozone loss. These ozone decreases in the stratosphere and troposphere will lead to increased surface UV fluxes and perhaps greater COS and CO production. A negative feedback is also possible, since increased UV could decrease biological productivity and hence decrease the source of DOM.

SUGGESTIONS FOR FURTHER RESEARCH

We believe that sufficient information is now available to make plausible global models of the air–sea fluxes of COS and CO. Field surveys and process-oriented studies have isolated many of the important mechanisms responsible for controlling the distributions of COS and CO in the surface ocean, and have related surface ocean COS and CO to variables that are in many cases available with global coverage and measurable from space. These variables include sea-surface temperature, wind speed, mixed layer depth, surface radiation, and chlorophyll concentration. Erickson (1989) and Erickson and Eaton (1993) have taken the first steps toward creating global models of the air–sea fluxes of CO and COS by considering the dependence of surface concentrations on light and chlorophyll concentration. We now propose that wind speed and mixed layer depth (through the effect on vertical mixing) and temperature (for COS hydrolysis) be incorporated into these models.

Many improvements can be made to our one-dimensional model as well. The diurnal cycle in mixing probably plays an important role in determining COS and CO concentrations and should be incorporated into our model. We also need to model the UV underwater light field with spectral resolution and to consider the scalar irradiance as opposed to the downwelling irradiance. All of these improvements are within reach, and we are in the process of incorporating these into global models of surface ocean COS and CO.

One of the major uncertainties in modeling COS and CO in surface waters is the surface UV flux. Under clear, aerosol-free skies, sufficiently accurate models have been developed; however, the presence of clouds and aerosols introduces great difficulty and uncertainty. Another uncertainty relates to the global representativeness of the action spectra published for CO and COS. No marine spectra are available for CO, and no open ocean action spectra are available for COS. Given the likely sensitivity of production to the action spectrum, good spatial and temporal coverage of action spectra measurements is needed.

Finally, there is an urgent need for vertical profiles of COS concentrations and a study of COS comparable in character and depth to the study by Conrad et al. (1982) for CO in the Atlantic Ocean. By gathering enough data on CO concentration with concurrent measurements of wind speed, light, and chlorophyll concentrations, and by conducting experiments on board ship, these investigators were able to quantify successfully many of the important processes governing the CO distribution in surface waters. Such a study for COS will be indispensable for evaluating models of COS in surface waters, such as ours, and suggesting areas for improvement. Specifically, an appropriately designed field study can evaluate our hypothesis that, beyond light and chlorophyll concentration, the important parameters determining surface ocean COS are wind speed, mixed layer depth, and temperature.

ACKNOWLEDGEMENTS

We are indebted to R. Zepp for providing us with unpublished data and for his encouragement and valuable discussions as well as to the reviewer, who made many helpful comments on an earlier version of this manuscript. Finally, we thank Tom Schnell for his programming assistance in developing the time-dependent COS model.

REFERENCES

Andreae, M.O., and R.J. Ferek. 1992. Photochemical production of carbonyl sulfide in seawater and its emission to the atmosphere. *Glob. Biogeochem. Cyc.* **6**:175–183.

Baker, K.S., and R.C. Smith. 1982. Bio-optical classification and model of natural waters. 2. *Limnol. Oceanogr.* **27**:500–509.

Behr, H.D. 1992. Net total and UV-B radiation at the sea surface. *J. Atmos. Chem.* **15**:299–314.

Cicerone, R. 1988. How has the atmospheric concentration of CO changed? In: The Changing Atmosphere, ed. F.S. Rowland and I.S.A. Isaksen, pp. 49–61. Dahlem Workshop Report PC 7. Chichester: Wiley.

Conrad, R., W. Seiler, G. Bunse, and H. Giehl. 1982. Carbon monoxide in seawater (Atlantic Ocean). *J. Geophys. Res.* **87**:8839–8852.

Elliot, S., E. Lu, and F.S. Rowland. 1987. Carbonyl sulfide hydrolysis as a source of hydrogen sulfide in open ocean seawater. *Geophys. Res. Lett.* **14**:131–134.

Erickson, D.J. III, 1989. Ocean to atmosphere carbon monoxide flux: Global inventory and climate implications. *Glob. Biogeochem. Cyc.* **3**:305–314.

Erickson, D.J., III, and B.E. Eaton. 1993. Global biogeochemical cycling estimates with CZCS data and general circulation models. *Geophys. Res. Lett.* **20**:683–686.

Erickson, D.J., III, and J.A. Taylor. 1992. 3-D tropospheric CO modeling: The possible influence of the ocean. *Geophys. Res. Lett.* **19**:1955–1958.

Ferek, R.J., and M.O. Andreae. 1984. Photochemical production of carbonyl sulphide in marine surface waters. *Nature* **307**:148–150.

Frouin, R., D.W. Lingner, C. Gautier, K.S. Baker, and R.C. Smith. 1989. A simple analytical formula to compute clear sky total and photosynthetically available solar irradiance at the ocean surface. *J. Geophys. Res.* **94**:9731–9742.

Johnson, J.E., and H. Harrison. 1986. Carbonyl sulfide concentrations in the surface waters and above the Pacific Ocean. *J. Geophys. Res.* **91**:7883–7888.

Jones, R.D. 1991. Carbon monoxide and methane distribution and consumption in the photic zone of the Sargasso Sea. *Deep-Sea Res.* **38**:625–635.

Khalil, M.A.K., and R.A. Rasmussen. 1984. Global sources, lifetimes and mass balances of carbonyl sulfide (COS) and carbon disulfide (CS_2) in the Earth's atmosphere. *Atmos. Environ.* **18**:1805–1813.

Large, W.G., J.C. McWilliams, and S.C. Doney. 1994. Oceanic vertical mixing: A review and a model with a non-local boundary layer parameterization. *Rev. Geophys.*, in press.

Li, Y.-H., T.-H. Peng, W.S. Broecker, and H.G. Ostlund. 1984. The average vertical mixing coefficient for the oceanic thermocline. *Tellus* **36B**:212–217.

Lumb, F.E. 1964. The influence of cloud on hourly amounts of total solar radiation at the sea surface. *Q. J. R. Meteorol. Soc.* **90**:43–56.

Madronich, S. 1992. Implications of recent total atmospheric ozone measurements for biologically active ultraviolet radiation reaching the earth's surface. *Geophys. Res. Lett.* **19**:37–40.

Mihalopoulos, N., B.C. Nguyen, J.P. Putaud, and S. Belviso. 1992. The oceanic source of carbonyl sulfide (COS). *Atmos. Environ.* **26A**:1382–1394.

Mihalopoulos, N., J.P. Putaud, B.C. Nguyen, and S. Belviso. 1991. Annual variation of atmospheric carbonyl sulfide in the marine atmosphere in the Southern Indian Ocean. *J. Atmos. Chem.* **13**:73–82.

Peng, T.-H., T. Takahashi, W.S. Broecker, and J. Olafsson. 1987. Seasonal variability of carbon dioxide, nutrients and oxygen in the northern North Atlantic surface water: Observations and a model. *Tellus* **39B**:439–458.

Price, J.F., R.A. Weller, C.M. Bowers, and M.G. Briscoe. 1987. Diurnal response of sea surface temperature observed at Long-Term Upper Ocean Study (34°N, 70°W) in the Sargasso Sea. *J. Geophys. Res.* **92**:14,480–14,490.

Reid, R.C., J.M. Prausnitz, and T.K. Sherwood. 1987. The Properties of Gases and Liquids, 3rd ed. New York: McGraw Hill.

Seiler, W., and R. Conrad. 1987. Contribution of tropical ecosystems to the global budget of trace gases, especially CH_4, H_2, CO and N_2O. In: The Geophysiology of Amazonia, ed. R.E. Dickinson, pp. 133–160. New York: Wiley.

Trenberth, K.E., W.G. Large, and J.G. Olson. 1989. The effective drag coefficient for evaluating wind stress over the ocean. *J. Climate* **2**:1507–1516.

Tyn, M.T., and W.F. Calus. 1975. Estimating liquid molal volume. *Processing* **21(4)**:16–17.

Valentine, R.L., and R.G. Zepp. 1993. Formation of carbon monoxide from the photodegradation of terrestrial dissolved organic carbon in natural waters. *Environ. Sci. Technol.* **27**:409–412.

Wanninkhof, R. 1992. Relationship between wind speed and gas exchange over the ocean. *J. Geophys. Res.* **97**:7373–7382.

Wilke, C.R., and P. Chang. 1955. Correlation of diffusion coefficients in dilute solutions. *AIChE J.* **1**:264–270.

Wilson, D.F., J.W. Swinnerton, and R.A. Lamontagne. 1970. Production of carbon monoxide and gaseous hydrocarbons in seawater: Relation to dissolved organic carbon. *Science* **168**:1577–1579.

Yoder, J.A., C.R. McClain, G.C. Feldman, and W.E. Esaias. 1993. Annual cycles of phytoplankton chlorophyll concentrations in the global ocean: A satellite view. *Glob. Biogeochem. Cyc.* **7**:181–194.

R.G. Najjar et al.

Zepp, R.G. 1982. Photochemical transformations induced by solar ultraviolet radiation in marine ecosystems. In: The Role of Ultraviolet Radiation in Marine Ecosystems, ed. J. Calkins, pp. 293–307. New York: Plenum.

Zepp, R.G., and M.O. Andreae. 1990. Photosensitized formation of carbonyl sulfide in sea water. In: Effects of Solar Ultraviolet Radiation on Biogeochemical Dynamics in Aquatic Environments, ed. N.V. Blough and R.G. Zepp, p. 180. Woods Hole Oceanogr. Inst. Tech. Rept. WHOI–90–09. Woods Hole, MA: Woods Hole Oceanogr. Inst.

9

Balancing the Global Carbon Cycle with Terrestrial Ecosystems

R.A. HOUGHTON

The Woods Hole Research Center, P.O. Box 296, Woods Hole,
Massachusetts 02543, U.S.A.

ABSTRACT

The long-term (1850–1990) net flux of carbon between terrestrial ecosystems and the atmosphere, based on geophysical analyses (Sarmiento et al. 1992), indicates a net release of 25 Pg C to the atmosphere, most of which was released prior to 1940. In contrast, changes in land use have resulted in a release that totals 120 Pg C, most of which occurred *after* 1940. By difference, undisturbed terrestrial ecosystems seem to have accumulated about 95 Pg C since 1850, presumably as a result of changing environmental conditions.

If the difference between the geophysical and land-use estimates represents changes in terrestrial ecosystems in response to a changing global environment, the implications for future climatic change are ambiguous: a global warming seems likely to release carbon from soil, but either the increased mineralization of soil nitrogen or the elevated concentrations of CO_2 in the atmosphere might counter this release. The net direction of the feedbacks is difficult to predict.

On the other hand, if undisturbed ecosystems have been accumulating carbon over the last 50 years or so, the future of this accumulation is less ambiguous. Deforestation in the Tropics is about 15×10^6 ha (or almost 1%) yr^{-1} currently and the rate seems to be increasing. The net effect of converting forests to cleared land will be a reduction in the capacity of ecosystems to accumulate carbon in the future even if they have in the past.

This long-term accumulation of carbon in undisturbed ecosystems is calculated by difference. It has not been directly observed. Measured rates of tree growth suggest that the accumulation is not in vegetation. On the basis of accumulation rates of organic matter, it appears not to be in soils either (Schlesinger 1990). It is impossible to rule out

Role of Nonliving Organic Matter in the Earth's Carbon Cycle
Edited by R.G. Zepp and Ch. Sonntag © 1995 John Wiley & Sons Ltd.

the oceans as an additional sink or to be certain that the imbalance in the global carbon equation is not the result of errors in one or more of the other terms.

INTRODUCTION

Are terrestrial ecosystems releasing carbon to the atmosphere, or are they withdrawing it from the atmosphere and accumulating it in vegetation and soils? The answer is almost as uncertain today as it was more than twenty years ago (Woodwell and Pecan 1973). Direct analyses based on changes in land use have consistently shown terrestrial ecosystems to be releasing carbon to the atmosphere (Houghton et al. 1983, 1985, 1987; Detwiler and Hall 1988; Hall and Uhlig 1991; Houghton 1991). Indirect estimates based on geophysical constraints from atmospheric and oceanic data show that terrestrial ecosystems must be accumulating carbon (Keeling et al. 1989; Tans et al. 1990; Sarmiento et al. 1992). The controversy pertains particularly to the last two or three decades. For the decade of the 1980s, the following "equation" applies (units are Pg carbon; 1 Pg $= 10^{15}$ g):

$$
\begin{array}{ccccccc}
\text{Fossil fuel} & = & \text{Atmospheric} & + & \text{Oceanic} & + & \text{Terrestrial} \\
\text{emissions} & & \text{increase} & & \text{uptake} & & \text{uptake} \qquad (9.1) \\
5.4\ (\pm 0.5) & & 3.2\ (\pm 0.1) & & 2.0\ (\pm 0.5) & & ??
\end{array}
$$

If terrestrial uptake is calculated by difference, terrestrial ecosystems are apparently either in balance with respect to carbon or accumulating about $1\,\mathrm{Pg\,C\,yr^{-1}}$, and the carbon budget is balanced. On the other hand, if the terrestrial term is determined from changes in land use, terrestrial ecosystems are releasing about $2\,\mathrm{Pg\,C\,yr^{-1}}$ to the atmosphere, and the terms of the global carbon equation are not balanced.

The purpose of this review is to offer an explanation for reconciling the two conflicting estimates of flux, those based on changes in land use and those based on analyses using geophysical data and models. Although nonliving organic matter (NLOM) in soils or in the sea may play a role in resolving the global carbon imbalance, a resolution is not limited to these pools of carbon. Indeed, neither the geographic location nor the form of the "missing carbon" is known.

CHANGES IN TERRESTRIAL CARBON RESULTING FROM CHANGES IN LAND USE

Clearing forests for new agricultural land causes a release of carbon to the atmosphere. The carbon initially held in trees and other vegetation is released through burning or through decomposition of above- and belowground plant material left in the soil at the time of clearing. Some of the carbon originally held in forest soil is also released to the atmosphere as a result of cultivation. The reverse is true when agricultural lands are abandoned and allowed to return to forests. With forest growth, carbon is withdrawn from the atmosphere and accumulated in biomass and soil. Determining the net flux of carbon over any interval of time is dependent upon two types of data: (a) rates of land-use change and (b) changes in carbon per ha in the vegetation and soil affected by the land-use change. These two types of data (change in area and change in carbon per unit area) have been assembled for the major ecosystems of the world and have been used in a number of analyses to compute regional and global changes in terrestrial carbon (Houghton et al. 1983, 1985, 1987; Detwiler and Hall 1988; Hall and Uhlig 1991; Houghton 1991; Melillo et al. 1988). Because forests contain so much more carbon than the lands that replace them, studies of change in terrestrial carbon have emphasized forests and rates of deforestation and reforestation. Before 1980, however, there were no estimates of deforestation for much of the Earth, and analyses of long-term trends in carbon storage relied on changes in the area of agricultural lands, rates of logging, and other changes in land use to determine changes in the area of forests. Beginning in 1980, estimates of deforestation provided a more direct appraisal of change.

Effects of Land-use Change on Carbon in Vegetation and Soil

Vegetation

The net flux of carbon between terrestrial ecosystems and the atmosphere depends upon the amount of carbon held in vegetation and soil. The biomass of forests is variable, however, and not well known. Estimates of tropical forest biomass based on surveys of wood volumes by foresters are substantially lower than estimates based on direct measurements of biomass. Because the lower estimates are based on a much larger sampling of area, they are generally believed to be more accurate than the higher estimates and more appropriate for calculation of carbon flux (Brown et al. 1989, 1991; Detwiler and Hall 1988). Recent evidence suggests, however, that the use of low estimates of biomass to calculate flux underestimates both the current and long-term (> 10 yr) flux. The evidence comes from the observed, widespread reduction in biomass of tropical forests over time (Uhl and Vieira 1989; Woods 1989; Gajanseni and Jordan 1990; Brown et al. 1991; Flint and Richards 1991). If the current biomass of forests is lower than it was formerly, analyses based on current estimates will underestimate past losses.

Use of current (low) estimates of biomass also underestimates the current release of carbon if the calculations are based on deforestation alone and do not include the carbon lost as a result of reduction of biomass *within* forests (degradation). The actual net flux includes that due to deforestation and reforestation (changes in area) as well as that due to degradation and regrowth within forests. Most analyses have not considered the loss of carbon from within forests, although some have approached the issue by modeling logging. Logging reduces biomass if rates of harvest are greater than rates of regrowth. The net flux of carbon calculated from logging has generally been small, however, while the flux calculated from degradation has been large. The most likely reason for the difference is that much of the extraction of timber and fuelwood from tropical forests is illicit and, hence, not reported. Reported rates are underestimates.

Soil

Estimates of the organic carbon in the top 1 m of soil have been summarized by Eswaran et al. (1993) for the major soil types, by Schlesinger (1984, 1986) and Zinke et al. (1986) for major types of vegetation, and by Brown and Lugo (1982), Post et al. (1982), and Zinke et al. (1986) for life zones. Life zones are classes of potential vegetation determined from climatic variables. On average, the soils of the world contain 1400–1500 Pg C, about three times more organic carbon than is contained in vegetation, 500–600 Pg C. Although the carbon content of soils greater than 1 m in depth is generally low, the great depths to which organic carbon may be found (Nepstad et al. 1991) indicates that the total organic carbon of soils may be considerably greater than 1500 Pg C. If this carbon is old and refractory, however, its mass is largely irrelevant in exchanges with the atmosphere.

Types of Land-use Change

The major types of land-use change and their effects on carbon storage are described below.

Cultivated land. When forests and woodlands are cleared for cultivated land, much of the aboveground biomass is burned and released immediately to the atmosphere as CO_2. Some of the wood may be harvested for products and oxidized more slowly. The rest of the aboveground and belowground material decays, as does the organic matter of newly cultivated soil (Figure 9.1). A small fraction of the organic matter burned may be converted to elemental carbon or charcoal that resists decay.

Disturbance, and particularly cultivation, of forest soils generally results in a loss of organic carbon. The variability is large, but, on average, 25% to 30% of the carbon in the surface meter seems to be lost to the atmosphere when forest soils are cleared of vegetation and cultivated (Detwiler 1986; Schlesinger 1986; Davidson and Ackerman 1993). Most of the loss occurs rapidly, within the first five years of clearing. It is important to note that the 25% to 30% loss is a net loss. It includes a relatively greater loss of carbon from the soil of the initial

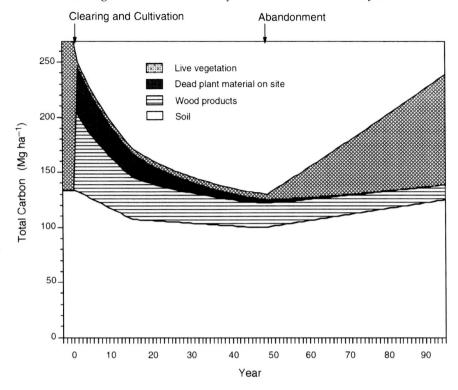

Figure 9.1   Changes in the initial stocks of carbon (Mg C ha^{-1}) following clearing and cultivation of a temperate deciduous forest (year 0) and following abandonment of agriculture (year 50).

ecosystem and an increase in soil carbon from the decay of newly planted crops. Thus, the labile component of NLOM in soils is greater than 25% to 30% and may be considerably greater.

Pastures. Because pastures are generally not cultivated, the loss of carbon from forest soils converted to pastures is less than the loss associated with cultivation. Under some conditions there appears to be no loss or even an increase in soil carbon. Nevertheless, most studies show a loss, sometimes as much as 40% of the carbon originally in the forest soil. In the estimates reported below, conversion of tropical forests to pastures was assumed to release to the atmosphere 12.5% of the organic matter contained in the top 1 m of soil (Detwiler 1986).

Shifting cultivation. Deforestation for shifting cultivation releases less carbon to the atmosphere than deforestation for permanently cleared land because of the partial recovery of the forests during fallow. Similarly, less soil organic matter is oxidized during the shifting cultivation cycle than during continuous cultivation (Detwiler 1986; Schlesinger 1986).

Logging. Although logging is not generally considered deforestation, it causes a release of carbon to the atmosphere from the mortality and decay of trees damaged in harvest operations, from decay of logging debris and from oxidation of wood products. It also withdraws carbon from the atmosphere if logged forests are allowed to regrow. Carbon may be lost from soils immediately after logging, but this loss is variable and often not observed at all (Johnson 1992). However, if carbon is lost from soils immediately after logging, the loss is generally balanced over time by the accumulation of soil carbon in regrowing of forests. Thus, calculations of the net flux of carbon from logging and regrowth, unlike those for clearing and cultivation, are relatively insensitive to assumptions about the loss of carbon from soils. For example, the difference between an assumed 20% loss (Houghton et al. 1987) and an assumed 0% loss (Melillo et al. 1988) of soil carbon following logging contributed a difference of less than $0.05\,Pg\,C\,yr^{-1}$ to the calculated flux of carbon from the combination of temperate and boreal forests in North America, Europe, and the former Soviet Union.

Afforestation and reforestation. Tree plantations may simply replace existing forests or may be established on lands that were cleared years ago. Depending on the carbon content of the soils planted, organic matter may either accumulate or be lost from newly established plantations.

Models Used to Calculate Regional and Global Changes in Terrestrial Carbon

The types of land-use change described above affect the carbon held in vegetation and soil to varying degrees. Studies of land-use change have defined these changes for different types of land use and different types of ecosystems. Because of the variety of ecosystems and land uses, and because calculations of carbon flux require accounting for cohorts of different ages, bookkeeping models have been used for the calculations (Houghton et al. 1983, 1985; Detwiler and Hall 1988). Simulation of annual changes in the different reservoirs of carbon (live vegetation, soils, debris left on site, and wood products) determines the annual net flux of carbon between the land and atmosphere.

The Net Flux of Carbon between Terrestrial Ecosystems and the Atmosphere

Long-term Flux of Carbon from Changes in Land Use

The long-term (1850–1990) flux of carbon to the atmosphere from global changes in land use is estimated to have been about 120 Pg C (Houghton 1994) (Figure 9.2). The annual flux increased from about $0.4\,Pg\,C\,yr^{-1}$ in 1850 to about $1.7\,Pg\,C\,yr^{-1}$ in 1990. Until about 1940, the region with the

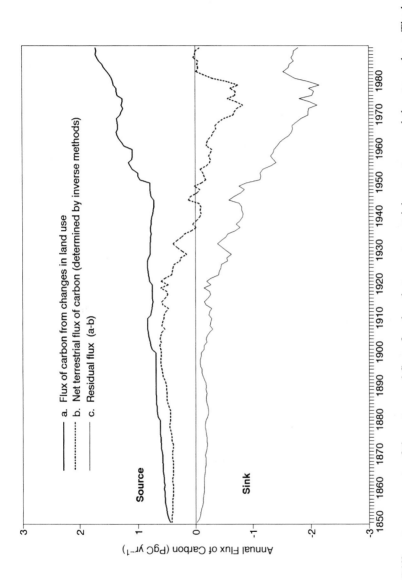

Figure 9.2 Different components of the net annual flux of carbon between terrestrial ecosystems and the atmosphere. The heavy solid curve is the flux of carbon from changes in land use. The heavy dashed curve is the net terrestrial flux, determined from inversion of an ocean model (Sarmiento et al. 1992). The thin solid curve is the residual flux of carbon, the difference between the net flux and the flux from land-use change. The negative residual flux is assumed here to indicate an accumulation of carbon in undisturbed ecosystems as a result of changes in the environment.

greatest flux was the temperate zone. Since 1950, the Tropics have been increasingly important. The current emissions from northern temperate and boreal zones are thought to be close to zero, with releases from decaying wood products approximately balanced by annual accumulations of carbon in regrowing forests (Houghton et al. 1987; Melillo et al. 1988; Houghton 1993).

About two-thirds of the total long-term flux, or 80 Pg C, was from oxidation of plant material, either burned or decayed. About one-third, or 40 Pg C, was from oxidation of soil carbon, largely from cultivation. About 23 Pg C were lost from tropical soils and about 17 Pg C were lost from soils in the temperate zone, mostly from grassland or prairie soils.

The ratio of soil loss to biomass loss (1:2 in units of carbon) is somewhat surprising given the ratio of total stocks (1450:600, or approximately 2.4:1). The difference is explained by the fact that, on average, only 25% to 30% of the soil carbon is lost with cultivation, with a smaller fraction lost when forests are converted to pastures or to shifting cultivation. In contrast, about 100% of the biomass is often oxidized and released as CO_2 to the atmosphere. Assuming that 25% of soil carbon is lost, the effective ratio of soil carbon to biomass carbon becomes about 350:600, or approximately 1:2, in agreement with the ratio determined in the analysis. The ratio varies with ecosystem as well as with land use. As long as the soil carbon below 1 m is refractory, for example, the ratio may be less than 1:5 when tropical forests are cleared for agriculture. The ratio may be higher than 1:1 for temperate zone grasslands. If some of the carbon in deep tropical soils is actively cycled, the net release associated with deforestation may be considerably greater than previously thought. Preliminary work with isotopes in Amazonian soils shows that about one-third of the carbon released after conversion of forest to pasture is from depths greater than 1 m (Davidson et al. 1993).

Current Flux of Carbon from Changes in Land Use

Recent published estimates of the flux of carbon from changes in land use, taken almost entirely from the Tropics, vary between 0.6 and 2.5 for 1980 (Houghton et al. 1987; Hall and Uhlig 1991; Houghton 1991) and between 1.1 and 3.6 Pg C for 1990 (Houghton 1991). The high end of this range now seems too high because rates of deforestation in Brazil are lower than initially thought. If recent estimates of Brazilian deforestation (Fearnside et al. 1990) are used to recalculate emissions of carbon from Latin America, the estimate of flux is reduced from about 0.7 Pg C yr^{-1} (Houghton et al. 1991) to about 0.5 Pg C yr^{-1} in 1980, and from about 0.9 to 0.7 Pg C yr^{-1} in 1990. If the still more recent estimates of Skole and Tucker (1993) are used, the calculated net flux is reduced by another 0.1 Pg C yr^{-1}.

Recent estimates of biomass and deforestation in tropical Asia and Africa require new estimates of carbon flux for those regions as well. Rates of deforestation in Southeast Asia were revised upward recently, and historical rates of degradation in the region were reassessed. These revisions give an estimate of flux for South and Southeast Asia that is about 0.7 Pg C in 1990 (Houghton and Hackler 1994). A reanalysis of land-use change in Africa, using improved estimates of biomass (Brown et al. 1989), gives an estimate of flux there of 0.35 Pg C yr^{-1} in 1990.

Globally, the annual flux of carbon from changes in land use is now estimated to have been about 1.4 Pg C in 1980 (1.3 Pg from the Tropics and 0.1 Pg from outside the Tropics) and 1.7 Pg C in 1990 (essentially all of it from the Tropics). This estimate for all the Tropics in 1980 is about 45% higher than the recent estimate by Hall and Uhlig (1991), which did not explicitly consider reduction of biomass within forests. The average annual flux over the decade of the 1980s is estimated to have been 1.6 ± 1.0 Pg C yr^{-1}.

CHANGES IN TERRESTRIAL METABOLISM

Indirect Estimates of Change in Terrestrial Carbon Based on Geophysical Constraints

The above estimate of 120 Pg C lost from terrestrial ecosystems since 1850 is in contrast to the estimate obtained independently from analyses based on atmospheric CO_2 data and models of oceanic uptake. Geophysical analyses suggest a net terrestrial release of 25 to 50 Pg C (Siegenthaler and Oeschger 1987; Keeling et al. 1989; Sarmiento et al. 1992). The approach is based on an inverse technique, sometimes referred to as deconvolution. Normally, ocean models of carbon uptake are used to calculate atmospheric concentrations of CO_2 from specified annual emissions of carbon from fossil fuels and deforestation. The models can be inverted, however. If the historic pattern of atmospheric concentrations of CO_2 is known, the same models can be used to calculate the annual emissions and accumulations required to generate that pattern. Subtraction of the annual emissions of fossil-fuel carbon from the total emissions calculated from these inversions yields a residual flux, a non-fossil flux of carbon. The residual flux need not be a terrestrial flux, but it is often assumed to be because the other terms in the budget are specified. Thus, the terrestrial term is determined by difference, and, indeed, short-term evidence from ^{13}C:^{12}C ratios suggests that the changes are of terrestrial rather than oceanic origin (Siegenthaler and Oeschger 1987; Keeling et al. 1989).

A recent inversion of atmospheric CO_2 data with an ocean model (Sarmiento et al. 1992) shows the residual (presumably terrestrial) flux to have been a cumulative net release of about 25 Pg C to the atmosphere over the period 1850 to 1990 (Figure 9.2). The net flux was positive before 1940 and largely negative after. The estimate is about half that calculated by Siegenthaler and

Oeschger (1987) with a different oceanic model, but the temporal patterns of the two estimates are almost identical.

A Possible Reconciliation of the Two Estimates

The estimate obtained indirectly from geophysical data (a release of 25 Pg C) is considerably less than the estimate based on changes in land use (120 Pg C over the period 1850–1990), and the temporal patterns of the two estimates do not bear much resemblance (Figure 9.2). The disagreement, however, may not be the result of errors in one or the other of the approaches. The difference between these two estimates of flux may represent an actual flux of carbon between terrestrial ecosystems and the atmosphere, a flux that is not related to changes in land use but to changes in the global environment. Analyses of flux based on changes in land use assume that ecosystems not directly affected by land-use change are in steady state with respect to carbon. This assumption is probably not valid. The amount of carbon stored in vegetation and soils is changed not only by deliberate human activity, but by inadvertent changes in climate, CO_2 concentrations, nutrient deposition, or pollution (Table 9.1).

It is useful to distinguish between these two causes or types of change: deliberate and inadvertent. Deliberate changes result directly from management, for example, deforestation or reforestation. Inadvertent changes occur through changes in the metabolism of ecosystems, more specifically through a change in the ratio of the rates of primary production and respiration. Clearly, management practices affect metabolism, and the distinction between deliberate and inadvertent effects is blurred. Nevertheless, the distinction is a useful one because changes in carbon stocks that result from deliberate human activity are more easily documented than changes resulting from metabolic changes alone. Deliberate changes usually involve a well-defined area; and the changes in carbon density ($Mg C ha^{-1}$) associated with land-use change are generally large: forests hold 20 to 50 times more carbon in their vegetation than the lands that replace them following deforestation. In contrast to these large and well-defined changes associated with management, changes that occur as a result of metabolic changes are generally subtle. They occur slowly over poorly defined areas and are difficult to measure against the large background levels of carbon in vegetation and soils and against the natural variability of diurnal, seasonal, and annual metabolic rates.

Assuming that the entire flux obtained through inversion of ocean models is terrestrial, the inverse approach provides an estimate of the total *net* terrestrial flux, including both a flux from land-use change and a flux due to other changes in terrestrial ecosystems. The flux of carbon from land-use change, on the other hand, includes only those lands directly and deliberately managed. The difference between them may thus define the flux of carbon from terrestrial ecosystems not directly modified by humans. The estimate is determined by

Table 9.1   Factors included and not included in analyses of the flux of carbon from changes in land use.

Included (Deliberate changes)	Not included (Inadvertent changes)
Change in the area of forests	Unmanaged, undisturbed ecosystems
Deforestation	assumed to be in steady state
croplands	
pastures	
degradation of land	
Reforestation	
abandonment of agriculture	
afforestation	Forest decline or enhancement
Change in biomass within forests	Eutrophication (including CO_2 fertilization)
Logging and regrowth	Frequency of fires
Shifting cultivation	Climatic change
Degradation of forests (reduction of	Desertification
biomass)	

difference, and the errors involved in calculating a flux from the difference between the inversion technique and the land-use approach are potentially large. Nevertheless, the gross trends in the residual flux may be correlated with global environmental factors and may suggest the direction and magnitude of terrestrial feedbacks between global carbon and global climate systems. Arguments are made, on one hand, that terrestrial ecosystems will amplify a global warming (positive feedback) by releasing additional carbon to the atmosphere as a result of a warming-enhanced increase in respiration rates; on the other hand, it is postulated that they will reduce the warming (negative feedback) by accumulating carbon in response to increasing concentrations of CO_2 in the atmosphere. The historical variation in the residual flux, calculated here, allows one to ask whether the variation is related to temperature, atmospheric CO_2, or some other environmental factor. Correlations do not demonstrate a cause–effect relationship, of course; however, they may reveal which environmental factors were important in the past and may suggest the direction of response of terrestrial ecosystems to global change in the future. Will terrestrial ecosystems act as a net source or sink for carbon?

The residual (inadvertent) flux, or the difference between the net biotic flux (from inversion) and the land-use flux, is shown in Figure 9.2 (thin solid line). The flux has always been negative and is interpreted here to mean that some terrestrial ecosystems were accumulating carbon independent of land-use change. Positive differences, if they occurred, would indicate that some terrestrial ecosystems were releasing carbon to the atmosphere in addition to that released from changes in land use. The pattern of this residual flux over time exhibits three features: (a) a period before 1920 showing a terrestrial flux not different from zero, (b) a period between about 1920 and 1975 when some of the world's terrestrial ecosystems were apparently accumulating carbon at an increasing rate, and (c) a period since the mid-1970s in which the rate of accumulation has decreased.

Possible Mechanisms

The amount of carbon held in terrestrial ecosystems is determined by the balance between photosynthesis and respiration, the latter including decay of soil organic matter. Both of these metabolic processes are affected by many environmental factors; however, the progressive increase in the residual flux of carbon after 1920 (Figure 9.2) parallels the trend in industrial activity, and one cause of an increasing accumulation of carbon on land may have been increasing concentrations of CO_2 in the atmosphere. Elevated concentrations of CO_2 have been shown to enhance photosynthesis and growth in many plants and may increase the storage of carbon in terrestrial ecosystems if the enhanced growth is not respired. Annual concentrations of CO_2 are highly correlated with variation in the residual flux over the period 1850 to 1990 (Figure 9.3) (Houghton 1994). About $0.035 \, Pg \, C \, yr^{-1}$ are stored per ppmv (parts per million by volume) increase in CO_2. The correlation breaks down after the late 1970s, when the annual storage declined despite a continuing increase in the concentration of CO_2.

The role of temperature in affecting metabolism and, hence, in affecting the storage of carbon in plants and soils is also important, but the net long-term effect on ecosystems is unclear. On one hand, rates of respiration (including decomposition) are temperature sensitive, and higher temperatures should increase these rates, releasing more carbon to the atmosphere (Woodwell 1989; Raich and Schlesinger 1992). On the other hand, respiration (mineralization) not only releases CO_2, it also releases nitrogen and other nutrients. Because the ratio of C:N is higher in woody tissue than in soil organic matter, a temperature-enhanced mineralization may have led to a net accumulation of carbon in wood. For such an accumulation to have occurred, nitrogen released during decomposition would have to have been taken up by plants. If the processes of mineralization and growth were decoupled in time, the mineralized nitrogen might have been lost from the system.

If the net effect of warmer temperatures was to release carbon, the pattern of the residual flux should parallel the trend in temperature deviations (Figure 9.3): a source of carbon between 1880 and 1940, an increasing sink between 1940 and 1970, and then an abruptly decreasing sink again after the late 1970s. If the net effect of warming was to store carbon (changing its form from soil to wood), the expected flux would be reversed. Comparison of the temperature trend with the missing flux is ambiguous (Figure 9.3). The flux shows no

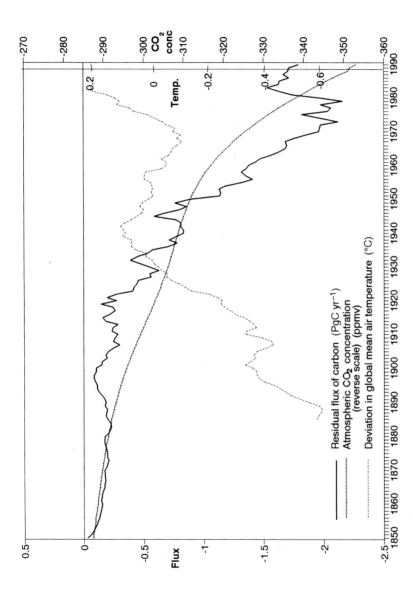

Figure 9.3 The residual flux of terrestrial carbon (heavy solid line); the concentration of CO_2 in the atmosphere (dotted line) (plotted with the ordinate reversed so that the overall trend in concentration parallels the trend in the residual flux of carbon); and mean global land surface temperature deviations from the 1951–1980 baseline (thin dashed line).

Residual flux of carbon ($PgC\ yr^{-1}$)
Atmospheric CO_2 concentration (reverse scale) (ppmv)
Deviation in global mean air temperature (°C)

response to the early warming between 1880 and 1940. After 1940, however, the increasing sink of carbon matches the decrease in global temperature until the 1970s; and the decreased sink after the 1970s matches the recent warming. Results over the last fifty years suggest a warming-caused release of carbon, a positive feedback.

Regression coefficients determined from an analysis of the post-1940 residual flux against global mean temperature suggest a Q_{10} of 2.5 to 3.2, if temperature is assumed to have affected only the rate of heterotrophic (largely soil) respiration (Houghton 1994). If the change in temperature is assumed to affect also the rate of plant respiration, the values of Q_{10} range between 1.6 to 2.1 (Table 9.2). These values of Q_{10} assume that only respiration is affected by temperature. If rates of photosynthesis have also been increased by warmer temperatures, the Q_{10} for respiration alone is higher.

Unfortunately, the relative importance of CO_2 concentration and temperature in explaining variation in the residual flux is ambiguous (Houghton 1994). With a regression equation that allowed for a 7-year lag between a change in temperature and flux, the effect of warming was to release 6.4 Pg C/°C, and the effect of CO_2 was to store 0.004 Pg C/ppmv. Without a lag, the effects were 3.4 Pg C/°C and 0.043 Pg C/ppmv. In the different regressions, the effect of temperature varied by a factor of 2; the effect of CO_2 concentration varied by a factor of 10. Considering both temperature and CO_2, together, feedback effects for a future world with a CO_2 concentration of 700 ppmv and a mean surface temperature 2.5°C warmer than today could range from a net annual release of 13 Pg C yr^{-1} (regression 1) to a net storage of 10 Pg C yr^{-1} (regression 2) (Table 9.2). These crude extrapolations demonstrate the ambiguity of the results to date. Even the net direction of the feedbacks is uncertain, to say nothing of the magnitudes possible under environmental conditions very different from those of the present day.

On the other hand, the effects of future changes in land-use change seem less ambiguous. Deforestation in the Tropics is about 15×10^6 ha (almost 1%) yr^{-1} at present and is increasing (FAO 1993). The net effect of this conversion will be a reduction in the capacity of ecosystems to accumulate carbon in the future even if they have in the past.

These analyses have not considered changes in precipitation, which could well be the dominating influence on the storage of terrestrial carbon. A recent analysis by Dai and Fung (1993), for example, shows that year-to-year variations in temperature and precipitation can account for about half of the imbalance in the global carbon budget over the period 1940–1988.

The abrupt decline (about 0.5 Pg C yr^{-1}) in the global (terrestrial?) sink for carbon since the end of the 1970s is, perhaps, important. It has been observed by others (Keeling et al. 1989; Sarmiento et al. 1992), but the cause remains uncertain. The release of carbon from tropical deforestation has increased over the last decade. From the analysis reported here, it appears that an increased

Table 9.2 Results of regression analyses.

Independent variable	R^2	Effect	Biotic* growth factor	Q_{10}**	Net effect of doubled CO_2 and 2.5°C increase ($Pg\,C\,yr^{-1}$)
1. CO_2 (no lag) temp. dev. (7-yr lag)	0.94	sink: 0.004 $Pg\,C$/ppmv source: 6.4 $Pg\,C$/°C	0.026	2.1	source: 13
2. CO_2 (no lag) temp. dev. (no lag)	0.84	sink: 0.043 $Pg\,C$/ppmv source: 3.4 $Pg\,C$/°C	0.14	1.6	sink: 10

*Biotic growth factor defined as: $Beta = \dfrac{NPP}{NPP_0} - 1/\ln (C/C_0)$,

where NPP is net primary production, C is atmospheric CO_2 concentration, and $_0$ refers to preindustrial time. These biotic growth factors assume that all of the increase in NPP is stored. If only a fraction of an increase in NPP is stored, the calculated biotic growth factor would be larger.
**These values for Q_{10} are calculated for a global respiration of $100\,Pg\,C\,yr^{-1}$ (autotrophic and heterotrophic respiration). If temperature affects only heterotrophic respiration, the resulting Q_{10}'s are 2.5 and 3.2 for regressions 2 and 1, respectively. In both cases, temperature is assumed to have no effect on photosynthesis. If higher temperatures increase the rate of photosynthesis as well as the rate of respiration, Q_{10}'s for respiration alone will be higher.

release (or reduced accumulation) has occurred elsewhere, perhaps as a result of a warming-enhanced increase in respiration and decomposition without a compensating storage of carbon in biomass.

WHERE IS CARBON ACCUMULATING?

The current imbalance in the global carbon budget is $1.5–2.0\,Pg\,C\,yr^{-1}$ (the difference between a net flux of $0\ Pg\,C\,yr^{-1}$ and a release from tropical defor-estation of $1.5–2.0\,Pg\,C\,yr^{-1}$). Based on geographically based estimates of industrial emissions, atmospheric data, and a model of atmospheric trans-port, the study by Tans et al. (1990) suggests that the missing carbon was accumulating in terrestrial ecosystems of the Northern Hemisphere's temper-ate zone. Two recent analyses of forest growth in this region have reported an accumulation of carbon that approaches $1\,Pg\,C\,yr^{-1}$ (Kauppi et al. 1992; Sedjo 1992), and, based on these two studies and the one by Tans et al. (1990), one might conclude that the carbon budget is close to balanced. These recent analyses are, however, incomplete. They included the accumulation of carbon in regrowing forests logged during the last several decades but not the fate of plant material initially held in those forests before logging. The rates of carbon accumulation calculated by Kauppi et al. (1992) and Sedjo (1992) are similar to rates obtained earlier by Houghton et al. (1987) and Melillo et al. (1988). The new estimates of carbon accumulation do not change those calculated earlier by more than a few tenths of a Pg C (Table 9.3).

The major difference between the earlier (Houghton et al. 1987; Melillo et al. 1988) and later (Kauppi et al. 1992; Sedjo 1992) studies arises not from estimates of forest growth rates, but from the methods used to account for the rest of the carbon originally held in the vegetation and soil of forests logged, cleared, or otherwise disturbed. The "rest of the carbon" includes plant material left to decay on site, plant material burned, wood products removed from the forest but gradually oxidized, and soil carbon released to the atmosphere as a result of oxidation following clearing of forests for agriculture (Figure 9.1). When these components are included in carbon budgets, the net flux of carbon from changes in land use is quite different from that calculated on the basis of forest regrowth alone (Table 9.3). Had Kauppi et al. (1992) or Sedjo (1992) accounted for all the initial carbon held in the disturbed ecosystems, they, too, might have found a net flux close to zero.

If the recent studies confirm that the accumulation of carbon in northern forests is at a rate expected from past and current logging, what does this imply for the missing carbon? If additional carbon were accumulating in the northern temperate zone, could we measure it? The global carbon imbalance is 1.5 to

Table 9.3   Annual rates of accumulation (+) and loss (–) of carbon from changes in the major forested regions of the northern temperate zone (Tg C yr^{-1}).

Sources	Former USSR	North America	Europe	Total
A. Gross Accumulation of Carbon in Regrowing Forests				
Kauppi et al. (1992)			85–120	850–1200**
Sedjo (1992)	416	179	91	686
Melillo et al. (1988)	154	407		786
B. Gross Releases of Carbon Melillo et al. (1988)				
Decay of logging debris	–82	–301	–143	–526
Burning*	–24	–6	–16	–46
Oxidation of wood products	–46	–96	–51	–193
Expansion and abandonment of agricultural lands	–39	4	26	9
Net (A. plus B.) Meilillo et al. (1988)	–37	8	41	12***

*Burning associated with logging and replanting, and with burning of fuelwood. Emissions do not include those resulting from prescribed burning or wildfires.
**The rest of the world's northern temperate and boreal forests is assumed (unrealistically) to be accumulating carbon at a rate similar to that of European forests.
***Houghton et al. (1987) calculated a net release of 32 Tg C yr^{-1} for these regions. The major difference between the analyses by Houghton et al. and Melillo et al. was that the former assumed a 20% reduction in soil carbon with logging followed by recovery during regrowth. Thus, the uncertainty associated with the response of soils to logging contributes an uncertainty of about 50 Tg C yr^{-1} to the net flux.

$2.0\,\mathrm{Pg\,C\,yr^{-1}}$. If all of that carbon were accumulating in the trees of the northern temperate zone (area of temperate zone forests $= 600 \times 10^6$ ha), the accumulation would be about $2.5\,\mathrm{Mg\,C\,ha^{-1}\,yr^{-1}}$. In comparison, the accumulation of carbon in European forests was 0.4 to 0.6 Mg C $\mathrm{ha^{-1}}$ $\mathrm{yr^{-1}}$ (Kauppi et al. 1992). Thus, the accumulation of missing carbon would be about 5 times higher than observed. It would be difficult to miss in data on forest growth rates. Including boreal forests in the calculation lowers the accumulation rate to 1.0 $\mathrm{Mg\,C\,ha^{-1}}$ $\mathrm{yr^{-1}}$, but it is still high relative to observed growth rates. If the missing carbon were accumulating throughout the temperate zone, not just in forests but in woodlands, grasslands, and cultivated lands, the average accumulation rate would be about 0.5 Mg C $\mathrm{ha^{-1}}$ $\mathrm{yr^{-1}}$, about the rate observed in European forests. This rate of accumulation in the vegetation of grasslands and cultivated lands is high relative to the standing stocks of carbon there to begin with and would be obvious. Finally, if the missing carbon were accumulating in trees throughout the world, the rate would, again, average about $0.5\,\mathrm{Mg\,C\,ha^{-1}\,yr^{-1}}$. However, this is the rate observed for regrowing European forests, forests with a long history of intensive use. That an *additional* accumulation of this magnitude, representing the missing sink, could be in European trees, is unlikely. That such an accumulation could be occurring in the rest of the forests of the world, many of which have not been logged and, hence, are not regrowing (for example, large regions of Canada, central and eastern Russia, and Brazil) seems equally unlikely. Regrowing forests are young; mortality is low and, as a result, little carbon is lost. Even if mature forests around the world are not in steady state, they are unlikely to be accumulating carbon at rates that approximate rates of regrowth immediately after logging. Overall, it seems unlikely that carbon is accumulating in vegetation anywhere at the rate required to balance the global carbon budget.

Perhaps the missing carbon is accumulating in temperate zone soils. If so, the chances of measuring it are small. The standing stock of carbon in soils is about 1500 ± 300 Pg C, and the global carbon imbalance is two to three orders of magnitude smaller than this uncertainty. However, observed rates of carbon accumulation in undisturbed soils are too low to account for 1.5–2.0 Pg C $\mathrm{yr^{-1}}$ globally (Schlesinger 1990). The more rapid rates associated with the regrowth of forests on abandoned lands are already included in the land-use term (Table 9.1). Furthermore, if the accumulation is in soils, what is its pathway? Have increasing concentrations of CO_2 enhanced belowground production (Norby et al. 1992), not measured in forestry surveys of growth? Perhaps. But it is unlikely that an increase in root material would not eventually lead to an increase in aboveground growth as well. Yet, analyses of ring widths have not consistently shown a trend of enhanced growth. The long-term residual flux of 100 Pg C represents about 20% of global terrestrial biomass. If stored in three rings, it should be observable. If stored in soils, it might not be observable, but the mechanism of such an accumulation in the absence of an observed increase in primary production remains problematic.

ACKNOWLEDGEMENTS

The research was supported by the U.S. Department of Energy, Office of Energy Research, Office of Health and Environmental Research, Carbon Dioxide Research Program (Grant Number DE–FG02–90ER61079).

REFERENCES

Brown, S., A.J.R. Gillespie, and A.E. Lugo. 1989. Biomass estimation methods for tropical forests with applications to forest inventory data. *For. Sci.* **35**:881–902.

Brown, S., A.J.R. Gillespie, and A.E. Lugo. 1991. Biomass of tropical forests of South and Southeast Asia. *Can. J. For. Res.* **21**:111–117.

Brown, S., and A.E. Lugo. 1982. The storage and production of organic matter in tropical forests and their role in the global carbon cycle. *Biotropica* **14(3)**:161–187.

Dai, A., and I.Y. Fung. 1993. Can climate variability contribute to the "missing" CO_2 sink? *Glob. Biogeochem. Cyc.* **7**:599–609.

Davidson, E.A., and I.L. Ackerman. 1993. Changes in soil carbon inventories following cultivation of previously untilled soils. *Biogeochemistry* **20**:161–193.

Davidson, E.A., D.C. Nepstad, and S.E. Trumbore. 1993. Soil carbon dynamics in pastures and forests of the eastern Amazon. *Bull. Ecol. Soc. Am.* **74(Suppl.)**:208.

Detwiler, R.P. 1986. Land-use change and the global carbon cycle: The role of tropical soils. *Biogeochemistry* **2**:67–93.

Detwiler, R.P., and C.A.S. Hall. 1988. Tropical forests and the global carbon cycle. *Science* **239**:42–47.

Eswaran, H., E. Van Den Berg, and P.F. Reich. 1993. Organic carbon in soils of the world. *Soil Sci. Soc. Am. J.* **57**:192–195.

FAO. 1993. Forest Resources Assessment 1990 Project. Tropical Forests. FAO Forestry Paper 112. Rome: FAO.

Fearnside, P.M., A.T. Tardin, and L.G.M. Filho. 1990. Deforestation rate in Brazilian Amazonia. Brazilia: Natl. Secretariat of Sci.& Tech., 8 pp.

Flint, E.P., and J.F. Richards. 1991. Historical analysis of changes in land use and carbon stock of vegetation in south and southeast Asia. *Can. J. For. Res.* **21**:91–110.

Gajaseni, J., and C.F. Jordan. 1990. Decline of teak yield in northern Thailand: Effects of selective logging on forest structure. *Biotropica* **22**:114–118.

Hall, C.A.S., and J. Uhlig. 1991. Refining estimates of carbon released from tropical land-use change. *Can. J. For. Res.* **21**:118–131.

Houghton, R.A. 1991. Tropical deforestation and atmospheric carbon dioxide. *Clim. Change* **19**:99–118.

Houghton, R.A. 1993. Is carbon accumulating in the northern temperate zone? *Glob. Biogeochem. Cyc.* **7**:611–617.

Houghton, R.A. 1994. Effects of land-use change, surface temperature, and CO_2 concentration on terrestrial stores of carbon. In: Biotic Feedbacks in the Global Climate

System: Will the Warming Speed the Warming?, ed. G.M. Woodwell. Oxford: Oxford Univ. Press, in press.

Houghton, R.A., R.D. Boone, J.R. Fruci, J.E. Hobbie, J.M. Melillo, C.A. Palm, B.J. Peterson, G.R. Shaver, G.M. Woodwell, B. Moore, D.L. Skole, and N. Myers. 1987. The flux of carbon from terrestrial ecosystems to the atmosphere in 1980 due to changes in land use: Geographic distribution of the global flux. *Tellus* **39B**:122–139.

Houghton, R.A., R.D. Boone, J.M. Melillo, C.A. Palm, G.M. Woodwell, N. Myers, B. Moore, and D.L. Skole. 1985. Net flux of carbon dioxide from tropical forests in 1980. *Nature* **316**:617–620.

Houghton, R.A., and J.L. Hackler. 1994. The net flux of carbon from deforestation and degradation in South and Southeast Asia. In: Effects of Land Use Change on Atmospheric CO_2 Concentrations. South and Southeast Asia as a Case Study, ed. V.H. Dale, pp. 301–327. New York: Springer.

Houghton, R.A., J.E. Hobbie, J.M. Melillo, B. Moore, B.J. Peterson, G.R. Shaver, and G.M. Woodwell. 1983. Changes in the carbon content of terrestrial biota and soils between 1860 and 1980: A net release of CO_2 to the atmosphere. *Ecol. Monogr.* **53**:235–262.

Houghton, R.A., D.L. Skole, and D.S. Lefkowitz. 1991. Changes in the landscape of Latin America between 1850 and 1980. II. A net release of CO_2 to the atmosphere. *For. Ecol. Manag.* **38**:173–199.

Johnson, D.W. 1992. Effects of forest management on soil carbon storage. *Water Air Soil Poll.* **64**:83–120.

Kauppi, P.E., K. Mielikainen, and K. Kuusela. 1992. Biomass and carbon budget of European forests, 1971–1990. *Science* **256**:70–74.

Keeling, C.D., R.B. Bacastow, A.F. Carter, S.C. Piper, T.P. Whorf, M. Heimann, W.G. Mook, and H. Roeloffzen. 1989. A three-dimensional model of atmospheric CO_2 transport based on observed winds: 1. Analysis of observational data. In: Aspects of Climate Variability in the Pacific and the Western Americas, ed. D.H. Peterson, pp. 165–236. Geophysical Monograph 55. Washington, D.C.: Am. Geophysical Union.

Melillo, J.M., J.R. Fruci, R.A. Houghton, B. Moore, and D.L. Skole. 1988. Land use change in the Soviet Union between 1850 and 1980: Causes of a net release of CO_2 to the atmosphere. *Tellus* **40B**:116–128.

Nepstad, D.C., C. Uhl, and E.A.S. Serrao. 1991. Recuperation of a degraded Amazonian landscape: Forest recovery and agricultural restoration. *Ambio* **20**:248–255.

Norby, R.J., C.A. Gunderson, S.D. Wullschleger, E.G. O'Neill, and M.K. McCracken. 1992. Productivity and compensatory responses of yellow-poplar trees in elevated CO_2. *Nature* **357**:322–324.

Post, W.M., W.R. Emanuel, P.J. Zinke, and A.G. Stangenberger. 1982. Soil carbon pools and world life zones. *Nature* **298**:156–159.

Raich, J.W., and W.H. Schlesinger. 1992. The global carbon dioxide flux in soil respiration and its relationship to vegetation and climate. *Tellus* **44B**:81–99.

Sarmiento, J.L, J.C. Orr, and U. Siegenthaler. 1992. A perturbation simulation of CO_2 uptake in an ocean general circulation model. *J. Geophys. Res.* **97**:3621–3645.

Schlesinger, W.H. 1984. The world carbon pool in soil organic matter: A source of atmospheric CO_2. In: The Role of Terrestrial Vegetation in the Global Carbon Cycle: Measurement by Remote Sensing, ed. G.M. Woodwell, pp. 111–124. SCOPE 23. New York: Wiley.

Schlesinger, W.H. 1986. Changes in soil carbon storage and associated properties with disturbance and recovery. In: The Changing Carbon Cycle: A Global Analysis, ed. J.R. Trabalka and D.E. Reichle, pp. 194–220. New York: Springer.

Schlesinger, W.H. 1990. Evidence from chronosequence studies for a low carbon-storage potential of soils. *Nature* **348**:232–234.

Sedjo, R.A. 1992. Temperate forest ecosystems in the global carbon cycle. *Ambio* **21**:274–277.

Siegenthaler, U., and H. Oeschger. 1987. Biospheric CO_2 emissions during the past 200 years reconstructed by deconvolution of ice core data. *Tellus* **39B**:140–154.

Skole, D., and C. Tucker. 1993. Tropical deforestation and habitat fragmentation in the Amazon: Satellite data from 1978 to 1988. *Science* **260**:1905–1910.

Tans, P.P., I.Y. Fung, and T. Takahashi. 1990. Observational constraints on the global atmospheric CO_2 budget. *Science* **247**:1431–1438.

Uhl, C., and I.C.G. Vieira. 1989. Ecological impacts of selective logging in the Brazilian Amazon: A case study from the Paragominas region of the State of Para. *Biotropica* **21**:98–106.

Woods, P. 1989. Effects of logging, drought, and fire on structure and composition of tropical forests in Sabah, Malaysia. *Biotropica* **21**:290–298.

Woodwell, G.M. 1989. The warming of the industrialized middle latitudes 1985–2050: Causes and consequences. *Clim. Change* **15**:31–50.

Woodwell, G.M., and E.V. Pecan, eds. 1973. Carbon and the Biosphere. U.S. Atomic Energy Commission, Symposium Series 30. Springfield, VA: Natl. Technical Information Service.

Zinke, P.J., A.G. Stangenberger, W.M. Post, W.R. Emanuel, and J.S. Olson. 1986. Worldwide organic soil carbon and nitrogen data. ORNL/CDIC–18. Oak Ridge, TN: Oak Ridge Natl. Lab.

Standing, left to right:
Rein Vaikmäe, Darwin Anderson, Skee Houghton, Andreas Dahmke, Uli Neue, Ray Najjar,
Ray Lassiter
Seated, left to right:
Tom Pedersen, Alain Huc, Susan Trumbore, Mac Post

10

Group Report: What Is the Role of Nonliving Organic Matter Cycling on the Global Scale?

W.M. Post, Rapporteur
D.W. Anderson, A. Dahmke, R.A. Houghton, A.-Y. Huc,
R. Lassiter, R.G. Najjar, H.-U. Neue, T.F. Pedersen,
S.E. Trumbore, R. Vaikmäe

INTRODUCTION

The global reservoirs of nonliving organic matter (NLOM) as well as the flows into and out of these reservoirs are of substantial magnitude. They significantly determine the concentration of CO_2 in the atmosphere and its fluctuations over many time scales, from seasonal to geological. In this report, we review the present state of knowledge of these NLOM reservoirs and their global dynamics, and explore ideas that could lead us to a better understanding of the NLOM's role in the modern global carbon cycle. We first examine carbon cycle dynamics that are represented in geochemical models of the current global carbon cycle. Next we determine whether it is plausible for some rates to have changed enough to result in a net sequestering of 1–2 Pg carbon per year (1 Pg $= 10^{15}$ g; the present best-estimate of the "missing sink") in NLOM or as a result of NLOM dynamics. We then consider the magnitude of changes in the global carbon cycle of the Quaternary and geological past and whether they reveal any information about how the current carbon cycle may operate in sequestering additional carbon. Finally, we identify (a) the key processes that determine the formation, retention, and destruction of significant NLOM at global scales, and (b) the tools, methods, and observations needed to quantify better the changing role of NLOM in the global carbon cycle.

Role of Nonliving Organic Matter in the Earth's Carbon Cycle
Edited by R.G. Zepp and Ch. Sonntag © 1995 John Wiley & Sons Ltd.

CURRENT GLOBAL CARBON CYCLE AND
THE MISSING SINK

Currently, fossil-fuel burning releases $5.4\,Pg\,C\,yr^{-1}$ into the atmosphere (averaged over the period 1980 to 1989). Box diffusion and general circulation models of ocean carbon uptake estimate a net flux of $2\,Pg\,C\,yr^{-1}$ into the ocean. Direct measurements show an increase of $3.2\,Pg\,C\,yr^{-1}$ in the atmosphere. Model-derived estimates of carbon fluxes, resulting from human land use, contribute an additional $1.6\,Pg\,C\,yr^{-1}$ to the atmosphere. If these are the only fluxes, then there is a net imbalance of $1.8\,Pg\,C\,yr^{-1}$ (Table 10.1). Table 10.1 estimates the uncertainty range for each of these estimates. While it appears plausible that this imbalance could be accommodated within the uncertainty of the flux estimates, the probability is very low. Furthermore, the imbalance is much larger when we consider the entire time period of 1850 to the present, during which atmospheric CO_2 has been increasing. Over this time the residual imbalance has accumulated to 105 Pg, or nearly the magnitude of the atmospheric increase. Furthermore, there are temporal trends in this residual, as shown by Houghton (this volume). It is therefore necessary to examine whether we can improve our understanding of the carbon cycle, to better account for this imbalance more accurately. In particular, are there aspects concerning global NLOM-dynamics that could help?

There are several mechanisms by which net accumulation of carbon in NLOM pools may be occurring. Below we summarize these separately as processes affecting NLOM in terrestrial and oceanic reservoirs.

Terrestrial Ecosystems

The land-use model described by Houghton (this volume) takes into account changes in ecosystem carbon accumulation resulting from afforestation and

Table 10.1 Net fluxes of carbon between global carbon cycle components (see Houghton, this volume).

Time Period	Fossil-fuel Burning	Atmospheric Increase	Land Use	Ocean Uptake	Residual Imbalance
1980–1990 Annual average $(Pg\,C\,yr^{-1})$	5.4	3.2	1.6	2.0	1.8
Uncertainty Range $(Pg\,C\,yr^{-1})$	±0.5	±0.2	±1.0	±0.5	±1.2
1850–1990 Cumulative (Pg-C)	225	140	120	100	105

recovery of vegetation on land abandoned from agriculture. Any possible sequestration of carbon in the Earth's terrestrial system must therefore be a result of ecological processes, largely in land areas not subject to land-use changes and assumed in Houghton's model to remain at an equilibrium carbon content.

CO₂ Fertilization

From laboratory and field experiments, we know that rates of photosynthesis increase with increasing air concentration of CO_2. As a general average, doubling of atmospheric CO_2 from 300 µl CO_2 l^{-1} air to 600 µl l^{-1} leads to a plant growth rate that is higher by about 30% (Bazzaz 1990). There are, however, some questions as to whether we can consider this 30% increase, found experimentally by observing young tree seedlings, to be a reasonable number representing the increase of terrestrial primary production. If, moreover, linear interpolation is correct, the net primary production should have increased from 60–63 Pg C yr^{-1} due to the atmospheric CO_2 increase of 25% since the mid-1800s. If this increased production is largely retained and not shed as additional detritus, it is large enough to account for the present flux imbalance. However, if we assume that it accumulated in aboveground biomass, then terrestrial biomass should have increased from 560 to 665 Pg, or by approximately 20%. Yet certainly we would have noticed such increases, although no systematic observations have been conducted that would confirm such an augmentation. An increase of 105 Pg represents only a 7% increase in the estimated 1500 Pg of soil organic carbon and would be much more difficult to detect in this reservoir. If the C:N ratio of this NLOM is 15, then an additional 7000 Tg N (1 Tg = 10^{12} g) would be sequestered with this carbon, i.e., many times the estimated amount of N fixed by atmospheric processes, fertilizer production, and combustion processes and deposited on land over this interval of time. This imposes a serious constraint on sequestering this amount of NLOM in soil.

   If, however, the average C:N ratio of soil organic matter changed from 15 in the mid-1800s to 16 at the present time, then the 105 Pg of soil organic matter could have been added to the soil organic matter pool without requiring the sequestering of any additional nitrogen. Even if we assume that half of soil organic matter, or 750 Pg, is resistant or persistent (say with a C:N ratio of 10), and all of the 105 Pg carbon is potentially added to a more active soil organic matter component, then an increase from a C:N ratio of 30 to 34 is all that would be required to keep the total amount of soil organic nitrogen the same. Melillo et al. (see Table 11.4, this volume) presents some information on the difference in C:N ratio of maple litter between experimental treatments with different levels of atmospheric CO_2; their data indicate that such changes may be plausible. Tree seedlings grown at a CO_2 concentration of 300 µl l^{-1}

produced leaf litter with C:N ratios of 35 (*Acer rubrum*) and 24 (*Acer pensyl-vanicum*). When grown at 700 μl l^{-1}, however, seedlings of these species produced leaf litter with C:N ratios three times as large (106 for *A. rubrum*, and 86 for *A. pensylvanicum*). Other investigators have found similar changes in C:N ratios in leaf litter from seedlings grown with CO_2 fertilization (Norby et al. 1986, white oak; Coûteaux et al. 1991, sweet chestnut). We need to conduct more measurements on the effect of CO_2 fertilization on abscised leaf litter C:N ratios, especially for conifer species, before we can examine the generality of this response. We must also conduct experiments on the decomposition of this low nitrogen litter to determine whether it can alter the soil organic matter C:N over time scales of a few years to a few decades.

Temperature Increase

Over the past century, a small increase of approximately 0.5°C in average temperature has occurred. Temperature affects metabolism and hence storage of carbon in plants and soil. However, temperature effects on ecosystems are not straightforward. Temperature increases will enhance the soil organic matter decomposition rate, resulting in a decline of carbon storage if input rates remain constant. Net primary production may increase or decrease with increasing temperature. Tree-ring information shows that increased growth of trees may be attributed to temperature increases at the taiga/tundra boundary (Jacoby and D'Arrigo 1989). However, no systematic trends attributed to temperature have been reported for tree rings from other regions.

Temperature increases may also increase the mobilization of nutrients in NLOM. In an experiment, where only soil temperature was increased so that the forest trees were not directly affected by temperature change, decomposition of soil organic matter increased substantially (Melillo et al., this volume). Associated with the warming-enhanced increase in soil decomposition, more nitrogen was mineralized and made available. If this additional N causes an increase in plant growth, then net carbon sequestration will result. The explanation lies in the different N concentrations of soil and plant organic matter. The C:N ratio of forest soil is typically 15 while the C:N ratio of trees is typically over 60. Thus for every gram of soil organic matter C lost, 4 grams or more of tree-tissue C may be constructed from the N released, depending upon the types of materials formed (wood, leaves, or roots).

Warming experiments on the Alaska tundra indicate that species composition may shift from a grass-dominated vegetation to woody vegetation under elevated temperature. Since the C:N ratio in woody tissue is considerably higher than grass tissue, no additional nutrients may be required to support this increased biomass.

Both of these experiments seem to indicate that carbon can be sequestered in terrestrial biomass without increasing the amounts of nitrogen required in the ecosystem, but merely redistributing it. We can recklessly extrapolate the soil warming experiment to the globe using some recent model results. Using the *Terrestrial Ecosystem Model* (TEM; McGuire et al. 1992), Melillo et al. (1993) estimated a cumulative net release of 18–40 Pg C with a 1°C warming of soils. This release is based on an assumed steady state, which may occur over 10–100 years. If 0.5°C warming over the past 100 years scales to the TEM results, it will release approximately 10 Pg C from soils, which will provide enough N to potentially fix 40 Pg C in biomass. Since soils would not have reached steady state as a result of such a temperature perturbation, the release may possibly be only one-half this amount, i.e., 5 Pg soil organic carbon. When spread over 50 years, this number gives an additional release of $0.1 \, Pg \, C \, yr^{-1}$ from soils. Obviously, these calculations are highly speculative, and the transfer of soil N to plants would not be 100% efficient. Yet the calculations serve to indicate that this mechanism could provide sufficient N to support a $0.4–1.0 \, Pg \, C \, yr^{-1}$ growth in terrestrial biomass for a few decades.

Nitrogen Deposition

Air pollution has resulted in a deposition of approximately $18 \, Tg \, N \, yr^{-1}$ on terrestrial ecosystems. Depending upon how one assumes this deposition to be distributed between soil organic matter and vegetation, Peterson and Melillo (1985) show that this additional N could result in a sequestering of at most $0.2 \, Pg \, C \, yr^{-1}$. It is likely that a portion of this N is lost through denitrification (either locally, or after leaching, in riparian or aquatic systems), so these represent upper estimates. Furthermore, the deposition of N is not spatially uniform, so dramatic increases in high deposition regions should be observed if C sequestration is occurring. Increased forest growth has been observed in the managed forests of Europe. It is not clear, however, if this is due to N deposition alone. Some of this N also falls on agricultural land where it will result in small rates of carbon sequestration.

In addition to this inadvertent N input, via air pollution, approximately 77 Tg N yr^{-1} (IFA 1992) is applied to agricultural crops as fertilizer. Perhaps 50% of the fertilizer is used in intensive, frequently tilled agricultural systems where past experience has indicated little increase in NLOM with added fertilizer, especially where most production is removed from the field. The other 50%, applied to systems, with a better potential for conservation (grasslands, crops in rotation, dryland cereal production, zero-tillage systems), still results primarily in production removed from the fields; however, a proportion of the fertilizer is fixed in organic matter that remains in the field. If 20% of the N ends up in the soil over the medium to long term, this would result in $0.076 \, Pg \, C \, yr^{-1}$ fixed.

In the semiarid region of western Canada, the use of N fertilizer over about 25 years in cereal rotations has resulted in about 0.1% increase in organic C, from 1.6% to 1.7%. A series of calculations indicates that about 1 Mg (1 Mg $= 10^6$ g) of N fertilizer used (over 25 years) results in about 2 Mg of C fixed. Assuming, as before, that one-half of the fertilizer is used in "conserving" systems, we return to fertilizer N contributing perhaps less than 0.1 Pg of C yr^{-1}.

Note, however, that other situations may result in very significant organic C increases. P fertilizer additions to P-deficient grasslands in New Zealand can result in increased N fixation, much improved production, and considerable organic matter stored for each unit of P fertilizer.

The above discussion provides estimates of carbon sequestering that could result from processes associated with increasing net ecosystem production. There are also possible changes that could result in decreases in decomposition rate that would have the same effect in sequestering more carbon in soil.

Humification Ratio

Data from agricultural fields and forest ecosystems indicate that the interaction between soil moisture, pH, and type of plant material can result in changes in the proportion of plant material that is converted to humified organic matter rather than mineralized to CO_2 (as indicated by the fulvic acid:humic acid ratio). It is impossible to estimate the magnitude of such an effect, since regions where the humification ratio is either increasing or decreasing have not been identified or investigated. We do not completely understand all of the factors that affect the rates of humus formation. Melillo et al. (this volume) states that we may need to learn more about how global change affects the rate of soil humus formation at many levels, including biochemical, to understand and predict soil carbon storage.

Charcoal

The rates of charcoal formation from biomass burning are not well known. The yield of charcoal during any burning event varies and depends upon many factors, including the size distributions and water content of materials, temperature of the fire, and environmental conditions. Two percent of the biomass burned may remain as charcoal in the soil after a fire in semiarid regions. Using this rate as an example, let us assume that the entire biosphere burned on average every 200 years. The global rate of charcoal formation would then be 0.056 Pg C yr^{-1}. This is probably an extremely high estimate, since if this rate had held for the past 10,000 years there would be as much carbon as charcoal in soil as there is carbon in live vegetation. It is implausible that human land use could generate a change in charcoal formation large enough to be significant in accounting for any of the residual imbalance.

Deep (> 1 m) Soil Organic Matter

Tropical soils are often quite deep (> 10 m). The quantity of organic carbon contained in soil layers deeper than 1 meter, when added to the quantity of approximately $10 \, kg \, C \, m^{-2}$ in the upper soil layers 0–1 m thick as being commonly considered for estimates of soil organic matter (SOM), increase the total soil carbon pool to more than $20 \, kg \, C \, m^{-2}$. This deep soil carbon is not important if it is all inert; however, the presence of living deep roots, the ^{14}C content of deep soil organic matter, and observed production rates of CO_2 in deep Oxisols at Paragominas, Brazil, suggest that a portion (5–10%) of this deep soil organic matter turns over on time scales of a decade or less (Nepstad et al. 1994).

Changes in deep soil carbon pools may occur in response to vegetation changes and CO_2 fertilization (which may increase root biomass). Models that attempt to account for the sequestration of carbon following regrowth of forests after pasture abandonment may underestimate the net C sequestered in these deep rooting systems, potentially underestimating the amount of biospheric C uptake. For example, a 1.5 Pg C yr^{-1} net sink, spread over the area for tropical forests (15 × 10^6 km^2), requires a net sequestration of roughly $0.1 \, kg \, C \, m^{-2} \, yr^{-1}$ in increased root and SOM. While a net change in carbon inventory of the order of $0.1 \, kg \, C \, m^{-2} \, yr^{-1}$ might be difficult to detect, even after a decade of accumulation, the increased ^{14}C content of deep soil carbon reservoirs that would result from new carbon inputs of this magnitude should be detectable after several years' time.

Oceans

The estimate of the current oceanic uptake of anthropogenic CO_2 reported by Houghton (this volume) is based on ocean models. Even direct measurements of ocean uptake (Takahashi et al. 1983; Quay et al. 1992) must employ a model to extrapolate those measurements to global ocean estimates. The models used to estimate the marine uptake of anthropogenic CO_2 assume that there have been no changes in the cycling of carbon by the marine biota. Thus the uptake by the ocean is modeled as an inorganic chemical and physical process. Physical models have been evaluated using transient tracers, such as bomb-produced radiocarbon, bomb-produced tritium, and freons. The models, though by no means perfect, reproduce many of the main features of these transient tracers and lend support to the model-based estimate of anthropogenic CO_2 uptake.

There is no direct proof, however, that the marine biota has remained unchanged over the past two hundred years. It is possible that there is an additional term in the global atmospheric CO_2 budget that could be included due to transient effects of marine biota. If such changes have occurred, it is likely that NLOM has played an important role.

Dissolved Organic Carbon Pool

The size of the dissolved organic carbon (DOC) pool in the ocean may have changed in response to changes in primary production or remineralization. The DOC pool in the oceans contains 600 to 900 Pg C. The bulk of this DOC is material that is uniformly distributed throughout the ocean, with a mean age on the order of 6,000 years (Williams and Druffel 1987). This uniformly distributed, old DOC must be fairly resistant to decomposition. On average, every 1,000 years the general circulation of ocean waters should cause this DOC to pass through the surface ocean where, presumably, decomposition processes are most active (Najjar et al., this volume). On average, a resistant DOC molecule makes the transit through surface waters 6 times before mineralization, assuming a steady-state circulation over thousands of years. Assuming additional equilibrium conditions, resistant DOC would have a formation rate of $0.1 \, Pg \, C \, yr^{-1}$ balanced by an equivalent decomposition rate. One would have to postulate fairly dramatic changes in the ocean carbon cycle in order for the resistant DOC pool to be substantially involved in the missing sink question.

In the surface waters there is an additional labile DOC pool containing 20 to 30 Pg C with a decadal-scale turnover time (Druffel et al. 1992). It is known that decadal-scale variations in ocean circulation occur, and these may, in some yet unknown way, have affected primary production, labile DOC remineralization, or both. The size of the labile DOC pool in the surface ocean and its turnover time in the ocean is such that it is possible for the DOC pool to play some role in the missing sink question. Because of the methodological problem of measuring DOC in seawater, there is currently considerable uncertainty about the magnitude of the DOC pool and its role in ocean carbon dynamics.

New Production

The downward flux of NLOM in the form of sinking particulate organic carbon (POC) may also have changed over the past 200 years. Like DOC, the POC flux at the base of the euphotic zone, which is also called new production, may have changed in response to changes in primary production or remineralization rate within the euphotic zone. It is possible that these processes may have changed in response to climate change, ocean circulation, or both. Estimates of the sinking flux of POC at the base of the euphotic zone range from 5 to 20 Pg C yr^{-1}. Thus a modest fractional increase in this flux may account for some of the missing sink. Whether this could be sustained over a long period of time is questionable, since the nitrate required to fuel this additional production would eventually be lost from the euphotic zone. Utilization of all of the nitrate in the top 100 meters (which has an average concentration of $5 \, \mu mol \, kg^{-1}$) would result in a carbon loss of 10 Pg C, assuming a C:N ratio of 8.1 (130:16). To account for the cumulative 105 Pg C since preindustrial time, it seems that the marine carbon cycle would be required to change in a dramatic and implausible manner.

Biological Production

Changes in the rates of DOC formation, new production, and rate of biogenic sedimentary organic matter accumulation, however, depend ultimately upon changes in biological production. Biological production, over annual or decadal time scales, is largely determined by nutrient availability in surface waters. Macronutrients (N and P) largely come from mixing with nutrient-rich deep water through wind-driven and thermohaline circulation. Availability of micronutrients, such as iron, requires external inputs from the continents that are independent of ocean circulation. Direct evidence for recent change in ocean biological production is equivocal. A time series of Secchi disk measurements indicate increases in production in some places and decreases at other locations in the North Pacific (Falkowski and Wilson 1992). Other data, such as long-term sediment trap records, show no systematic changes in the direction required to explain the missing sink.

Biological production in some estuaries and adjacent coastal waters has increased as a result of river pollution. Berner (1992) estimates that a global total of $0.1–0.2$ Pg C yr^{-1} may be preserved on the continental shelf. Deposition of nitrogen resulting from atmospheric pollution, if it falls in oligotrophic gyres could increase biological production. It is estimated that 12 Tg N may be deposited from this source and that it could, using a Redfield C:N ratio of 8.1, result in 0.097 Pg C yr^{-1} of additional production. Recent changes in these fluxes resulting from pollution would therefore be quite small, falling far short of the residual imbalance.

PAST CHANGES IN GLOBAL NLOM DYNAMICS

The amounts of NLOM on land and oceans, rates of its formation and destruction, as well as the influences of NLOM on carbon cycle processes have changed over geological time. The magnitudes of the changes are as large as or even larger than those we considered above but occurred over much longer periods. Nevertheless, consideration of past change in the global carbon cycle is instructive in suggesting potentials and constraints in current global carbon cycle processes.

Geological Time NLOM Changes

The organic carbon in sedimentary rocks is a large pool, consisting of 12,000,000 Pg C. The sedimentary record supports the observation of spatial complexity on time scales ranging from thousands to tens of millions of years.

It is estimated that 60% of this organic matter accumulated during the middle to upper Cretaceous and upper Jurassic periods of the Earth's history. Such episodic carbon-burial events varied in length and timing from place to place.

While the processes that are responsible for producing these organic-rich rocks are probably quite similar to those operating today, the global situation was very different in the geological past than today. In particular, the continental configuration, the hypsographic distribution, sea level, types of terrestrial plants (angiosperms do not appear until the middle to upper Cretaceous), and hydrologic cycle were vastly different. As a result, very few present environments qualify as satisfactory analogues for the conditions under which ancient organic rich rocks were deposited.

For geologically old time periods it is difficult to relate variation in atmospheric CO_2 levels and rates of organic sediment formation. At these time scales, the rate of CO_2 production from volcanos and removal by formation of limestone and weathering of rocks also varied considerably.

Quaternary NLOM Changes

In the recent geological past, say the last million years, marine organic matter accumulation rates have varied on glacial–interglacial time scales. Such cyclicity fundamentally reflects climatically driven changes in new production or export of organic matter from the mixed layer of the upper ocean. This may be mediated by iron inputs to the ocean as well as by nutrient fluxes to the mixed layer. One of the more important areas of high new production is the Southern Ocean. This region is frequently invoked as a potentially critical zone for enhanced production (or nutrient utilization) during glacial stages. Such augmentation of the flux of settling organic matter could theoretically play a major role in drawing down the atmospheric CO_2 level during glacial time, as is seen in ice core records. However, recent interpretation based on five different paleogeochemical tracers (biogenic opal accumulation rates [Mortlock et al. 1991], particle fluxes determined by [230]Th and [231]Pa measurements [Kumar et al. 1993], Ba and organic carbon concentration and accumulation in sediments [Shimmield et al. 1994], and $\delta^{15}N$ in sedimentary organic matter [Francois and Altabet 1992; Shemesh et al. 1993]) all rule out the possibility of increased production south of the Antarctic polar front during the Last Glacial Maximum. It now appears that productivity variations in the glacial Southern Ocean are not critical to drawing down atmospheric CO_2 during glacial time. A glacial-period sink apparently must be elsewhere.

Figure 10.1 shows the pattern of atmospheric CO_2 concentration from the present to 150,000 years ago. Superimposed on this curve is a record of organic carbon accumulation rate in marine sediment compiled from low-latitude sediment cores. The temporal pattern of organic carbon accumulation rate suggests that variations in atmospheric CO_2 concentrations may be influenced

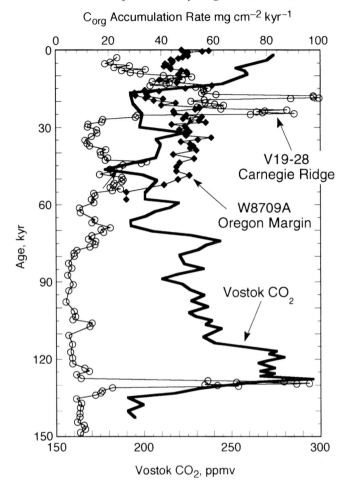

C_{org} Accumulation Rate mg cm^{-2} kyr^{-1}

Figure 10.1   Comparison of atmospheric CO_2 concentration (data after Barnola et al. 1987) and organic carbon accumulation rates in marine sediments from two locations in the Pacific Ocean (data from Lyle et al. 1988, 1992). Spatial differences point out the need to construct global organic-carbon sediment accumulation-rate averages to compare with atmospheric CO_2 rather than extrapolating from limited samples.

by burial of carbon in marine sediments. Also plotted in Figure 10.1, however, is a record of organic carbon accumulation rate from temperate zone continental-margin sediment cores, which show a completely different temporal pattern. This points out the difficulties associated with extrapolating such data to a global scale. Currently there is not enough known about the spatial variation in the rates of organic carbon accumulation to construct global budgets of sedimentary carbon during the Quaternary. As a result, it is impossible to relate marine carbon sequestration by sediments directly to atmospheric CO_2 concentration.

Evidence from ice cores shows that atmospheric dust concentrations were much higher globally during the last glacial periods. If iron availability limits open-ocean primary production, as is suggested by Martin et al. (1991), then it is possible that deposition of dust, which contains iron, to the nutrient-rich equatorial oceans during the past glacial maxima might have supported higher biological production and therefore increased rates of organic C accumulation in sediments. Figure 10.2 shows a strong relationship between ice core dust content and organic carbon accumulation rates from two equatorial Pacific ocean sediment cores. This evidence is not a verification of the iron fertilization hypothesis, rather it does not cause one to reject the hypothesis.

Evidence for changes in terrestrial ecosystems during glacial time to present is largely limited to regions where opportunities for preservation of pollen grains in layered sediments or peats are abundant. Harden et al. (1992) have used such information combined with records of glacier retreat to calculate the rate of organic carbon sequestration in newly exposed soils resulting from deglaciation of the Laurentide Ice Sheet in North America. The greatest increase in the rate of carbon sequestration by soils occurred from 8,000 to 4,000 years ago and continues today. Steady state has not been reached today. By extrapolating their results to the globe, Harden et al. estimate that 0.075 to 0.18 $Pg\,C\,yr^{-1}$ would have been sequestered on deglaciated land globally during preindustrial time.

Carbon isotopes have been used to determine shifts between vegetation dominated by C_3 plants under moist conditions and vegetation dominated by C_4 plants under drier conditions in several sites containing soil carbonate deposits (Quade et al. 1989). These historical records of vegetation show large changes in vegetation composition between the present and glacial times. Currently it is not possible to reconstruct completely from such records the distribution of ecosystem types globally over glacial–interglacial time. It is also difficult to interpret past vegetation records in terms of carbon stocks, in living plants and soil organic matter, and carbon fluxes. In many cases, there are not current ecosystems that are analogous to those in the past, adding additional difficulties to such interpretations. Nevertheless, several investigators (Prentice and Fung 1990; Adams et al. 1990; Smith et al. 1992) have made global estimates of carbon stored in terrestrial systems during glacial time. There is little correspondence between their results, which range from little change from the present total of 2,000 Pg C to more than double this amount 18,000 years ago. It is not currently useful to apply these results to global carbon cycle analyses.

Past changes in global carbon cycle dynamics indicate that biological processes in the terrestrial system and ocean can undergo large changes resulting in

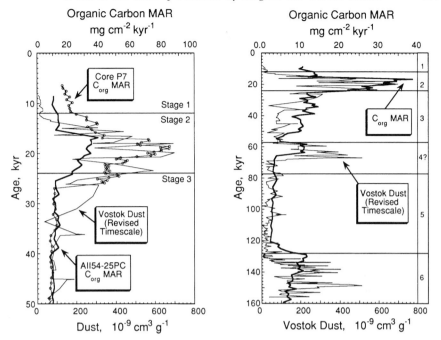

Figure 10.2 Comparison of the C_{org} MAR profiles from cores P7 (open circles) and AII54–25PC (bold line) and the Vostok dust record (thin line, plotted as volume of dust per gram of ice) as a function of time for the last 50 kyr (left) and the last 160 kyr (right). The Vostok dust data (Petit et al. 1990) are plotted on the revised time scale of Shackleton et al. (1992). From Pedersen et al. 1991; reprinted with permission from the American Geophysical Union.

considerable alteration of NLOM pools and associated fluxes. Caution, as pointed out by Sundquist et al. (1993), should be followed when applying insights from the past to current carbon cycle issues. First, the rates of change in global carbon pools and fluxes associated with observable events in the geological record (constrained by fine-scale dating problems and by spatial resolution due to compaction of sediments) occurred at rates an order of magnitude slower than rates of change during recent times needed to account for the fate of fossil-fuel derived CO_2. Second, the forces that are driving the carbon cycle changes are fundamentally different. Present changes in atmospheric CO_2 concentration are caused by a direct addition of CO_2 from a new source. The glacial–interglacial change in atmospheric CO_2 concentration is a consequence of change taking place in the terrestrial system and the ocean due to climate forcings that were further amplified or damped by CO_2 changes.

ADVANCING OUR KNOWLEDGE OF NLOM
AT GLOBAL SCALE

Atmospheric measurements of gases and the isotopic composition of these gases provide valuable information concerning processes related to the global carbon cycle. Because more measuring stations have been established and provided data, it has been possible to obtain more information about the spatial distribution of carbon cycle processes. So far, most gas sampling stations are in locations designed to collect from tropospheric air masses that are well mixed and free from terrestrial influences. As additional stations are added that are designed to collect from air masses reflecting exchanges with terrestrial systems or oceanic water masses, we will be able to refine our knowledge of the spatial variations in carbon exchange with the atmosphere.

In principle, combined measurements of the long-term trends in atmospheric CO_2 and O_2 will determine the relative roles of the oceans and biota in current carbon sequestration (Keeling and Shertz 1992). Spatial gradients in these gases and in $^{13}CO_2$ can provide additional information on the locations of carbon sources and sinks (Tans et al. 1990; Keeling et al. 1989; Keeling et al. 1993). Atmospheric transport makes it difficult to infer the spatial distribution of carbon and oxygen fluxes directly from gas measurements. To account for transport, atmospheric transport models and general circulation models are currently used. Although considerable work on modeling large-scale atmospheric transport still needs to be done, these models, combined with atmospheric gas measurements, can provide estimates of spatially distributed carbon fluxes between the atmosphere, terrestrial systems, and oceans. These indirect carbon flux estimates can provide important constraints on models of terrestrial and ocean exchange with the atmosphere. Terrestrial models and ocean models for directly estimating carbon fluxes have been developed independently. Considerable work on these two classes of models will continue in order to account more accurately for observations and processes represented within the models.

Other group reports discuss how detailed biological, chemical, and physical descriptions of organic matter pools can be incorporated into NLOM models to improve them as tools for exploring hypotheses about NLOM processes. As models incorporate more processes, they become more complex and require more inputs and parameters. For global CO_2 investigations, however, it may be possible to gloss over many details and represent processes more simply. Here we discuss what may be critical components that must be captured in models of NLOM dynamics in order to understand the fate of fossil-fuel-produced CO_2.

Terrestrial NLOM Models

Biological processes play a key role in NLOM decomposition. A simplified terrestrial model, therefore, can represent decomposition as a biochemical process acting on a heterogeneous substrate to meet the energy and nutritional demands of the decomposer organisms. This can roughly be divided into three phases. Fresh, carbon-rich detritus is attacked to meet the energy demands of a growing decomposer population. As the material declines in quality (the yield of energy declines as the organic matter changes in relative proportions of chemical constituents), the carbon to nutrient ratio declines and the organic matter enters a second phase, where further attack is less important in meeting the energy needs and more important in providing nutrients. After a while, the quality declines until it is no longer energetically feasible to attack the material, even for the remaining nutrients. The material then enters a third phase of decomposition, where it is relatively inert or chemically resistant. During the decomposition processes there are environmental factors influencing the decomposer activity rate. The decomposer activity rate is influenced by temperature and water. Low temperatures or low oxygen availability, due to water saturation, can inhibit decomposer activity, thus allowing organic matter to be persistent despite its chemical structure. Soil mineral particles also react with organic matter, which binds both with charged mineral surfaces and dissolved ions. This changes the organic matter to make it less available as a resource to decomposers.

While there is one model that represents these energetic and nutritional processes as a continuous decline of quality (Bosatta and Ågren 1991), most models use three discrete pools that are operationally defined as homogeneous, with different time constants associated with their decay rates. While there is a general correspondence between the three model pools and the three stages identified above, resulting from changes in substrate quality from the decomposer perspective, the environmental influences that result in organic matter persistence cause these operational pools to have various combinations of physically and chemically protected components. Therefore, there are no physical or chemical extraction methods that exactly correspond to these operational pools. Carbon isotope techniques are beginning to aid in identifying the sizes of organic matter pools in soils that correspond to these model concepts (Trumbore 1993; Balesdent et al. 1987). Other methods would be valuable, particularly for use in validating models. A first-order model, therefore, needs information on the rate of supply of fresh detritus and some measure of its initial quality and nutrient content. Often this measure is lignin and nitrogen content. Further research, however, may indicate that additional measures would be helpful. An estimate of the rate of decomposer activity for each pool, for example, is also necessary to determine the rate of mass loss as carbon is oxidized. Decomposer activity depends upon environmental characteristics. Soil temperature may be determined from air temperature and soil hydrologic regime. Soil hydrologic regime can be determined from soil factors such as structure, depth, topography, and climate factors of temperature and precipitation. Physical protection may be quantified simply as an index based on the mineralogy and particle size distribution of the soil. For a global analysis, these basic factors must be supplied with sufficient spatial resolution to be representative of a relatively large land unit for all land units on the globe.

Ocean NLOM Models

In oceans, plants do not produce a significant amount of structural material so carbon:nutrient ratios are low for primary producers. Distinguishing three pools for NLOM as for terrestrial systems is not as important. It is more convenient to define physical pools of NLOM that have different patterns of mobility in seawater. These include particles that are (a) small enough to remain and decompose in the surface ocean, (b) particles that are packed into pellets by grazers, are scavenged by other particles (including mineral matter), or simply large enough to fall below the surface mixed layer before being substantially decomposed (i.e., components of the so called new production), and (c) dissolved organic matter.

Models of ocean NLOM can be generally described as follows. First, the factors that control the rate of biological production must be specified. Important parameters include rate of nutrient supply to the surface ocean; efficiency of nutrient utilization, particularly in nutrient-rich regions (this may involve carbon:nutrient ratios in plant matter, iron availability, and ecological factors such as species composition and zooplankton grazing efficiency); and physical factors such as temperature and light penetration. Second, the relationships between gross biological production and the rate of new production, DOC, and the relationship between production and sedimentation rate must be quantitatively described. Equally important, but less well known and quantified are factors that result in destruction and/or production of DOC and particulate organic matter in sediments. Actually, each of these classes of organic material requires the development of submodels to aid in quantifying their dynamics. In the development of these submodels, multiple analytical tools must be applied simultaneously to answer specific questions. In studies of the sedimentary record, for example, such tools include stable isotopic analyses ($\delta^{13}C$, $\delta^{15}N$), inorganic chemical measurements (opal, $CaCO^{-3}$, Ba), and organic chemical determinations (biomarkers). Multidisciplinary collaboration is required to apply this suite of analysis.

Global ocean databases, particularly of nutrient and oxygen distributions, are key elements in evaluating marine carbon cycle models. In principle, these models can be used to distinguish the relative importance of the different NLOM pools in the marine carbon cycle (Najjar et al. 1992).

SUMMARY

As the increasing use of fossil fuel continues to add to the accelerating increase of CO_2 in the atmosphere, and irrefutable evidence accumulates concerning its effects on climate, policymakers will make increasingly more difficult demands on the scientific community to determine whether or not expensive carbon management policies should be mandated. We will be asked whether a 20% cutback in fossil-fuel emissions will result in noticeable declines in atmospheric CO_2 concentration. Was the CO_2 increase in 1991 smaller than expected because of a cold summer in Siberia, because of increased upwelling off the cost of Peru, or because of effects due to the Mt. Pinatubo eruption? We do not yet understand the global carbon cycle well enough to to answer such questions. The role of NLOM dynamics plays a pivotal role in carbon cycle questions concerning the potential of the Earth's biosphere to sequester or release carbon in response to global changes.

There are broad biogeochemical constraints on how responsive terrestrial ecosystems can be to global changes. It is conceivable that a combination of CO_2 fertilization and climatic effects are increasing the amount of carbon stored in vegetation and soil, particularly in unmanaged ecosystems. These increases will eventually reach a saturation point due to some combination of resource limitations. Land use is also increasing and, as a result, biotic potential is decreasing. If the missing sink is found to be a terrestrial phenomenon, its capacity is likely to decrease in the future. We need, however, to be more precise than this for our science to be useful.

For ocean uptake of CO_2 we can be a little more precise if it can be assumed that physical and chemical processes are of primary importance in CO_2 uptake and that ocean biological processes remain constant. This assumption, though, requires rigorous testing. Again, as for terrestrial systems, there are broad biogeochemical constraints on how much and how fast the ocean biology and NLOM dynamics can change. We are only beginning to explore the magnitudes of the pools and fluxes involved. It will be somewhat longer before we understand the nature of most interactions, quantify them, and understand the forces that can make them change.

REFERENCES

Adams, J.M., H. Faure, L. Faure-Denard, J.M. McGlade, and F.I. Woodward. 1990. Increases in terrestrial carbon storage from the last glacial maximum to the present. *Nature* **348**:711–714.

172   *W.M. Post et al.*

Balesdent, J., A. Mariotti, and B. Guillet. 1987. Natural ^{13}C abundance as a tracer for studies of soil organic matter dynamics. *Soil Biol. Biochem.* **19**:25–30.

Barnola, J.M., D. Raynaud, Y.S. Korotkevich, and C. Lorius. 1987. Vostok ice core provides 160,000-year record of atmospheric CO_2. *Nature* **329**:408–414.

Bazzaz, F.A. 1990. The response of natural ecosystems to the rising global CO_2 levels. *Ann. Rev. Ecol. Syst.* **21**:167–196.

Berner, R.A. 1992. Comments on the role of marine sediment burial as a repository for anthropogenic CO_2. *Glob. Biogeochem. Cyc.* **6**:1–2.

Bosatta, E., and G.I. Ågren. 1991. Dynamics of carbon and nitrogen in the organic matter of the soil: A generic theory. *Am. Nat.* **138**:227–245.

Coûteaux, M.-M., M. Mousseau, M.-L. Célérier and P. Bottner. 1991. Increased atmospheric CO_2 and litter quality: Decomposition of sweet chestnut litter with animal food webs of different complexities. *Oikos* **61**:54–64.

Druffel, E.R.M., P.M. Williams, J.E. Bauer, and J.R. Ertel. 1992. Cycling of dissolved and particulate organic matter in the open ocean. *J. Geophys. Res.* **97**:15,639–15,659.

Falkowski, P.G., and C. Wilson. 1992. Phytoplankton productivity in the North Pacific ocean since 1900 and implications for absorption of anthropogenic CO_2. *Nature* **358**:741–743.

Francois, R., and M. Altabet. 1992. Glacial to interglacial changes in surface nitrate utilization in the Indian sector of the Southern Ocean as recorded by sediment $\delta^{15}N$. *Paleoceanography* **7**:589–603.

Harden, J.W., E.T. Sundquist, R.F. Stallard, and R.K. Mark. 1992. Dynamics of soil carbon during deglaciation of the Laurentide ice sheet. *Science* **258**:1921–1924.

IFA 1992. Nitrogen fertilizer statistics 1986/87 to 1990/91. International Fertilizer Industry Association Report A/92/147, Nov. 1992. Netherlands.

Jacoby, G.C., and R. D'Arrigo. 1989. Reconstructed Northern Hemisphere annual temperature since 1671 based on high latitude tree-ring data. *Clim. Change* **14**:39–59.

Keeling, C.D., R.B. Bacastow, A.F. Carter, S.C. Piper, T.P. Whorf, M. Heimann, W.G. Mook, and H. Roeloffzen. 1989. A three-dimensional model of atmospheric CO_2 transport based on observed winds: 1. Analysis of observational data. In: Aspects of Climate Variability in the Pacific and Western Americas, ed. D.H. Peterson, Geophysical Monograph 55, pp. 165–236. Washington, D.C.: Am. Geophysical Union.

Keeling, R.F., R.P. Najjar, M.L. Bender, and P.P. Tans. 1993. What atmospheric oxygen measurement can tell us about the global carbon cycle. *Glob. Biogeochem. Cyc.* **7**:37–67.

Keeling, R.F., and S.R. Shertz. 1992. Seasonal and interannual variations in atmospheric oxygen and implications for the global carbon cycle. *Nature* **358**:723.

Kumar, N., R. Gwiazda, R.F. Anderson, and P.N. Froelich. 1993. $^{213}Pa/^{230}Th$ ratios in sediments as a proxy for past changes in Southern Ocean productivity. *Nature* **362**:45–48.

Lyle, M., D.W. Murray, B.P. Finney, J. Dymond, J.M. Robbins, and K. Brookforce. 1988. The record of late Pleistocene biogenic sedimentation in the eastern tropical Pacific Ocean. *Paleoceanography* **3**:39–60.

Lyle, M., R. Zahn, F. Prahl, J. Dymond, R. Collier, N. Pisias, and E. Suess. 1992. Paleoproductivity and carbon burial across the California Current: The multitracers transect, 42N. *Paleoceanography* **7**:251–272.

Martin, J.H., R.M. Gordon, and S.E. Fitzwater. 1991. The case for iron. *Limnol. Oceanogr.* **36**:1793–1802.

McGuire, A.D., J.M. Melillo, L.A. Joyce, D.W. Kicklighter, A.L. Grace, B. Moore, and C.J. Vorosmarty. 1992. Interactions between carbon and nitrogen dynamics in estimating net primary productivity for potential vegetation in North America. *Glob. Biogeochem. Cyc.* **6**:101–124.

Melillo, J.M., A.D. McGuire, D.W. Kicklighter, B. Moore, C.J. Vorosmarty, and A.L. Schloss. 1993. Global climate change and terrestrial net primary production. *Nature* **363**:234–240.

Mortlock, R.A., C.D. Charles, P.N. Froelich, M.A. Zibella, J. Saltzman, J.D. Hays, and L.H. Burkle. 1991. Evidence for lower productivity in the Antarctic Ocean during the last glaciation. *Nature* **351**:220–223.

Najjar, G.R., J.L. Sarmiento, and J.R. Toggweiler 1992. Downward transport and fate of organic matter in the ocean: Simulations with a general circulation model. *Glob. Biogeochem. Cyc.* **6**:45–76.

Nepstad, D.C., P. Jipp, E.A. Davidson, R. Reis, R. Vieira, and S. Trumbore. 1994. Deforestation affects linkages between carbon and hydrological cycles in deep soil of the eastern Amazon basin. *Science,* in press.

Norby, R.J., J. Pastor, and J.M. Melillo. 1986. Carbon–nitrogen interactions in CO_2-enriched white oak: Physiological and long-term perspectives. *Tree Physiol.* **2**:233–241.

Pedersen, T.F., B. Nielsen, and M. Pickering. 1991. Timing of late Quaternary productivity pulses in the Panama basin and implications for atmospheric CO_2. *Paleoceanography* **6**:657–677.

Peterson, B.J., and J.M. Melillo. 1985. The potential storage of carbon caused by eutrophication of the biosphere. *Tellus* **37B**:117–127.

Petit, J.R., L. Mounier, J. Jouzel, Y.S. Korotkevich, V.I. Kotlyakov, and C. Lorius. 1990. Palaeoclimatological and chronological implications of the Vostok core dust record. *Nature* **343**:56–58.

Prentice, K.C., and I.Y. Fung. 1990. The sensitivity of terrestrial carbon storage to climate change. *Nature* **346**:48–51.

Quade, J.T., E. Cerling, and J.R. Bowman. 1989. Development of Asian monsoon revealed by marked ecological shift during the latest Miocene in northern Pakistan. *Nature* **342**:163–166.

Quay, P.D., B. Tilbrook, and C.S. Wong. 1992. Oceanic uptake of fossil fuel CO_2: Carbon-13 evidence. *Science* **256**:74–79.

Shackleton, N.J., J. Le, A. Mix, and M.A. Hall. 1992. Carbon isotope records from Pacific surface waters and atmospheric carbon dioxide. In: Glacial–Interglacial Variability of the Climatic System, ed. D.Q. Bowen, pp. 387–400. New York: Pergamon.

Shemesh, A., S.A. Macko, C.D. Charles, and G.H. Rau. 1993. Isotopic evidence for reduced productivity in the glacial Southern Ocean. *Science* **262**:407–410.

Shimmield, G.B., S. Derrick, A. Mackensen, H. Grobe, and C. Pudsey. 1994. The history of barium, biogenic silica, and organic carbon accumulation in the Weddell Sea and Antarctic Ocean over the last 150,000 years. In: Carbon Cycling in the Glacial Ocean: Constraints on the Ocean's Role in Global Change, ed. M.K.R. Zahn, L.D. Labeyrie, and T.F. Pedersen. Heidelberg: Springer, in press.

Smith, T.M., R. Leemans, and H.H. Shugart. 1992. Sensitivity of terrestrial carbon storage to CO_2-induced climate change: Comparison of four scenarios based on general circulation models. *Clim. Change* **21**:367–384.

Sundquist, E. 1993. The global carbon dioxide budget. *Science* **259**:934–941.

Takahashi, T., D. Chipman, and T. Volk. 1983. Geographical, seasonal, and secular variations of the partial pressure of CO_2 in the surface waters of the North Atlantic Ocean. The results of the TTO Program. pp. 15–143. Proceeding: Carbon Dioxide, Science and Consensus. CONF 820970. Washington, D.C.: U.S. Dept. of Energy.

Tans, P.P., I.Y. Fung, and T. Takahashi. 1990. Observational constraints on the global atmospheric CO_2 budget. *Science* **247**:1431–1438.

Trumbore, S.E. 1993. Comparison of carbon dynamics in temperate and tropical soils using radiocarbon measurements. *Glob. Biogeochem. Cyc.* **7**:275–290.

Williams, P.M., and E.R.M. Druffel. 1987. Radiocarbon in dissolved organic matter in the central North Pacific Ocean. *Nature* **330**:246–248.

11

Global Change and Its Effects on Soil Organic Carbon Stocks

J.M. Melillo, D.W. Kicklighter, A.D. McGuire,
W.T. Peterjohn, and K.M. Newkirk
The Ecosystems Center, Marine Biological Laboratory, Woods Hole,
Massachusetts 02543, U.S.A.

INTRODUCTION

Soil plays a central role in the global carbon budget. It is the storehouse of an estimated 1300 to 1500 Pg (10^{15}g) carbon in nonliving organic matter (NLOM): an amount of carbon about twice as large as that in the atmosphere and almost three times as large as that in living plants (Watson et al. 1990). As scientists, resource managers, and politicians consider the possible responses of terrestrial ecosystems to global change, they have begun to ask a series of questions about the soil's organic carbon pool: How will this large carbon pool respond to global change? Will it increase in size and function as a net sink of atmospheric carbon? Or will it decrease in size and act as a net source of carbon to the atmosphere?

To help answer these critical questions, our research group in Woods Hole, in collaboration with colleagues at the University of New Hampshire, has been conducting a soil-warming experiment in a mid-latitude forest and building a coupled plant–soil model for the terrestrial part of the globe. Both the field experiment and the modeling activity point to the need for understanding carbon–nitrogen interactions at the whole ecosystem level in order to address global change questions about terrestrial ecosystem responses. In this chapter we explore some aspects of carbon–nitrogen interactions in the whole ecosystem context.

This chapter is divided into three parts. In the first part, we report the results to date of our soil-warming experiment, with special emphasis on carbon and

Role of Nonliving Organic Matter in the Earth's Carbon Cycle
Edited by R.G. Zepp and Ch. Sonntag © 1995 John Wiley & Sons Ltd.

nitrogen dynamics. We point to carbon–nitrogen linkages when we review the global application of our coupled plant–soil model in the second part and identify future research topics in the area of carbon–nitrogen linkages in the final section of the paper.

SOIL WARMING

Current models of climate change predict that the global mean temperature will increase about 3°C during the next century (Houghton et al. 1992). One result of global warming may be changes in the rates of temperature-dependent soil processes such as decomposition (Swift et al. 1979) and net nitrogen mineralization, and nitrification (Focht and Verstraete 1977). Here, we present the results from the first two years of our ongoing soil-warming experiment.

The purpose of this experiment is to determine the response of belowground processes in a mixed deciduous forest to elevated soil temperatures, with special emphasis on soil processes that could significantly alter ecosystem function, atmospheric chemistry, and global climate. It is being conducted at the Harvard Forest in central Massachusetts (42°30'N, 72°10'W). The research site is slightly more than 0.1 ha in an even-aged, mixed deciduous forest, where the dominant tree species are black oak (*Quercus velutina* Lam.) and red maple (*Acer rubrum* L.), with some paper birch (*Betula papyrifera* Marsh.) and striped maple (*Acer pensylvanicum* L.).

Soils are mainly of the Gloucester series (fine loamy, mixed, mesic Typic Dystrochrept) with a surface pH of 3.83 and subsurface pH of 4.85. The average bulk density of the upper 15 cm is 0.64 g cm^{-3}. A distinct Ap horizon indicates past cultivation, and historical records confirm that this old-field forest was established after abandonment at the turn of the century (pers. comm., David F. Foster, Director of Harvard Forest). There has been some cutting for firewood subsequent to abandonment. The climate is cool temperate and humid. The mean weekly air temperature varies from a high of approximately 20°C in July to a low of –6°C in January (Spurr 1957). Precipitation is distributed evenly throughout the year and annually averages approximately 108 cm (Spurr 1957).

At the research site we established eighteen 6 × 6 m plots in April 1991 (Figure 11.1). The plots are grouped into six blocks and the three plots within a block were randomly assigned to one of three treatments. The treatments are:

1. heated plots in which the average soil temperature is elevated 5°C above ambient using buried heating cables;

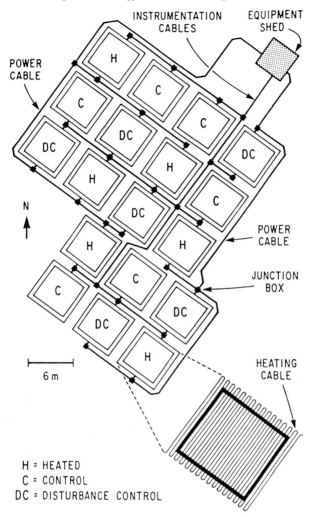

Figure 11.1   Plot design of soil warming experiment at the Harvard Forest.

2. disturbance control plots that are identical to heated plots except they receive no electrical power; and
3. undisturbed control plots that have been left in their natural state.

To warm a given plot, we buried heating cables at a depth of 10 cm in 6-m long rows spaced 20 cm apart. When supplied with 240 VAC, the heating cables have a power output of 13.6 W m^{-1} and produce a power density of approximately 77 W m^{-2}. This method of warming soil has been previously evaluated and performs well under a variety of moisture and temperature conditions (Peterjohn et al. 1993). Details of the system's operation are given in Peterjohn et al. (1994).

In a prototype soil-warming experiment (Peterjohn et al. 1993), we found that soil temperatures were consistently lower near the edges of a 6×6 m heated area. To avoid regions where the temperature is $< 5°C$ above ambient, all samples in the Harvard Forest experiment were taken from a 5×5 m experimental area that is nested within each 6×6 m plot.

We measured a variety of parameters including CO_2 emissions from the soil and indexes of N availability on two occasions every month. For a given month, nine plots (three blocks) were sampled on the first occasion and the remaining nine plots were sampled on the second. Thus, each plot was sampled once a month.

We measured the net flux of CO_2 between the soil and the atmosphere using previously published sampling and analytical techniques (Raich et al. 1990). Briefly, we made these measurements by placing static chambers over the surface of the soil for 30 minutes and sampling the headspace at 10-min intervals. The samples were analyzed for trace gas concentrations by gas chromatography and the changes in concentration were used to calculate net flux rates. On each sampling date, fluxes were measured in one chamber per plot at 0600, 1000, 1400, and 1800 hrs. Adjacent to each chamber, we also measured soil temperatures at depths of 2 and 4 cm. Throughout the experiment the locations of the chambers never changed.

Net N mineralization and nitrification were measured for the forest floor and mineral soils using the buried bag technique (Eno 1960; Westermann and Crothers 1980; Pastor et al. 1984). Subsamples of two soil cores from each plot were extracted for 48 hrs with 2 N KCl; the remaining soil was used to measure soil moisture gravimetrically. Soil cores adjacent to the initial cores were incubated *in situ* in polyethylene bags for 22–36 days before determining inorganic N concentrations and calculating the net change in extractable NH_4^+ and NO_3^-. Soil extracts were analyzed for NH_4^+ and NO_3^- using continuous flow colorimetry (Technicon Methods 780–86T and 782–86T). After sampling, all holes were filled with soil from outside the research site and identified to avoid resampling.

Results from the first year of study (July 1991 through June 1992) at the Harvard Forest soil warming experiment indicated that heating increased emission of CO_2 and nitrogen mineralization. We estimated CO_2 fluxes to be approximately 7,100 kg C ha^{-1} yr^{-1}, 7,900 kg C ha^{-1} yr^{-1}, and 11,100 kg C ha^{-1} yr^{-1} in the control, disturbance control, and heated plots, respectively; this represents an increase due to warming of approximately 3,200 kg C ha^{-1} yr^{-1} (Figure 11.2). Nitrogen mineralization was doubled in the forest floor plus the top 10 cm of the mineral soil. During the first growing season, we observed that in the forest floor, heating increased the average net mineralization rates from

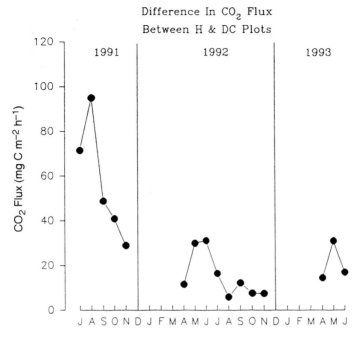

Figure 11.2 Difference in CO_2 flux between the heated and disturbed control plots (see text) at the warming experiment at the Harvard Forest. Two full years of data are presented. H = heated; DC = disturbance control.

1.02 to 2.47 mg N kg^{-1} d^{-1}. In the mineral soil, heating increased the average net mineralization rates from 0.09 to 0.19 mg N kg^{-1} d^{-1}. No nitrification was observed in either the forest floor or the mineral soil in any of the treatments.

Results from the second year of study (July 1992 through June 1993) indicated that warming had a much less dramatic effect on CO_2 flux, but a sustained dramatic effect on nitrogen mineralization. We estimated CO_2 fluxes in the second year to be approximately 6,800 kg C ha^{-1}, 7,600 kg C ha^{-1}, and 8,700 kg C ha^{-1} for control, disturbed control, and heated treatments, respectively. Nitrogen mineralization rates remained doubled. Again, no nitrification was observed in either the forest floor or the mineral soil.

One interpretation of the differences in the CO_2 response (heated versus disturbed control) between the first and second years versus the constant response in nitrogen mineralization is as follows: (a) there are two major NLOM pools—a fast and a slow pool; (b) the fast pool contains high C:N ratio material (e.g., recent litter)—the decay of fast pool material results in large CO_2 losses per unit of material processed and small net nitrogen release in the mineralization process; (c) the slow pool contains low C:N ratio material (e.g., metastable humus)—the decay of the slow pool material results in a smaller loss of CO_2 per unit of material processed and large net nitrogen release in the mineralization process. If this is realistic, it suggests that the fast pool is relatively unimportant for ecosystem response to climate change over the long term and that the slow pool is the key to ecosystem response to climate change over the long term. As elevated soil temperature increases the rate of decay in the slow pool, both carbon (as CO_2) and nitrogen (as inorganic N) are released. The nitrogen then becomes available to be taken up by plants. Since the C:N ratio of plant material is substantially larger than the C:N ratio of NLOM in the slow pool, warming may lead to increased carbon storage in the ecosystem. The magnitude of such an increase would, of course, depend on how plant carbon balance is affected by other components of climate change including the availability of water and the effect of increased temperature on both photosynthesis and respiration.

COUPLED PLANT–SOIL MODEL

Over the last few years, a number of research teams have begun to build simulation models for the purpose of predicting the responses of terrestrial ecosystems to global change. Our research group has developed one of these models and we call it TEM, an acronym for the Terrestrial Ecosystem Model (Raich et al. 1990; McGuire et al. 1992; Melillo et al. 1993). The model uses information on soils, vegetation, and climate variables such as temperature, precipitation, cloudiness, and solar radiation input to make monthly estimates of important transfers of carbon and nitrogen in ecosystems. The data sets used in TEM are currently organized into 0.5 latitude by 0.5 longitude parcels that cover the vegetated surface of the globe.

We applied the model to more than 56,000 of these parcels to develop estimates of plant and soil carbon stocks for current climate conditions and atmospheric CO_2 levels. Our model-based estimate of the total terrestrial carbon stocks is 2279 Pg (Table 11.1): 977 Pg C in the vegetation and 1301 Pg C in the soil.

Since we do not yet include a consideration of land-use change in the model, the estimates are for potential plant and soil carbon stocks, not present-day stocks. The conversion of natural ecosystems that have large woody plant components into agricultural systems throughout the world has certainly reduced the globe's plant carbon pool (Houghton et al. 1983; Melillo et al. 1988). This pool has also been reduced by firewood gathering, especially in poor tropical countries. In addition, the present-day soil carbon stock may be less than the potential stock because of the initial effects of converting natural ecosystems to agricultural systems that are tilled.

When both land area and carbon density are considered, tropical ecosystems account for about 47% of the total terrestrial carbon pool: 56% of the

Table 11.1 Response of carbon storage (10^{15} g C) by region for experiment involving two levels of atmospheric CO_2 and five levels of climate.

CO2 concentration	312.5 ppmv					625.0 ppmv				
Climates:	Contemporary[1]	GFDL 1	GFDL Q	GISS	OSU	Contemporary	GFDL 1	GFDL Q	GISS	OSU
Vegetation carbon										
Northern ecosystems[2]	140	+43	+34	+32	+31	+4	+71	+51	+42	+40
Temperate ecosystems[3]	290	−1	+5	+24	+13	+33	+67	+71	+80	+63
Tropical ecosystems[4]	547	−37	−40	−50	−74	+108	+116	+116	+110	+68
Total vegetation carbon	977	+5	−1	+6	−30	+145	+254	+238	+232	+171
Soil carbon										
Northern ecosystems	327	−42	−31	−17	−8	+11	−7	−6	+2	+8
Temperate ecosystems	441	−66	−67	−47	−34	+73	+28	+22	+36	+48
Tropical ecosystems	533	−103	−108	−140	−138	+89	+8	+5	−31	−37
Total soil carbon	1301	−211	−206	−204	−180	+173	+29	+21	+7	+19
Total carbon										
Northern ecosystems	466	+1	+3	+15	+22	+15	+64	+46	+44	+48
Temperate ecosystems	733	−67	−62	−23	−20	+105	+95	+92	+118	+112
Tropical ecosystems	1080	−140	−148	−190	−212	+198	+124	+121	+78	+30
Total terrestrial carbon	2279	−206	−207	−198	−210	+318	+283	+259	+240	+190

[1] Pool size that is used as the baseline for responses.
[2] Northern ecosystems include polar desert, alpine tundra, wet/moist tundra, boreal woodland, and boreal forest. These ecosystems occupy a combined areas of 28.2 million km^2.
[3] Temperate ecosystems include forests, savannas, grasslands, shrublands and deserts. These ecosystems occupy a combined area of 56.7 million km^2.
[4] Tropical ecosystems include forests, savannas and shrublands. These ecosystems occupy a combined area of 42.5 million km^2.

vegetation carbon and 41% of the soil carbon. With most of the world's population growth expected to occur in the Tropics over the next century, these large carbon stocks are likely to be reduced through land-use change activities and perhaps other aspects of global change.

To simulate changes in terrestrial carbon stocks under various climatic conditions, we used the output from four global climate models to run TEM (Melillo et al. 1993). The four climate models, more correctly referred to as general circulation models (GCMs), were: the Goddard Institute of Space Studies GCM (GISS); the Oregon State University GCM, (OSU); and two GCM simulations from the Geophysical Fluid Dynamics Laboratory (GFDL 1 and GFDL Q). Among the GCMs, mean global temperature increases between 2.8°C and 4.2°C, global precipitation increases between 7.8% and 11.0%, and cloudiness decreases between 0.4% and 3.4%. These forecasts describe climate under conditions of doubled atmospheric CO_2.

Changes in climate with no change in CO_2 concentration are predicted by the model to result in substantial losses of carbon from terrestrial ecosystem (Table 11.1). For all four GCM climates, the predicted carbon storage reductions are about 200 Pg and they come mostly from soils. This represents a 15% loss from the soil carbon pool.

Tropical ecosystems account for much of the global response. With the GFDL climates they account for about 75% of the carbon losses and with the GISS and OSU climates they account for virtually all of the losses.

In contrast to tropical systems, our model predicted that northern ecosystems respond to climate change alone by increasing their carbon storage, albeit by relatively small amounts (Table 11.1). The increase in carbon storage occurred because the increase in vegetation carbon was larger than the decrease in soil carbon. The sequence of processes changes leading to this increased carbon storage in the vegetation at the expense of the soil is thought to be as follows: as temperature increases, the soil loses carbon, and nitrogen release from the soil is accelerated (see previous section on soil warming); this results in a plant fertilization response and an increase in plant productivity and plant carbon storage. This sequence represents an important carbon–nitrogen interaction at the ecosystem level.

For doubled CO_2 with no climate change, TEM predicts an increase in global terrestrial carbon storage of 318 Pg, which is a 14% increase in terrestrial carbon storage relative to the estimated contemporary condition (Table 11.1). In absolute terms, tropical systems are the most responsive to CO_2 doubling. Our model predicts that they will account for almost two-thirds of the increased carbon storage from CO_2 doubling. The carbon storage in tropical ecosystems under this scenario is predicted to be almost equally divided between vegetation and soils. Northern ecosystems, including tundra, boreal woodlands and boreal forests, are predicted to be the least responsive. The small response occurs because TEM predicts that productivity in nitrogen-limited ecosystems does not substantially respond to elevated CO_2 alone.

Our model predicts substantial increases in terrestrial carbon stocks in response to changes in both CO_2 and climate, although the magnitude of the increase varies according to GCM climate (Table 11.1). The OSU climate combined with a CO_2 doubling resulted in the smallest increase in terrestrial carbon storage globally—190 Pg (an 8% increase)—while the GFDL 1 climate combined with a CO_2 doubling yielded the largest increase from all the GCM climates—283 Pg (a 12% increase). Changes in vegetation carbon stocks accounted for about 90% of the increased storage.

We are in the early stages of working with TEM and are continuing to explore how its parameterization and structure affect its behavior. Recently, we examined what would happen if we changed the depth of the soil profile from 1 meter to 20 centimeters and then applied a 1°C temperature increase globally. The difference in the soil response was dramatic (Table 11.2). With the 1-meter soil depth parameterization, the soil carbon loss at $1 \times CO_2$ with a 1°C increase was 42.7 Pg, whereas with a 20-cm soil depth parameterization, the loss was 18.8 Pg C.

The results from the 20-cm parameterization can be compared to the results of Schimel et al. (1994), who used both USDA field data and the CENTURY model to estimate soil carbon losses from the upper 20 centimeters of soil globally for a 1°C increase. Using the field data, they estimated a 14.1 Pg C decrease (25% smaller than the decrease estimated with TEM). With CENTURY they estimated an 11.1 Pg C decrease. In the TEM estimate, tropical soils accounted for most of the carbon losses, while in the USDA and CENTURY estimates, extratropical soils accounted for most of the soil carbon losses. The difference between TEM's results and the results of the two approaches reported by Schimel et al. (1994) may be related to the one compartment conceptualization of the soil carbon pool in TEM. We are currently exploring a two compartment conceptualization of soil in TEM to enable us to capture the soil carbon and nitrogen dynamics we have observed in the soil warming experiment at the Harvard Forest.

In another recent modeling exercise, we compared predictions of changes in carbon stocks for the two soil depth parameterizations of TEM when the model was driven by the $2 \times CO_2$ climate predicted by the GFDL 1 GCM. For climate change alone, we found that TEM predicted greater soil carbon losses with the deep soil profile parameterization (Table 11.3); that is, a 211 Pg C loss with the 1-m soil depth parameterization, and a 128 Pg C loss with the 20-cm parameterization. When both climate and CO_2 were changed, the soils acted as a small carbon sink for both parameterizations: 29 Pg C for the 1-m parameterization and 5 Pg C for the 20-cm parameterization.

Table 11.2 Response of carbon storage and annual NPP (10^{15} gC) to +1°C at $1 \times CO_2$ and $2 \times CO_2$ by region.

CO_2 concentration	(parameterized for "1 m" soil carbon)			(parameterized for 20 cm soil carbon)		
Climates:	312.5 ppmv Contemporary (pool size)	312.5 ppmv +1°C (response)	625.0 ppmv +1°C (response)	312.5 ppmv Contemporary (pool size)	312.5 ppmv +1°C (response)	625.0 ppmv +1°C (response)
Vegetation C						
Northern ecosystems	139.5	+9.8	+14.5	139.5	+9.8	+14.5
Temperate ecosystems	290.4	+6.7	+46.0	285.5	+6.3	+44.6
Tropical ecosystems	547.3	−5.8	+121.5	548.3	−5.2	+97.4
Total vegetation C	977.2	+10.7	+182.0	973.3	+10.9	+156.5
Soil C						
Northern ecosystems	327.2	−0.1	+13.0	327.2	−0.1	+13.0
Temperate ecosystems	441.3	−11.4	+65.3	288.5	−5.4	+30.2
Tropical ecosystems	533.0	−31.2	+68.4	221.5	−13.3	+25.8
Total soil C	1301.5	−42.7	+146.7	837.2	−18.8	+69.0
Total C						
Northern ecosystems	466.7	+9.7	+27.5	466.7	+9.7	+27.5
Temperate ecosystems	731.7	−4.7	+111.3	574.0	+0.9	+74.8
Tropical ecosystems	1080.3	−37.0	+189.9	769.8	−18.5	+123.2
Total terrestrial C	2278.7	−32.0	+328.7	1810.5	−7.9	+225.5
NPP						
Northern ecosystems	5.0	+0.2	+0.5	5.0	+0.2	+0.5
Temperate ecosystems	16.1	+0.2	+2.8	16.0	+0.1	+2.7
Tropical ecosystems	30.0	−0.2	+6.7	29.9	0.0	+5.7
Total terrestrial NPP	51.1	+0.2	+10.0	50.9	+0.3	+8.9

Table 11.3 Response of carbon storage and annual NPP (10^{15} gC) to GFDL 1 at $1 \times CO_2$ and $2 \times CO_2$ by region.

CO$_2$ concentration	(parameterized for "1 m" soil carbon)			(parameterized for 20 cm soil carbon)		
Climates:	312.5 ppmv Contemporary (pool size)	312.5 ppmv GFDL 1 (response)	625.0 ppmv GFDL 1 (response)	312.5 ppmv Contemporary (pool size)	312.5 ppmv GFDL 1 (response)	625.0 ppmv GFDL 1 (response)
Vegetation C						
Northern ecosystems	139.5	+43.4	+70.6	139.5	+43.4	+70.6
Temperate ecosystems	290.4	−0.6	+67.5	285.5	−0.4	+66.0
Tropical ecosystems	547.3	−37.4	+115.5	548.3	−36.4	+100.5
Total vegetation C	977.2	+5.4	+253.6	973.3	+6.6	+237.1
Soil C						
Northern ecosystems	327.2	−41.7	−6.9	327.2	−41.7	−6.9
Temperate ecosystems	441.3	−66.4	+27.7	288.5	−41.2	+10.5
Tropical ecosystems	533.0	−103.0	+8.4	221.5	−44.6	+1.4
Total soil C	1301.5	−211.1	+29.2	837.2	−127.5	+5.0
Total C						
Northern ecosystems	466.7	+1.7	+63.7	466.7	+1.7	+63.7
Temperate ecosystems	731.7	−67.0	+95.2	574.0	−41.6	+76.5
Tropical ecosystems	1080.3	−140.4	+123.9	769.8	−81.0	+101.9
Total terrestrial C	2278.7	−205.7	+282.8	1810.5	−120.9	+242.1
NPP						
Northern ecosystems	5.0	+1.4	+2.3	5.0	+1.3	+2.3
Temperate ecosystems	16.1	+0.5	+4.3	16.0	+0.4	+4.3
Tropical ecosystems	30.0	−1.8	+6.4	29.9	−1.6	+5.8
Total terrestrial NPP	51.1	+0.1	+13.0	50.9	+0.1	+12.4

FUTURE RESEARCH

Many basic questions remain unanswered about how terrestrial ecosystems will respond to climate change. Some of the most important ones have to do with plant–soil couplings and carbon–nitrogen interactions. One key question is: Will global change, including an increase in the atmospheric CO_2 concentration, affect plant litter quality and soil processes such as carbon storage and nitrogen mineralization rates? Plants grown under elevated CO_2 often show dramatically increased C:N ratios and sometimes more subtle changes in carbon-fraction chemistry. We have seen examples of this for deciduous tree seedlings (Tables 11.4 and 11.5). Largely because of the changes in C:N ratio of the litter, we see significant changes in the litter's lignin:N ratio (Table 11.4), an index we have used to predict leaf litter decay rates (Melillo et al. 1982). This leads us to the prediction that litter from plants grown in elevated CO_2 will decompose more slowly than those grown at ambient CO_2 (Table 11.4).

Table 11.4 Nitrogen and lignin chemistry, decay indexes and predicted decomposition rates of leaf litter from plants grown at low nutrient availability and two levels of CO_2: 350 ppmv and 700 ppmv.

Species	CO_2 level	Concentrations		Decomposition Indexes		Predicted Decomposition Rate (k)[*]
		% N	% lignin	C:N	lignin/N	
Red maple	350	1.28	11.52	34.82	9.00	0.96
	700	0.43	13.59	106.33	31.61	0.39
Striped maple	350	1.80	10.93	23.88	6.07	1.04
	700	0.51	11.55	85.61	22.65	0.61

[*]Predicted decay rates based on a simple statistical decomposition model developed for leaf litter material decaying at the Harvard Forest (Aber et al. 1990).

Table 11.5 Carbon chemistry of leaf litter from plants grown at low nutrient availability and two levels of CO_2: 350 ppmv and 700 ppmv.

Species	CO_2 level	%NPE	%WS	%acidsol	%lignin	%soluble phenolics[*]
Red maple	350	4.47	49.10	34.91	11.51	16.43
	700	3.73	52.72	29.96	13.59	27.92
Striped maple	350	3.62	50.05	35.40	10.93	7.07
	700	2.52	57.24	28.69	11.55	21.22

[*]A component of the water soluble fraction.

Besides affecting litter decay rates, changes in litter quality may affect the rate of humus formation. We still do not completely understand the pathway(s) by which soil humus is formed. Stevenson (1982) has proposed three major pathways: a lignin modification pathway, a polyphenol-quione pathway, and a sugar-amine pathway (Figure 11.3). It is possible that all three pathways operate in all systems, but to different degrees. The relative mix of carbon compounds in the litter may affect which pathway is dominant, but other environmental factors may also be involved. Stevenson (1982) has made the following speculations. A lignin pathway may be the dominant one in poorly drained soils. In many forests, polyphenols derived from microbial processing of leaf litter may be of considerable importance. In addition, the frequent and sharp fluctuations in temperature and moisture in terrestrial surface soils under harsh continental climate may favor humus synthesis by sugar-amine condensation. If the pathways have different conversion efficiencies of litter carbon into humus, changes in initial litter quality could affect the rate of humus formation. An important question to consider is: Do we need to know more about how global change may affect the rate of litter decay and the rate of soil humus formation at the biochemical level if we want to build predictive models of soil carbon storage in response to global change?

One of the most critical issues in terrestrial global change research is: Will plant and soil processes acclimate to climate change such that after a short period of transient response, the process rates return to "predisturbance"

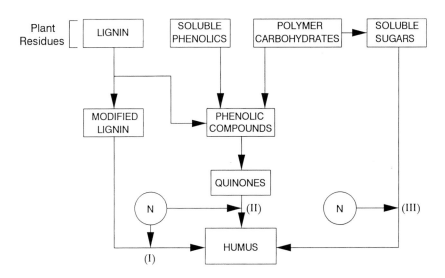

Figure 11.3 Major pathways of soil humus formation (modified from Stevenson 1982).

levels? It is their inability to handle acclimation that makes the predictions of global models of land ecosystems very suspect.

Another critical question is: How will ecosystems respond to rapid changes in climate? If the rapid changes cause ecosystems to "fall apart," with major plant die-back and an increased incidence of fire, the future state of these ecosystems will be very difficult to predict. In this rapid change scenario, soils play a critical role. How will they respond to rapid climate change? How will soil changes affect plant species migrations in the long term? Clearly, the role of NLOM, particularly soil carbon and nitrogen dynamics, will be central to ecosystem responses to rapid global change. Many of these questions will have to be pursued through the use of well-crafted models that capture important ecosystem (coupled plant–soil) dynamics without being too complex. Perhaps one of the most important issues is: What is the minimum amount of information about soil processes that we need to include in coupled plant–soil models so that they can be used in global research?

ACKNOWLEDGEMENTS

This research is being supported by funds from the United States Environmental Protection Agency (Cooperative agreement CR817734–010, Athens Environmental Research Laboratory), the Department of Energy's National Institute for Global Change (Northeast Regional Center, subagreement 901214–HAR), and the National Aeronautics and Space Administration (contract NAGW–1825, subcontract 91–14).

REFERENCES

Aber, J.D., J.M. Melillo, and C.A. McClaugherty. 1990. Predicting long-term patterns of mass loss, nitrogen dynamics, and soil organic matter formation from initial fine litter chemistry in temperate forest ecosystems. *Can. J. Bot.* **68**:2201–2208.
Eno, C.F. 1960. Nitrate production in the field by incubating the soil in polyethylene bags. *Soil Sci. Soc. Am. J.* **24**:277–279.
Focht, D.D., and W. Verstraete. 1977. Biochemical ecology of nitrification and denitrification. In: Advances in Microbial Ecology, ed. M. Alexander, pp. 135–214. New York: Plenum.
Houghton, J.T., B.A. Callander, and S.K. Varney. 1992. Climate Change 1992: The Supplementary Report to the IPCC Scientific Assessment. Cambridge: Cambridge Univ. Press.
Houghton, R.A., J.E. Hobbie, J.M. Melillo, B. Moore, B.J. Peterson, G.R. Shaver, and G.M. Woodwell. 1983. Changes in the carbon content of terrestrial biota and soils between 1860 and 1980: A net release of CO_2 to the atmosphere. *Ecol. Monogr.* **53**:235–262.

McGuire, A.D., J.M. Melillo, L.A. Joyce, D.W. Kicklighter, A.L. Grace, B. Moore III, and C.J. Vorosmarty. 1992. Interactions between carbon and nitrogen dynamics in estimating net primary productivity for potential vegetation in North America. *Glob. Biogeochem. Cyc.* **6**:101–134.

Melillo, J.M., J.D. Aber, and J.F. Muratore. 1982. The influence of substrate quality on leaf litter decay in a northern hardwood forest. *Ecology* **63**:621–626.

Melillo, J.M., J.R. Fruci, R.A. Houghton, B. Moore III, and D.L. Skole. 1988. Land-use change in the Soviet Union between 1850 and 1980: Causes of a net release of CO_2 to the atmosphere. *Tellus* **40B**:116–128.

Melillo, J.M., A.D. McGuire, D.W. Kicklighter, B. Moore III, C.J. Vorosmarty, and A.L. Schloss. 1993. Global climate change and terrestrial net primary production. *Nature* **363**:234–240.

Pastor, J., J.D. Aber, G.A. McClaugherty, and J.M. Melillo. 1984. Aboveground production and N and P cycling along a nitrogen mineralization gradient on Blackhawk Island, Wisconsin. *Ecology* **65**:256–268.

Peterjohn, W.T., J.M. Melillo, F.P. Bowles, and P.A. Steudler. 1993. Soil warming and trace gas fluxes: Experimental design and preliminary flux results. *Oecologia* **93**:18–24.

Peterjohn, W.T., J.M. Melillo, P.A. Steudler, K.N. Newkirk, F.P. Bowles, and J.D. Aber. 1994. The response of trace gas fluxes and N availability to experimentally elevated soil temperatures in an eastern U.S. deciduous forest. *Ecol. Appl.* **4(3)**:617–625.

Raich, J.W., R.D. Bowden, and P.A. Steudler. 1990. Comparison of two static chamber techniques for determining carbon dioxide efflux from forest soils. *Soil Sci. Soc. Am. J.* **54**:1754–1757.

Schimel, D.S., B.H. Braswell, Jr., E.A. Holland, R. McKeown, D.S. Ojima, T.H. Painter, W.J. Parton, and A.R. Townsend. 1993. Climatic, edaphic and biotic controls over storage and turnover of carbon in soils. *Glob. Biogeochem. Cyc.* **8**:279–293.

Spurr, S.H. 1957. Local climate in the Harvard Forest. *Ecology* **38**:37–46.

Stevenson, F.J. 1982. Humus Chemistry. New York: Wiley.

Swift, M.J., O.W. Heal, and J.M. Anderson. 1979. Decomposition in Terrestrial Ecosystems. Berkeley: Univ. of California Press.

Watson, R.H., H. Rodhe, H. Oeschger, and U. Siegenthaler. 1990. Greenhouse gases and aerosols. In: Climate Change: The IPCC Scientific Assessment, ed. J.T. Houghton, G.J. Jenkins, and J.J. Ephraums, pp. 5–40. Cambridge: Cambridge Univ. Press.

Westermann, D.T., and S.E. Crothers. 1980. Measuring soil nitrogen mineralization under field conditions. *Agr. J.* **72**:1009–1012.

12

The Influence of Nonliving Organic Matter on the Availability and Cycling of Plant Nutrients in Seawater

W.G. SUNDA

Beaufort Laboratory, NMFS/NOAA, 101 Pivers Island Road,
Beaufort, North Carolina 18516, U.S.A.

ABSTRACT

Nonliving organic matter (NLOM) has a profound influence on the supply of limiting nutrients to phytoplankton in the ocean and, consequently, on marine productivity and the biogeochemical cycling of nutrients. It plays several essential roles in facilitating and stabilizing nutrient supply. The interconversion of nitrogen and phosphorus between simple available molecular species, such as nitrate, ammonia, and phosphate, and complex biologically refractory organic molecules effectively buffers the supply of these nutrients and thereby helps to stabilize productivity. Dynamic equilibria between organic chelates and dissolved inorganic metal species provide a similar function for the supply of limiting micronutrient metals such as zinc. Association of Fe(III) with organic ligands promotes the solubility and retention of this critical, but sparingly soluble nutrient metal, and thermal- and photoreduction of Fe(III) and Mn(III&IV) oxides enhance the availabilities of iron and manganese. Finally, the release of dimethyl sulfide by phytoplankton and its oxidation to sulfuric acid in the atmosphere solubilizes iron and manganese in mineral aerosols and mediates their delivery to the ocean's surface by facilitating cloud droplet and rain formation. Since the various pools of NLOM that participate in the above functions are either directly or indirectly derived from the activities of phytoplankton, intricate feedback loops exist among the biological release of NLOM, nutrient supply, and algal production of organic carbon in the ocean.

Role of Nonliving Organic Matter in the Earth's Carbon Cycle
Edited by R.G. Zepp and Ch. Sonntag © 1995 John Wiley & Sons Ltd.

INTRODUCTION

Nonliving organic matter (NLOM) has a significant influence on the biological availability and elemental cycling of primary plant macro- and micronutrients in the ocean. Nitrogen and phosphorus represent the primary macronutrients affected by organic cycling while the micronutrients affected include a variety of trace metals (Fe, Mn, Zn, Cu, Co, and Ni).

We can identify four major roles of NLOM in elemental nutrient cycling. The first is the sequestration of available nitrogen and phosphorus into complex mixtures of organic nitrogen- and phosphate-containing compounds possessing varying degrees of biological availability. The rates of interconversion between nonavailable forms (such as complex polymeric compounds) and available forms (such as nitrate, ammonia, and urea) can regulate the overall productivity in many marine systems.

The second role of NLOM is as complexing agents for trace metal nutrients. This role is significant since the availability of trace metals is generally controlled by the concentration of free aquated metal ions or of dissolved inorganic species, and therefore, the complexation of metals by organic chelators can decrease the immediate availability of metals; in doing so, however, chelators act as metal buffering agents and provide long-term stability to metal availability. In the case of iron, organic chelators can provide an additional beneficial effect as solubilizing agents.

A third role of NLOM is to provide a source of electrons for the reduction and solubilization of iron and manganese. The lower oxidation states of both of these metals [Mn(II) and Fe(II)] are soluble in seawater, whereas the thermodynamically stable higher oxidation states [Fe(III) and Mn(IV)] are insoluble or only sparingly soluble. Oxides of these metals undergo thermal and photochemical reduction by NLOM resulting in the release of soluble Mn(II) and Fe(II). This solubilization enhances the availability of these metals and decreases their loss from surface seawater via settling of particulate oxides.

A final role of NLOM is the enhancement of iron and nitrogen availability via the acidification of the marine atmosphere. This acidification occurs through the production of dimethyl sulfide (DMS) by phytoplankton and the diffusion of this volatile compound to the atmosphere where it is oxidized to sulfuric acid. The resulting H_2SO_4 acts as an acidic hygroscopic leaching agent for iron and other nutrients present in continental mineral aerosols and as an acidic trapping agent for ammonia. It also facilitates the delivery of leached iron and ammonium salts to the ocean via rain by providing nucleation sites for cloud droplet formation.

In this chapter, I will discuss all four roles beginning with that of inorganic/organic interconversions of nitrogen and phosphorus.

THE ROLE OF ORGANIC NITROGEN AND PHOSPHORUS IN THE FERTILITY OF SEAWATER

Nitrate and phosphate represent the predominant forms of biologically utilizable nitrogen and phosphorus compounds in seawater. This is evidenced by their abundance in deep seawater and by the fact that the relative change in their concentrations with depth (dNO$_3$/dPO$_4$ $\cong$ 16:1) approximates the ratio of these nutrients in phytoplankton (Redfield et al. 1963). When deep, nutrient-laden seawater is advected into near-surface sunlit regions, phytoplankton take up nitrate and phosphate in the approximate ratio of 16:1 and use them to synthesize the various nitrogenous and phosphate compounds needed for essential cell functions, such as structural support, metabolism, transport, and replication. The most abundant of these compounds include amino acids, proteins, amino sugars, purines, pyrimidines, nucleic acids, phospholipids, and adenylates. Phytoplankton are known to release many of these compounds (along with simple sugars, polysaccharides, and carboxylic acids) into their external environment, although the reasons for this release are uncertain (Hellebust 1974). Other compounds are released when phytoplankton or other organisms die or are eaten, and there is an active release of urea and ammonia as waste products of zooplankton metabolism. The compounds released back into seawater are composed largely of labile plant nutrients such as ammonia, urea, and phosphate, which are readily taken up by phytoplankton, and biologically labile organic compounds such as amino acids and proteins, which are rapidly utilized by bacteria. For example, radiotracer studies indicate that turnover times of dissolved free amino acids in near-surface seawater, with respect to their release and subsequent bacterial uptake, range from only a few hours in productive coastal waters to a day or so in oceanic regions (Ferguson and Sunda 1984). Because of this effective recycling by bacteria, these and other labile compounds are held to very low concentrations (Fuhrman and Ferguson 1986).

However, not all organic nitrogen and phosphorus compounds released into seawater are biologically labile. Some biological polymers have complex structures and are less readily broken down by microbial enzymes. Also, some intracellular biomolecules become damaged during metabolism by oxidative free radical reactions. If the damage is too severe or randomized, as often occurs during lipid peroxidation, these molecules cannot be reprocessed by the cell and accumulate as waste products. Typically they are packaged into vesicles (lysosomes), which are often ejected (exocytosed) from the cell. Many biological molecules released into seawater are subject to similar oxidative free radical attack, such as occurs during photooxidation or thermal oxidation by transition metals [e.g., Mn(III)]. They may undergo free radical condensation reactions, producing large, randomized and highly oxidized structures, typical of "humic" compounds (Harvey et al. 1983). Complex mixtures of biologically refractory polymeric compounds may be also be produced from existing bio-molecules via phenol condensation, Maillard type reactions between amino acids and carbonyls, or mellanoidin reactions between amino acids and sugars (Ishiwatari 1992).

The biologically labile nitrogen and phosphorus compounds are recycled by the biota, while those whose structures are too complex or randomized to be readily used as biological substrates accumulate in the water column. The net result is a progressive loss of nitrate and phosphate from near-surface water and an accumulation of complex mixtures of biologically refractory organic nitrogen and phosphorus compounds.

Evidence for the above scenario can be seen in the vertical distributions and seasonal variations of nitrogen and phosphorus compounds. As one ascends in the ocean from aphotic depths towards the surface, there is a decrease in nitrate and phosphate due to biological utilization and a concomitant increase in organic nitrogen and phosphorus compounds (Lara et al. 1993) (Figure 12.1). The overall concentration of total dissolved inorganic (minus N_2) plus organic nitrogen and phosphorus also decrease towards the surface due to their removal via settling of biogenic particulates, such as zooplankton fecal pellets. While labile nitrogenous nutrients (nitrate, ammonia, urea, and amino acids) drop to exceedingly low concentrations (that are often below analytical detec-tion limits) in oligotrophic surface waters, the concentrations of dissolved organic nitrogen (DON) remain at moderate levels of about 4–6 μmol l^{-1}. Shipboard incubation experiments demonstrate that this material is not imme-diately available for phytoplankton growth (Thomas et al. 1971). If it were, a

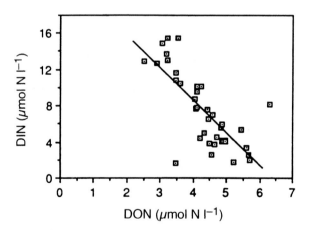

Figure 12.1 Negative correlation between dissolved inorganic nitrogen (DIN) (NO_3^-, NO_2^-, and NH_4^+), and dissolved organic nitrogen (DON) in water sampled with depth in the Greenland Sea. High DON and low DIN values are found in near-surface biologically productive waters. Data are from Lara et al. (1993).

concentration of 5 μmol l^{-1} would be sufficient to support the nitrogen needs for growth of phytoplankton at a typical oceanic productivity of 2 μg C l^{-1} d^{-1} for a period of 200 days without any nitrogen recycling. With a typical oceanic ratio of 10:1 for recycled use vs. loss via particulate settling, the same surface DON concentration could support the above algal growth rate for up to 5.5 years.

Although biologically unavailable in an immediate sense, refractory DON should be available on longer time scales since photolysis of NLOM facilitates its slow degradation to simpler, biologically labile inorganic and organic compounds (Mopper et al. 1991). Thus, the complex mixtures of nitrogen- and phosphate-containing organic compounds can be viewed as a kind of nutrient bank into which some portion of the available nutrients are deposited during periods of high availability and growth (e.g., during upwelling or springtime stratification) and from which they are slowly withdrawn during periods of low external supply (e.g., summer stratification). Thus, the "refractory organic nutrient pool" acts as a kinetic buffer (Jackson and Williams 1985) and functions to stabilize nutrient supply, and thereby to stabilize productivity over periods of months to years. This effect may be important in the overall stability of oceanic planktonic communities.

CHELATION OF TRACE METAL NUTRIENTS AND TOXICANTS BY NLOM

Numerous studies with phytoplankton and bacteria have shown that chelation of trace metals by both synthetic and natural organic ligands decreases their biological availability. Studies in chemically well-defined systems have shown that the biological uptake of trace metals is generally controlled by the free metal ion concentration, or in some cases, by the concentration of kinetically labile inorganic species (Sunda 1988/1989). Thus, chelation decreases metal availability by decreasing free ion concentrations. The importance of free metal ions or labile inorganic species in controlling biological availability appears to be a widespread phenomenon. There are, however, notable exceptions, such as the uptake of iron-siderophore chelates or of vitamin B_{12} (which contains cobalt) by specific microbial transport systems.

Although the importance of free metal ion concentrations in controlling biological availability has been recognized for some time, only recently have we been able to quantify the extent to which trace metals are complexed by organic ligands in seawater. Historically, it has been extremely difficult to find analytical techniques that were sufficiently sensitive and selective to measure the speciation of trace metals in seawater. A number of improved speciation techniques have been developed, including differential pulse anodic stripping voltammetry (DPASV) (Coale and Bruland 1988) and several competitive ligand equilibration methods coupled to cathodic stripping voltammetry (van den Berg 1984), solvent extraction (Moffett and Zika 1987), C_{18} SEP-PAK adsorption (Sunda and Hanson 1987), and Cu-chemiluminescence analysis (Sunda and Huntsman 1991).

Most studies with these new techniques have been with copper, a toxic metal that is also an essential micronutrient. These studies have shown that the overwhelming majority of copper (98–99.99%) is highly chelated in near-surface seawater by unidentified organic ligands. Complexometric titrations indicate that the copper is complexed by at least two separate ligands or classes of ligands: a first, present at low (ca. nanomolar) concentrations, which possesses extremely high conditional stability constants for binding copper (ca. 10^{13} mol l^{-1} at pH 8.1) and a second, present at higher concentrations with lower conditional constants (10^9 to 10^{10} mol l^{-1}).[1] Most of the copper in near-surface seawater is complexed by the first class of ligands, and the distribution of this ligand class controls free cupric ion concentrations.

The presence of copper-complexing ligands in seawater is beneficial to biological productivity. In their presence $[Cu^{2+}]$ is maintained within a biologically optimal range of approximately $10^{-13.5}$ to $10^{-12.5}$ mol l^{-1}, high enough to meet algal nutritional requirements, but low enough to avoid copper toxicity (Sunda 1988/1989). In the absence of organic complexation, free cupric ion concentrations would be three orders of magnitude higher (ca. $10^{-10.6}$ to $10^{-9.6}$ mol l^{-1} at typical seawater copper concentrations of 0.5 to 5 nmol l^{-1} and pH 8.1), sufficient to be toxic to many phytoplankton species.

There is evidence that the strong copper-binding ligand (or ligands) in seawater is produced biologically. *Synechococcus* sp., an oceanic cyanobacterium that is particularly sensitive to copper toxicity, has been shown to produce a ligand in culture, whose conditional constant for copper ($10^{12.8}$ mol l^{-1}) is consistent with values for the strong binding ligand in seawater (Moffett et al. 1990). In the North Pacific, the concentration of strong ligand is relatively invariant with location and time of year in the euphotic zone; however, at greater depths, the ligand disappears from all locations (Coale and Bruland 1988, 1990) (Figure 12.2). Thus, it is present only in waters that have recently experienced phytoplankton growth, again suggesting a biological source. These observations also indicate the existence of unknown degradation mechanisms operating below the euphotic zone.

An additional intriguing observation is the relative constancy of free cupric ion concentrations in near-surface seawater, irrespective of the location or the total copper concentration (Sunda and Huntsman 1991). This situation occurs despite the extremely steep relationships between $[Cu^{2+}]$ and total copper

---

[1]The conditional stability constant, K_c, is defined by the equation: $K_c = [CuL]/([Cu^{2+}][L'])$, where [CuL] is the concentration of copper-ligand complex and [L'] is the concentration of total ligand minus [CuL] (see Coale and Bruland 1988).

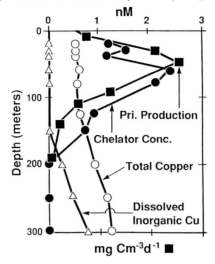

Figure 12.2 Vertical distribution of primary production and concentrations of strong copper-binding chelator, total copper, and dissolved inorganic copper (Cu^{2+} plus inorganic complexes) in the upper 300 m of the central North Pacific. In the productive surface waters, copper is strongly chelated, resulting in extremely low, biologically favorable free cupric ion concentrations (ca. 10^{-13} mol l^{-1}). Below productive surface waters, the strong ligand disappears and, as a consequence, $[Cu^{2+}]$ increases 1000-fold to toxic levels. Data are from Coale and Bruland (1988).

concentration within any single water sample. The concentration of strong complexing ligand always appears to adjust itself to a level just sufficient to yield a free cupric ion concentration of about 10^{-13} mol l^{-1}. It has been speculated that the ligand concentration is biologically regulated by one or more groups of microorganisms (e.g., *Synechococcus*) in order to maintain biologically favorable free cupric ion concentrations (Figure 12.3) (Coale and Bruland 1990; Sunda 1988/1989).

Data are also available on the chelation of zinc in seawater. Recent DPASV measurements of zinc speciation in the North Pacific indicate that this metal is highly (98–99%) chelated in near-surface seawater by an unidentified strong organic ligand present at virtually the same concentration (ca. 1.2 nmol l^{-1}) from the surface to 400 m, well below the vernal mixed layer (Bruland 1989). The ligand has an extremely high conditional stability constant for binding zinc of approximately $10^{11.0}$ mol l^{-1}. The ligand is not the same as the one that binds copper since it has a distinctly different depth distribution and because additions of copper in slight excess of the ligand concentration have no effect on zinc complexation. The conservative distribution of the ligand indicates that it has a greater stability with respect to microbial or chemical degradation than the strong organic ligand that binds copper.

W.G. Sunda

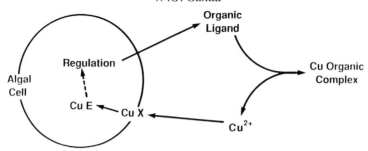

Figure 12.3 Conceptual model for the regulation of free cupric ion concentration in seawater via algal release of strong copper-binding organic ligands. By binding with copper, the released ligands reduce free cupric ion concentration and thereby reduce cellular copper uptake and toxicity.

At first glance it would appear that chelation of zinc in seawater would have a net adverse effect on biological productivity by decreasing the availability of this essential nutrient. However, the presence of the strong zinc-binding ligand actually has a beneficial effect on the long-term productivity of oceanic regions by preventing zinc from being completely stripped from surface waters by the extremely efficient zinc transport systems possessed by marine algae (see Sunda and Huntsman 1992). In this respect the strong zinc ligand acts to buffer the biological supply of zinc, similar to action of biologically refractory organic compounds in buffering the supply of nitrogen and phosphorus.

Much less is known about the chelation of other trace metal micronutrients, such as iron, manganese, cobalt, and nickel. Soluble Mn(II) is probably not chelated to any appreciable extent in seawater because of its low reactivity toward complex formation (Sunda 1988/1989). Cobalt and nickel have higher complexation reactivities than manganese, but little is known about their organic complexation in seawater. There is also little firm data on organic complexation of iron, a subject that will be discussed further in the following section.

THE ROLE OF ORGANIC INTERACTIONS IN MODULATING THE AVAILABILITY OF IRON AND MANGANESE

Iron is the fourth most abundant element in the Earth's crust and manganese the eleventh. Despite their crustal abundances, both of these essential elements are present at subnanomolar concentrations in remote oceanic regions such as the mid-Pacific and Southern Ocean (Martin et al. 1989, 1990). Iron limitation appears to be particularly severe, and there is mounting evidence that an insufficient supply of this element limits growth and controls trophic dynamics of many oceanic systems (Martin et al. 1989, 1991; Morel et al. 1991).

The low concentrations and low biological availability of iron and manganese result from the exceedingly low solubilities of their thermodynamically stable (or metastable) oxidation phases in seawater: Fe(III) oxides and hydroxides and Mn(III&IV) oxides. Iron and manganese present in these solid phases is not directly biologically available and tends to be lost from the water column via particulate settling. By contrast, the lower (2+) oxidation states of these metals are far more soluble, and any reduction of iron or manganese to the 2+ state can greatly increase their retention in seawater.

In seawater, there is evidence that NLOM thermally and/or photochemically reduces iron(III) and Mn(III&IV) oxides to their soluble 2+ states, and thereby increases their availability to phytoplankton. I will discuss each of these metals separately since there are important differences in their respective marine chemistries and biogeochemical cycling.

Although thermodynamically unstable with respect to oxidation by molecular oxygen, manganese(II) can, nonetheless, persist for some period of time in oxygenated seawater because of its slow oxidation kinetics (Sung and Morgan 1981). The oxidation rate of this metal in seawater, however, is increased by several orders of magnitude by the ubiquitous presence of bacteria that catalyze Mn(II) oxidation and precipitate the oxides on their surfaces (Emerson et al. 1982; Cowen and Bruland 1985). Because of bacterial catalysis, the turnover time of Mn(II) with respect to oxidation ranges from days to weeks in coastal waters to weeks or months in oceanic waters (Sunda and Huntsman 1987, 1988); in the absence of such catalysis, the oxidation of Mn(II) in oceanic waters would take several hundred years to reach half completion (Morgan 1967). Once formed, the Mn oxides avidly adsorb NLOM (e.g., humic compounds), which reductively dissolves them by thermal and photochemical pathways (Sunda et al. 1983; Waite et al. 1988) and returns Mn(II) to solution where it is available for uptake by phytoplankton. A pronounced photochemical reduction of Mn oxides is believed to be important in contributing to the surface maxima in soluble Mn(II) concentrations observed in the world's oceans, thereby increasing the supply of Mn to phytoplankton (Sunda and Huntsman 1988). The higher reductive dissolution rates of Mn oxides observed in surface waters, however, may also be partially due to higher thermal rates of reduction resulting from higher near-surface NLOM concentrations (Sunda and Huntsman 1988).

Although there is good evidence that the thermal reduction of Mn oxides in seawater is mediated by external NLOM, the source of electrons and mechanism responsible for the observed high specific rate of photoreduction of natural Mn oxides (ca. 0.07 to 0.12 h^{-1} in full sunlight) is uncertain (Sunda and Huntsman 1994). In laboratory experiments, the addition of marine humates markedly increased the thermal reductive dissolution of natural Mn oxides in seawater, but it did not stimulate photoreductive dissolution. The photodissolution appears to be largely independent of the external concentration of natural organics, and may involve transfer of electrons from the extracellular bacterial polymers onto which the oxides are precipitated (Sunda and Huntsman 1994).

There is evidence that iron is photochemically reduced by NLOM in seawater and that this photoreduction enhances iron availability to phytoplankton (Waite and Morel 1984; Sunda 1988/1989; Miller 1990; Wells and Mayer 1991) (Figures 12.4, 12.5). Like Mn oxides, iron oxides can avidly adsorb

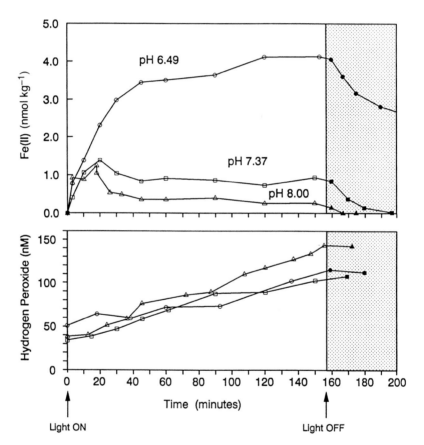

Figure 12.4 Experimental data (Miller 1990) showing the simultaneous photoreduction of Fe(III) to Fe(II) and O_2 to H_2O_2 by NLOM in filtered (0.2 μm) Narragansett Bay water containing 28 nM total iron. The photoproduced H_2O_2 rapidly oxidizes Fe(II) resulting in rapid photoredox cycling of iron, as depicted in Figure 12.5. The higher steady state Fe(II) concentrations observed at low pH probably result primarily from lower Fe(II) oxidation rates.

various NLOM molecules to form surface complex species. The adsorbed NLOM can undergo photochemically catalyzed electron transfer reactions to yield reduced Fe(II) and oxidized organics. Soluble iron chelates can undergo similar photochemical reactions. Unlike Mn(II), which is kinetically meta-stable, Fe(II) is unstable at the pH of surface seawater (ca. 8.1) and is rapidly oxidized back to Fe(III) by reaction with O_2 or photochemically produced H_2O_2. The resultant Fe(III) then undergoes rapid hydrolysis, followed by slow polymerization, dehydration, and molecular rearrangement reactions to form iron hydroxides, and then progressively less soluble and kinetically unreactive iron (hydroxy)oxides. The rapid photochemical cycling of iron that is mediated by NLOM acts to increase iron availability by substantially increasing the steady-state concentrations of kinetically labile, biologically available, soluble inorganic species of Fe(II) and Fe(III) (Figure 12.5).

Association with NLOM can also increase the long-term retention of iron in seawater through the formation of soluble chelates or through the stabilization of colloidal iron hydroxides and oxides. Although there is little direct evidence for such associations, indirect evidence is provided by the high concentrations of filterable (<0.4 µm) iron in many coastal waters. Concentrations can reach micromolar levels, several orders of magnitude above the solubility of iron hydroxides (see Table 1 in Sunda 1988/1989). It is not known whether the

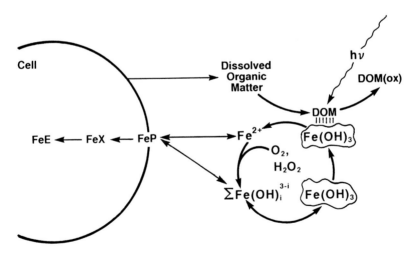

Figure 12.5 Conceptual model for the photoredox cycle of iron in seawater. In this cycle dissolved organic matter, derived from algal productivity, adsorbs onto iron hydroxides (and oxides) and facilitates photoreduction of Fe(III) to soluble Fe(II). The Fe(II) is rapidly reoxidized to dissolved Fe(III) which then hydrolyzes and slowly polymerizes to reform iron hydroxides. The photoredox cycle increases the concentration of soluble inorganic Fe(II) and Fe(III) species, and thereby, increases the supply of iron to phytoplankton.

high filterable iron concentrations result from soluble chelate formation or from adsorption of NLOM onto colloidal iron hydroxides and oxides, and resulting peptization (stabilization with respect to coagulation) of the colloids. In a practical sense, it may not matter since both organic associations will retard the formation of large iron-containing particles (which would then be lost via sedimentation) and both associations can undergo photochemical reduction of iron.

On a final note, it should be pointed out that numerous terrestrial and freshwater organisms, including bacteria, fungi, and higher plants, typically access iron in their environments via the release of specialized chelators called siderophores (Nielands 1981). These chelators have extremely high stabilities for Fe(III) complexation, which allow them to out-compete most other ligands for binding iron and to solubilize iron oxides. The iron siderophore chelates that form in the external environment are taken into the cell by specific membrane transport proteins, and the iron is then released for use in metabolism. Although this is a common mode of iron acquisition in soils, its relevance to the biological uptake of iron in seawater is at present unknown.

THE ROLE OF DMS IN THE ATMOSPHERIC DELIVERY OF AVAILABLE IRON, MANGANESE, AND NITROGEN

Recent evidence indicates that DMS, a biologically produced volatile organic molecule, has a profound influence on the chemistry and physics of the marine atmosphere, and ultimately on the atmospheric supply of available iron and manganese to marine phytoplankton (Duce and Tindale 1991). Unlike most other nutrients (e.g., phosphorus, nitrogen, and zinc) that are primarily supplied to surface waters by vertical advection and turbulent mixing of nutrient-rich deeper waters, the supply of iron and manganese in many oceanic situations appears to be derived predominantly from atmospheric deposition of continental dust (Duce and Tindale 1991). The iron and manganese present in continental materials, however, usually occur within stable oxides or other minerals and are, therefore, not readily available for uptake by phytoplankton. This is where the DMS enters the picture.

DMS is derived from dimethylsulfoniopropionate, which occurs in large amounts in certain groups of phytoplankton, where it is thought to function primarily as a osmoregulant (Keller et al. 1989). It is released into seawater upon the death of the phytoplankton (e.g., during zooplankton grazing) and diffuses into the atmosphere where it is oxidized to sulfuric acid by as yet unresolved pathways (Charlson et al. 1987; Yin et al. 1990) (Figure 12.6). The resulting sulfuric acid has a profound influence on the acidity of the marine atmosphere. It is surface-active and highly hygroscopic and, therefore, adsorbs onto mineral aerosol surfaces, especially wetted surfaces. As a

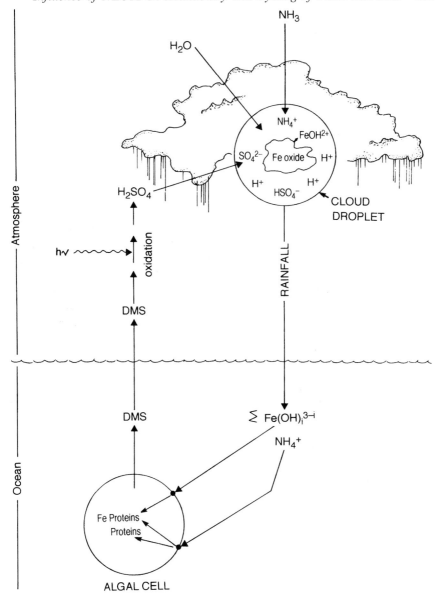

Figure 12.6 Schematic diagram for the increased supply of atmospheric iron and ammonia to phytoplankton mediated by dimethyl sulfide (DMS). DMS released by phytoplankton diffuses into the atmosphere where it is oxidized to sulfuric acid. The sulfuric acid increases iron availability via acid solubilization of iron in mineral aerosols (e.g., iron oxides) and provides an effective trapping agent for ammonium ions. Aerosols composed of acidic ammonium sulfates function as cloud nucleation sites, facilitating rainfall which delivers soluble iron and ammonium ions to surface seawater.

result, the aerosols become bathed in a sulfuric acid solution that effectively leaches and solubilizes iron and manganese from mineral phases. The resulting low pH also favors the thermal reduction and/or photoreduction of Fe(III) and Mn(III&IV) to Fe(II) and Mn(II), and retards their reoxidation, which further promotes solubilization (Behra and Sigg 1990; Zhuang et al. 1992). Finally, the sulfuric acid acts as a trapping agent for gaseous ammonia and results in its retention in the aerosols as soluble ammonium bisulfate, or upon desiccation, as ammonium salts.

The hygroscopic sulfuric acid and ammonium bisulfate aerosols derived from DMS oxidation provide an additional important function in the supply of biologically available iron, manganese, and ammonia by acting as effective nucleation sites for the formation of cloud droplets (Charlson et al. 1987). By doing so, they generate rainfall, and the rain provides an effective vehicle for the delivery of soluble Fe, Mn, and NH_4^+ to the ocean. Indeed, deposition via rain is known to be the primary mechanism for delivery of atmospheric Fe (and probably Mn) to the ocean (Duce and Tindale 1991), and as mentioned above, atmospheric deposition appears to be a major source of metabolic Fe and Mn to oceanic phytoplankton. The atmospherically supplied ammonium ions, although important during episodic events, are generally less important to oceanic productivity than available nitrogen supplied from advection and vertical mixing of nutrient-rich deeper water (Donaghay et al. 1991).

The above scenario represents a major feedback mechanism whereby release of DMS from marine phytoplankton enhances the supply of important limiting nutrients (iron, manganese, and nitrogen) by mediating large-scale changes in the chemistry and physics of the atmosphere (Duce and Tindale 1991). That such a system exists is not really in question since there is firm evidence for each piece of the scenario. A major unanswered question remains as to how such large-scale, beneficial feedback systems evolved.

CONCLUSIONS

It is clear from the preceding discussion that NLOM plays several critical roles in enhancing and regulating the supply of nutrients needed for marine productivity. It acts as a buffering system for the supply of major nutrients (N and P) and micronutrient metals (Zn, Cu, and possibly Fe) and thereby helps to maintain an adequate supply of these nutrients during periods of minimal external supply. It also ensures an adequate supply of iron and manganese by fulfilling several critical functions: (a) solubilizing Fe(III) or maintaining it in stable colloids through the formation of solution or surface complexes, (b) solubilizing and increasing the kinetic lability of Fe and Mn through (photo)reduction of Fe and Mn oxides and soluble Fe present in chelates, and (c) enhancing the supply, solubility, and kinetic lability of Fe and Mn from atmospheric sources via the release of DMS. The NLOM that carries out these critical functions is ultimately derived from phytoplankton productivity, and thus, positive feedback loops are established in which the activities of the biological community promote future productivity through the release of some portion of their organic carbon as a kind of productivity subsidy. How such beneficial feedback loops evolve, involving entire ecosystems or the Earth's system as a whole, remains uncertain.

ACKNOWLEDGEMENTS

I would like to thank the Office of Naval Research for financial support and Susan Huntsman and Elizabeth Lowrey for editorial assistance during preparation of this background paper.

REFERENCES

Behra, P., and L. Sigg. 1990. Evidence for redox cycling of iron in atmospheric water droplets. *Nature* **344**:419–421.

Bruland, K.W. 1989. Complexation of zinc by natural organic ligands in the central North Pacific. *Limnol. Oceanogr.* **34**:269–285.

Charlson, R.J., J.E. Lovelock, M.O. Andreae, and S.G. Warren. 1987. Oceanic phytoplankton, atmospheric sulfur, cloud albedo, and climate. *Nature* **326**:655–661.

Coale, K.H., and K.W. Bruland. 1988. Copper complexation in the northeast Pacific. *Limnol. Oceanogr.* **33**:1084–1101.

Coale, K.H., and K.W. Bruland. 1990. Spatial and temporal variability in copper complexation in the North Pacific. *Deep-Sea Res.* **34**:317–336.

Cowen, J.P., and K.W. Bruland. 1985. Metal deposits associated with bacteria: Implications for Fe and Mn marine biogeochemistry. *Deep-Sea Res.* **32**:253–272.

Donaghay, P.L., P.S. Liss, R.A. Duce, D.R. Kester, A.K. Hanson, T. Villareal, N.W. Tindale, and D.J. Gifford. 1991. The role of episodic atmospheric nutrient inputs in the chemical and biological dynamics of oceanic ecosystems. *Oceanogr.* **4**:62–70.

Duce, R.A., and N.W. Tindale. 1991. Atmospheric transport of iron and its deposition in the ocean. *Limnol. Oceanogr.* **36**:1715–1726.

Emerson, S., S. Kalhorn, L. Jacobs, B.M. Tebo, K.H. Nealson, and R.A. Rosson. 1982. Environmental oxidation rate of manganese(II): Bacterial catalysis. *Geochim. Cosmochim. Acta.* **46**:1073–1079.

Ferguson, R.L., and W.G. Sunda. 1984. Utilization of amino acids by planktonic marine bacteria: Importance of clean technique and low substrate additions. *Limnol. Oceanogr.* **29**:258–274.

Fuhrman, J.A., and R.L. Ferguson. 1986. Nanomolar concentrations and rapid turnover of dissolved free amino acids in seawater: Agreement between chemical and microbiological measurements. *Mar. Ecol. Prog. Ser.* **33**:237–242.

Harvey, G.R., D.A. Boran, L.A. Chesal, and J.M. Tokar. 1983. The structure of marine fulvic and humic acids. *Mar. Chem.* **12**:119–132.

Hellebust, J.A. 1974. Extracellular products. In: Algal Physiology and Biochemistry, ed. W.D.P. Stewart, pp. 838–863. Berkeley: Univ. of California Press.

Ishiwatari, R. 1992. Macromolecular material (humic substance) in the water column and sediments. *Mar. Chem.* **39**:151–166.

Jackson, G.A., and P.M. Williams. 1985. Importance of dissolved organic nitrogen and phosphorus to biological nutrient cycling. *Deep-Sea Res.* **32**:223–235.

Keller, M.D., W.K. Bellows, and R.R.L. Guillard. 1989. Dimethyl sulfide production in marine phytoplankton. In: Biogenic Sulfur in the Environment, ed. E.S. Saltzman and W.J. Cooper, pp. 167–182. ACS Symposium Series 393. Washington, D.C.: Am. Chemical Society.

Lara, R.J., U. Hubberten, and G. Kattner. 1993. Contribution of humic substances to the dissolved nitrogen pool in the Greenland Sea. *Mar. Chem.* **41**:327–336.

Martin, J.H., R.M. Gordon, and S.E. Fitzwater. 1990. Iron in Antarctic waters. *Nature* **345**:156–158.

Martin, J.H., R.M. Gordon, and S.E. Fitzwater. 1991. The case for iron. *Limnol. Oceanogr.* **36**:1793–1802.

Martin, J.H., R.M. Gordon, S.E. Fitzwater, and W.W. Broenkow. 1989. VERTEX: Phytoplankton/iron studies in the Gulf of Alaska. *Deep-Sea Res.* **36**:649–680.

Miller, W.L. 1990. An investigation of peroxide, iron, and iron bioavailability in irradiated marine waters. Ph.D thesis, Univ. of Rhode Island, Kingston, RI, 400 pp.

Moffett, J.W., L.E. Brand, and R.G. Zika. 1990. Distribution and potential sources and sinks of copper chelators in the Sargasso Sea. *Deep-Sea Res.* **37**:27–36.

Moffett, J.W., and R.G. Zika. 1987. Solvent extraction of copper acetylacetonate in studies of copper(II) speciation in seawater. *Mar. Chem.* **21**:301–313.

Mopper, K., X. Zhou, R.J. Kieber, D.J. Kieber, R.J. Sikowski, and R.D. Jones. 1991. Photochemical degradation of dissolved organic carbon and its impact on the oceanic carbon cycle. *Nature* **353**:60–62.

Morel, F.M.M., R.J.M. Hudson, and N.M. Price. 1991. Limitation of productivity for the biogeochemical control of new production. *Limnol. Oceanogr.* **36**:1742–1755.

Morgan, J.J. 1967. Chemical equilibria and kinetic properties of manganese in natural waters. In: Principles and Applications of Water Chemistry, ed. S.D. Faust and J.V. Hunter, pp. 561–620. Chichester: Wiley.

Nielands, J.B. 1981. Iron absorption and transport in microorganisms. *Ann. Rev. Nutr.* **1**:27–46.

Redfield, A.C., B.H. Ketchum, and F.A. Richards. 1963. The influence of organisms on the composition of seawater. In: The Sea, vol. 2, ed. M.N. Hill, pp. 26–77. Chichester: Wiley.

Sunda, W.G. 1988/1989. Trace metal interactions with marine phytoplankton. *Biol. Oceanogr.* **6**:411–442.

Sunda, W.G., and A.K. Hanson. 1987. Measurement of free cupric ion concentration in seawater by a ligand competition technique involving copper sorption onto C_{18} SEP-PAK cartridges. *Limnol. Oceanogr.* **32**:537–551.

Sunda, W.G., and S.A. Huntsman. 1987. Microbial oxidation of manganese in a North Carolina estuary. *Limnol. Oceanogr.* **32**:552–564.

Sunda, W.G., and S.A. Huntsman. 1988. Effect of sunlight on redox cycles of manganese in the southwestern Sargasso Sea. *Deep-Sea Res.* **35**:1297–1317.

Sunda, W.G., and S.A. Huntsman. 1991. The use of chemiluminescence and ligand competition with EDTA to measure copper concentration and speciation in seawater. *Mar. Chem.* **36**:137–163.

Sunda, W.G., and S.A. Huntsman. 1992. Feedback interactions between zinc and phytoplankton in seawater. *Limnol. Oceanogr.* **37**:25–40.

Sunda, W.G., and S.A. Huntsman. 1994. Photoreduction of manganese oxides in seawater. *Mar. Chem.* **46**:133–152.

Sunda, W.G., S.A. Huntsman, and G.R. Harvey. 1983. Photoreduction of manganese oxides in seawater and its geochemical and biological implications. *Nature* **301**:234–236.

Sung, W., and J.J. Morgan. 1981. Oxidative removal of Mn(II) from solution by the [γ-FeOOH (lepidocrocite) surface. *Geochim. Cosmochim. Acta* **45**:2377–2383.

Thomas, W.H., E.H. Renger, and A.N. Dodson. 1971. Near-surface organic nitrogen in the eastern Tropical Pacific Ocean. *Deep-Sea Res.* **18**:65–71.

van den Berg, C.M.G. 1984. Determining the complexing capacity and conditional stability constants of complexes of copper(II) in seawater by cathodic stripping voltammetry of copper catechol complex ions. *Mar. Chem.* **15**:1–18.

Waite, T.D., and F.M.M. Morel. 1984. Photoreductive dissolution of colloidal iron oxides in natural waters. *Environ. Sci. Technol.* **18**:860–868.

Waite, T.D., I.C. Wrigley, and R. Szymczak. 1988. Photoassisted dissolution of a colloidal manganese oxide in the presence of fulvic acid. *Eviron. Sci. Technol.* **22**:778–785.

Wells, M.L., and L.M. Mayer 1991. The photoconversion of colloidal iron oxyhydroxides in seawater. *Deep-Sea Res.* **38**:1379–1395.

Yin, E.-D., D. Grosjean, and D. Seinfeld. 1990. Photooxidation of dimethyl sulfide and dimethyl disulfide. I. Mechanism development. *J. Atmos. Chem.* **11**:309–364.

Zhuang, G., Y. Zhen, R.A. Duce, and P.R. Brown. 1992. Link between iron and sulfur cycles suggested by detection of Fe(II) in remote marine aerosols. *Nature* **355**:537–539.

13

The Effect of Land Use on Nonliving Organic Matter in the Soil

R.J. SCHOLES[1] and M.C. SCHOLES[2]

[1]Division of Forest Science and Technology, CSIR, Box 395,
Pretoria 0001, South Africa
[2]Botany Department, University of the Witwatersrand, Private Bag 3,
Wits 2050, South Africa

ABSTRACT

Changes in land-use practice, including conversion of natural vegetation to cropland and cropland to urban use and more extensive but less dramatic alterations in the management of seminatural vegetation, have a multifaceted impact on nonliving organic matter (NLOM) in the soil. Conversion to short-duration crops usually, but not necessarily, results in a 20% to 50% loss in soil NLOM over a period of several decades. This is mostly due to decreased organic inputs to the soil and, to a lesser extent, to the impact of tillage on the NLOM decomposition rate. There are agricultural practices that result in the conservation or accumulation of NLOM, and the land-use trend in many developed countries is towards reversion of cropland to pasture or forest. Changes in the management regime of seminatural vegetation, such as fire, herbivory, and harvesting, affect soil NLOM more subtly, but also more extensively. The current conceptual and numerical models of soil NLOM are able to predict the observed trends following changes in land use, despite large gaps in the understanding of the chemistry of soil NLOM.

INTRODUCTION

The impact of changing land use, particularly the conversion of natural vegetation into croplands, is second only to the burning of fossil fuels as a current human perturbation to the global carbon cycle (Houghton et al. 1990) and in

Role of Nonliving Organic Matter in the Earth's Carbon Cycle
Edited by R.G. Zepp and Ch. Sonntag © 1995 John Wiley & Sons Ltd.

terms of its historical effect may even exceed fossil-fuel combustion (Houghton, this volume). The nonliving organic matter (NLOM) in the soil is the larger of the two principal terrestrial carbon pools affected by land use within a decadal time frame; the other is the plant biomass (Goudriaan and Ketner 1984; Schlesinger 1993). In this chapter, rather than reviewing the experimental data in detail, we present current ideas on the effects human management has on NLOM in the soil and conceptual models used to interpret the observed trends in NLOM.

A CONCEPTUAL MODEL OF SOIL ORGANIC MATTER

In this chapter, we base our arguments on two key concepts regarding soil organic matter (SOM): one relates to the amount of SOM in a given environment, and the other to its behavior in time. The first concept proposes that SOM is an equilibrium system, which means that the SOM content tends towards a balance between organic inputs and carbon loss due to decomposition. The second concept proposes that SOM can be represented as a series of interconnected pools, each with a characteristic range of turnover rates. The predicted effects of land-use change on SOM follow directly from the acceptance of these two concepts.

Neither of these concepts is new, but they have evolved significantly over the past two decades to a stage where they form a useful and widely accepted basis for prediction and experimentation. At a purely empirical level, it is not essential that this model of SOM is entirely correct, as long as it predicts the direction and rate of SOM change with acceptable accuracy. However, global change is likely to lead to completely novel combinations of land-use, climate, and organic inputs. Since these will be outside the empirical envelope, it would be comforting to have a SOM model that is grounded in measurable entities and mechanistic cause-and-effect relationships, and could thus be expected to be robust under reasonable extrapolation.

The Concept of Equilibrium in SOM

Half a century ago, Hans Jenny (1930) conducted pioneering work in this field when he attempted to explain the observed geographical variation in SOM content. Figure 13.1 illustrates the basic form of the model. SOM is fed by inputs of above- and belowground litter. In a steady-state situation, these inputs would be equal to the net primary productivity (NPP) of the system, which is largely under control of the climate, with some influence of soil fertility status, vegetation type, and ecological factors such as fire and herbivory. SOM is lost primarily by decomposition, as CO_2 resulting from aerobic respiration, and CH_4 under anaerobic conditions. The decomposition rate is

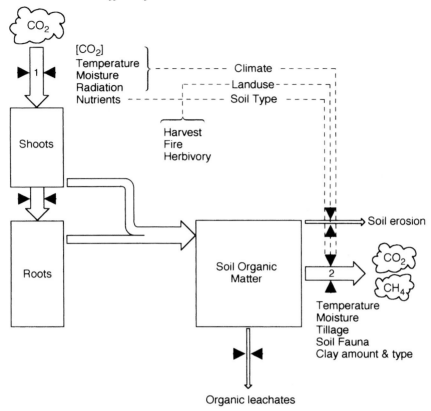

Figure 13.1 The equilibrium model for soil organic matter. The solid arrows are transfers of matter, the dashed lines are feedback controls. The two key processes are (1) primary production and (2) decomposition.

also a function of climate; however, the form of this relationship is different to that for primary production. Soil properties, such as texture and mineralogy, influence decomposition rate. Most importantly, it is directly related to the size of the SOM pool. This linkage causes the amount of SOM to tend towards an equilibrium where decomposition losses equal NPP gains. This level will be a function of climate, soil, and ecosystem type.

If a change in land use results in a change in the quantity or type of organic inputs to the soil, or the soil microclimate, or alters the decomposition rate, then the SOM will adjust to a new equilibrium level. In effect, this means that all land-use changes will alter SOM, but the change need not be negative.

Soil water availability and soil temperature affect the input and output sides of SOM differently, resulting in a strong effect of soil climate on SOM. Plant growth, which drives the input side, effectively stops below a soil water potential of $-1.5\,MPa$, while microbial respiration continues down to $-9\,MPa$ (Scholes et al. 1994). Thus in arid soils, decay is favored over input, and the SOM content is low. In wet soils, anaerobic conditions reduce the decomposition rate more than they reduce plant production, leading to the accumulation of organic matter.

The temperature optimum for plant production is typically around 25–30°C, while the optimum for soil respiration is 35–40°C (Scholes et al. 1994). Therefore cool climate soils accumulate SOM, while very hot climates ($> 30°C$) have low SOM soils. Organic inputs to the soil are mostly influenced by the air temperature, whereas decay rates are controlled by the soil temperature. In very cold climates, such as the tundra, the soil temperatures remain low throughout the year, while air temperatures rise sufficiently in summer to allow plant production. Consequently, these soils store a substantial part of the biospheric carbon.

In natural ecosystems there is an important feedback from SOM on the rate of plant production. Nitrogen, sulfur, and sometimes phosphorus, which are all essential plant nutrients, are mostly found in the SOM, but can only be taken up in the inorganic form. Since plant tissue must have a minimum concentration of these elements, in steady-state systems, the long-term rate of uptake by plants (and thus plant production) must equal their rate of release through decay of SOM. Due to this *ecosystem stoichiometry,* the total amount of carbon in an ecosystem is constrained by the supply of other essential nutrients in the system. The rate of nutrient input constrains the rate of new carbon sequestration in ecosystems that were at carbon equilibrium. The SOM carbon:nutrient ratio is typically substantially lower than the ratio in plant biomass as a whole, especially woody plants. Therefore, in a nutrient-limited nonleaky wooded ecosystem, the consequences of SOM decay will be enhanced plant growth and net ecosystem carbon sequestration.

This simple model can successfully account for most of the observed geographical patterns of SOM distribution in space and can predict the direction of SOM change following perturbation; however, it cannot predict the rate of change very well. This requires the second concept.

The Concept of Multiple SOM Pools

A single-pool model of SOM is mathematically incompatible with the observed dynamics. While a first-order equation, such as

$$SOM_t = SOM_0\, e^{kt} , \qquad (13.1)$$

where SOM_0 and SOM_t are the initial SOM content and content at time t respectively, and k is a constant, can be calibrated to approximate either the short- or long-term dynamics, it never fits well over the entire range. An adequate fit to, for instance, long-term ^{14}C-labeled decomposition experiments requires at least a three-term equation such as

$$\text{SOM}_t = \text{SOM}_1\, e^{kt} + \text{SOM}_2\, e^{lt} + \text{SOM}_3\, e^{mt}, \qquad (13.2)$$

where SOM_1 to SOM_3 are interpreted as SOM fractions with different turnover rates, as indexed by the decay constants k, l, and m (Paul and van Veen 1978; Paul 1984). Defined this way, fractions of SOM are called *functional pools*. Typically, there is a small pool (1–5%) with a rapid turnover (weeks to years) and two larger pools with a slow (decades) and very slow (centuries) turnover rate, respectively.

Almost all contemporary SOM models have at least three pools (although their names, definitions, and interrelationships vary), and most perform an adequate job of predicting both short- and long-term dynamics. The pool structure of two widely cited models is illustrated in Figure 13.2. There is little data unequivocally supporting one model structure over another: they are all abstractions to varying degrees. Most workers in this field agree that SOM is, in reality, a continuum of materials, with continuously varying turnover rates and no absolutely discrete pools. On the other hand, the pools are not entirely arbitrary either and probably correspond to some real differentiation in decay rate related to chemical or physical factors. The multiple-pool concept is a powerful pragmatic simplification for modeling purposes. However, unless the "pools" can be related to some measurable soil property, the models need to be empirically calibrated against long-term SOM dynamics data, which is scarce and expensive to collect. A major current research thrust is aimed at identification, characterization, and indexing of the pools.

The standard agronomic measure of SOM lumps all organic compounds into a single value, either the percentage organic carbon or organic matter in the soil. For nearly two centuries, researchers have been struggling to analyze SOM in terms of its components and have had limited success. This is because SOM consists of a large number of compounds, of which only a few have known chemical structures. A small fraction consists of familiar biological compounds such as carbohydrates, sugars, proteins, and nucleic acids. Bodies of microscopic living soil organisms, such as bacteria and fungi, make up another small fraction. While this fraction is strictly speaking not "nonliving organic matter," it is logical (and practical) to treat it as part of the SOM, but as a distinct subcategory. There are relatively simple analytical methods to measure it.

A large part of SOM consists of partly decomposed plant and animal tissues; the rest is amorphous, colloidal, high molecular weight organic compounds known collectively as "humic substances," the product of soil microbial processes.

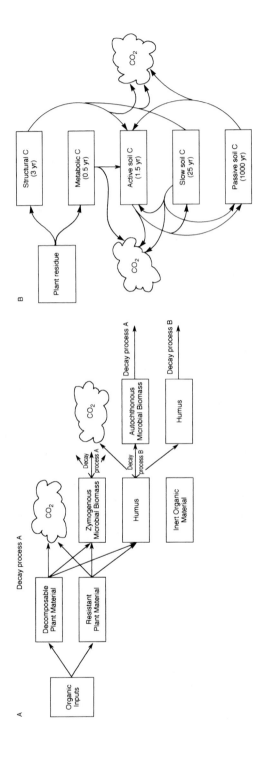

Figure 13.2 Two model structures which have been used to represent multiple-pool SOM relationships. A: Jenkinson and Rayner (1977); B: Parton et al. (1988).

The extraction and chemical characterization of humic substances has diverted soil chemists for over a century, but their structure and origin still remains largely a mystery. The well-known fractions, such as humic and fulvic acids, have not proved to be ecologically predictive. In particular, they do not have a consistent turnover time that would allow them to be related to the "functional" pools defined above. Humic substances tend to have ^{14}C ages that would put them in the slowly decaying pools.

There is consensus that the most rapidly turning over SOM pool, sometimes referred to as the "active" fraction, consists mainly of living microbial biomass, plus some readily metabolizable organic compounds leached out of litter, roots and recently dead microbes. Due to its rapid turnover, the active fraction contributes far more to short-term plant nutrition than its 1% to 5% contribution to SOM would suggest. It is also the fraction most responsive to land-use practices. Its impact on carbon sequestration is minor.

There is growing evidence that the principal difference between pools with different turnover rates may be less chemical than physical (Schulten and Hempfling 1992). The very slow turnover pool seems mainly to be associated with the fine-earth particle fractions (silt plus clay, < 50 μm). Two possible explanations, which are not mutually exclusive, have been advanced for this observation: (a) organic matter in the small pores between small particles is inaccessible to the microorganisms responsible for decomposition, and (b) the organic matter may form a complex with the clay minerals that protects it from attack. Whatever the reasons, in all soil types, climates, and ecosystems, fine-earth content and organic matter content are highly correlated in undisturbed soils. Furthermore, when cultivated, soils with a high fine-earth content show a slower rate of SOM loss. The "high-activity" clays, such as allophanes and smectites, accumulate and protect more SOM per unit mass than the "low-activity" clay minerals such as kaolinite (Jones 1973). The close adhesion of the SOM to the mineral particles results in the mineral-organic matter complex having a combined density intermediate between that of silicate minerals (about $2.5\,Mg\,m^{-3}$) and organic matter (about $0.9\,Mg\,m^{-3}$). This pool is therefore sometimes known as the "heavy fraction" and can be separated from the "light fraction" by flotation of the latter in a dense fluid.

SOM that is not in the fine-earth fraction is mostly particulate in structure. Under electron microscopy, it is revealed as partly decomposed resistant plant material, with a high lignin to cellulose ratio (Cambardella and Elliott 1992). It is suggested that when plant material is added to the soil, the soluble non-structural parts pass immediately into the active fraction, while the lignified structural parts pass into the medium-turnover rate pool. The particulate organic matter is easily quantified by dispersing the soil and passing it through a 50 μm sieve (Cambardella and Elliott 1992). Yet it is the SOM fraction in which the largest absolute changes occur following cultivation, thus making particulate organic matter a promising candidate as an index of the slow pool.

The above discussion relates to the relationship between SOM and soil primary particles. In undisturbed soils, the primary particles form aggregates. This process affects the stability and proportions of SOM fractions. The size of the aggregate influences its suitability as a microbial habitat, as well as the diffusion rate of water and oxygen to the sites of decomposition. Much of the rapid-turnover SOM is associated with macro-aggregates (> 50 μm) and the spaces between micro-aggregates, while the slow-turnover SOM is located within micro-aggregates.

CONVERSION OF NATURAL VEGETATION TO CROPLAND

The cultivation of formerly undisturbed soils has multiple effects on the SOM, and usually, but not necessarily, results in a decline in SOM content (Johnson 1992; see also Table 13.1). Organic inputs to the soil are typically reduced under crop agriculture. Net primary production is often increased, especially with fertilization, but a large fraction is removed as harvest. The fate of this harvest is discussed later (see section on Industrial and Urban Land Use). The high yields of modern cultivars are achieved by altering allocation within the plant, away from the roots and into the harvestable parts. This is made possible by fertilizing and irrigating the plants, and thus reducing their need for extensive root systems. Since most organic inputs to the soil come via the roots, either through root turnover or through root exudates, upward reallocation results in reduced SOM.

Tilling the soil serves three main purposes: it buries weeds, conserves water, and creates a penetrable and well-aerated seed-bed for the crop. Tillage disrupts macro-aggregates to a greater degree than micro-aggregates; therefore SOM associated with the macro-aggregates is lost first. The largest changes occur in the particulate SOM greater than 50 μm in diameter, which primarily consists of finely fragmented soil litter. Tillage also homogenizes the distribution of SOM in the ploughed layer and reduces the soil bulk density. Therefore decreases in SOM content of the top 10 cm are much more dramatic than the changes in SOM in the entire profile. Tillage and cropping also alter the soil water and temperature balance. Tillage-induced changes in soil climate are hard to disentangle from changes in soil structure. Overall, tillage practice probably contributes less to declining SOM than does decreased inputs of organic matter (Table 13.2).

A consequence of this conclusion is that the effects of crop agriculture are expressed throughout the rooting depth (which could be up to several meters) and not just in the ploughed layer. Carbon loss calculations based only on the top 30 cm are likely to be significant underestimates. Accurate measurements of changes in the soil carbon stock must also take into account the changes in the bulk density of the soil, dilution of topsoil with subsoil, and carbon losses through erosion.

Contemporary high-input agriculture depends upon the use of fertilizers and pesticides to maintain high yields. The "no-till" planting methods, which have been developed to reduce soil erosion and loss of soil structure, still require inorganic fertilizer and the application of herbicides to control weed competition. There is no evidence that the use of biocides at recommended dosage levels has either a negative or positive effect on SOM (Dick 1992). There is evidence, however, that sustained use of inorganic fertilizers reduces soil enzyme activity, but none to indicate a negative impact on the total amount of SOM. SOM frequently accumulates (or declines less rapidly) when fertilizer is applied, due to the stimulation of NPP.

Monocultures have widely replaced traditional multispecies cropping systems. This has resulted in soil biota being exposed to only one quality of organic input. This influences the composition and biomass of soil biota, especially through promoting the growth of pathogens; however, in the few known experiments, it has not had a detectable effect on the SOM content in the period of a few decades.

Alternative Agricultural Systems

Concerns about the high rate of soil loss, destruction of soil structure and loss of organic matter due to conventional agriculture have led to the development of a number of alternative techniques. No-tillage systems drop each seed into a small hole rather than into a ploughed furrow and control weeds using herbicides. There is little evidence that no-till systems increase the total profile SOM, but they tend to concentrate it in the upper layers of the soil. Accumulation of crop residues on the soil surface alters the soil microclimate.

Returned residue systems aim to maximize the organic matter inputs to the soil. Instead of burning the crop residues after harvest, or feeding them to livestock, they are incorporated into the soil. This practice significantly increases SOM relative to systems where the residue is not returned to the soil. The application of organic fertilizers, such as compost, farmyard manure, or green mulches, increases SOM and, in particular, microbial biomass.

Crop rotations are mainly directed at controlling soil pathogens or, in the case of rotations including legumes, at capturing nitrogen by biological fixation. Where the rotation includes a long-duration fallow, they can increase the SOM content as well.

Reversion of Cropland to Fallow

Not all land-use changes are in the direction of more cropland. Swidden agriculture, where periods of cropping alternate with long periods of vegetated fallow, is the norm in many parts of the world, as it was in Europe before the

Table 13.1 The effect of land-use practice on topsoil carbon content in recent publications reporting long-term experiments. For additional review of this subject, see Houghton et al. (1987) and Johnson (1992).

Location and crop	Treatments	Duration years	Depth cm	Initial C g kg^{-1}	Final C	Source
Northeast Iowa, U.S.A. Corn/soybean rotations	Moldboard plow	11	20	15.3	20.0	Karlen et al. (1991)
	Chisel plow			15.3	20.0	
	Ridge tillage			15.3	18.7	
	No-tillage			15.3	18.6	
Caterbury, New Zealand Pasture and intensive cultivation	"Wilderness"	?	4	40.0	40.0	Haynes & Williams (1992)
	Pasture	37	4	40.0?	41.0	
	Arable	11	4	40.0?	21.0	
Rostock, Germany	Continuous grass pot experiment	34		1.30	9.1	Schulten et al. (1992)
Costa Rica	Alleycropping	9	10	43.0	44.9	Mazzarino et al. (1993)
	• trees +N	9	10	43.0	41.5	
	• trees −N	9	10	43.0	40.3	
Puch, Bavaria	Fallow	39	10	7.7	7.7	Capriel et al. (1992)
	Single crop	39	10	7.7	9.6	
	Crop rotation	39	10	7.7	13.3	
	Grassland	39	10	7.7	17.6	
Tallassee, Alabama Pine community study	Herbaceous-hardwood-pine	9	15	8.99	7.09	Wood et al. (1992)
	Herbaceous-pine	9	15	8.99	7.29	
	Hardwood-pine	9	15	8.99	7.27	
	Pine only	9	15	8.99	5.89	
Pernambuco, NE Brazil Shifting cultivation	Climax vegetation	12	20(?)	12.0	12.0	Tiessen et al. (1992)
	Cassava/cotton/fallow		20(?)	12.0	8.0	
	Cassava/beans/4-yr fallow		20(?)	12.0	9.4	
	Cassava/8-yr fallow		20(?)	12.0	10.0	
	Cassava/10-yr fallow		20(?)	12.0	11.0	
	Continuous cropping				9.0	

(continued)

Table 13.1 (*continued*)

Location and crop	Treatments	Duration years	Depth cm	Initial C $g\,kg^{-1}$	Final C	Source
North Dakota, U.S.A.	Relict grassland	25	46	22.4	22.4	Bauer & Black (1992)
	Grazed	25(?)	46	22.4	18.8	
	Stubble mulched		46	22.4	15.0	
	Convential till		46	22.4	13.2	
New Zealand						
Taita hill soil	Native beech	120	20	39	39	Sparling (1992)
	Radiata pine	17	20	39	30	
	Gorse	15	20	39	45	
	Unimproved pasture	35	20	39	31	
	Fertilized pasture	35	20	39	36	
Rotoiti, NZ						
Agroforestry trial	Ryegrass/clover	25	20	44.9	44.9	
	Ryegrass/clover with radiata pine	17	20	44.9	39.9	
Manawatu, Kairanga,						
Moutu	Pasture	17	20	71.2	71.2	
NZ (meaned)	Maize	12	20	71.2	60.3	
Piracicaba, SE Brazil	Native forest		20	21.0	21.0	Vitorello et al. (1989)
	12-yr sugar cane	12	20	21.0	22.0	
	50-yr sugar cane	50	20		15.0	
Pacific Northwest, U.S.A.	Grass pasture	56	20	22.2	22.2	Collins et al. (1992)
Long-term cultivation	Stubble incorporated	56	20	22.2	11.6	
	Stubble burned	56	20	22.2	10.5	
	Wheat	56	20	22.2	15.5	
Iowa, U.S.A.	Uncultivated	120	10	65.3?	65.3	Zhang et al. (1988)
Tama	Cultivated		14	65.3?	21.8	
Fayette	Uncultivated	120	10	35.1?	35.1	
	Cultivated		16	35.1?	15.1	

Table 13.2 The effect of different tillage practices on carbon content of the soil, as reported in recent papers concerning long-term experiments. For further reviews, see Johnson (1992).

Experiment	Duration years	No till	Conventional g kg^{-1}	Depth cm	Source
Wheat/lupin rotation Wagga Wagga, Australia	10	24.2	16.8	100	Chan et al. (1992)
Continuous corn/rotation corn/rotation soybean Northeast Iowa, U.S.A.	11	18.6	19.6	20	Karlen et al. (1991)
Long term tillage/rotation Crossville, U.S.A.	10	86.0	58.0	20	Wood & Edwards (1992)
Long term tillage/rotation, Eastern Kansas, U.S.A.	12	21.8	19.9	30	Havlin et al. (1990)
Continuous cropping, New Zealand	12	71.2	60.3	20	Sparling (1992)

Industrial Revolution. The classical view of swidden agriculture is that declining soil fertility causes the farmer to clear new land. After a fallow period of a decade or two, fertility is regained allowing the farmer to return. In tropical regions, however, the farmer must often abandon the cleared land, well before nutrient deficiency sets in, due to weed pressure.

Shifting agriculture can be a carbon-neutral, sustainable practice, provided that the interval between cropping is sufficient to allow the SOM to return to its previous level. However, increasing land pressure in most parts of the world has resulted in a shortening of the fallow period and a sporadic net decline in SOM.

Agricultural overproduction in the industrialized world, coupled with the need to reduce farm subsidies in accordance with international trade agreements, has increased the amount of fallow land in most developed countries. In eastern North America, this trend has already led to the reversion of much cropland to hardwood forest. In the West, farmers receive subsidies to keep their fields as grass fallows. In Europe it is estimated that the forest area will increase by 25% in the next decade. All of these activities will lead to increased SOM. In the case of grasslands, new soil NLOM equilibria may be reached in a matter of decades. Forest soils may require centuries to attain equilibrium.

CHANGES IN THE MANAGEMENT REGIME OF NATURAL VEGETATION

While the clearing of tropical forests captures the public and scientific imagination, four-fifths of the Earth's land surface remains under natural vegetation. This does not mean that soil NLOM beneath it is not subject to land-use change. These changes are less rapid and dramatic than those caused by crop agriculture, largely because belowground processes are less affected. When they are more extensive, however, their cumulative effect can be large.

Herbivory

Herbivory can result in either an increase or a decrease in the amount of organic matter reaching the soil, depending upon its intensity. Light grazing does not reduce NPP, and may even mildly stimulate it; repeated heavy defoliation, however, reduces organic inputs to the soil. In general, mammalian herbivory in Africa, Australia, and South America is greater now than before European colonization. Loss of SOM, litter, and vegetation cover, plus trampling by large animal herds, leads to soil erosion, which carries large amounts of SOM into fluvial and marine sediments. Therefore, although overgrazing reduces terrestrial NLOM, not all of the carbon is necessarily released directly into the atmosphere.

Herbivory also changes the quality of the organic matter returning to the soil and its spatial distribution. Although dung consists primarily of a more-recalcitrant fraction ("fiber"), it also has a higher concentration of readily-available nitrogen than the original plant material and therefore decomposes faster. Herbivores, especially in nutrient-poor environments, tend to concentrate their dung in localized spots, thereby introducing patchiness of SOM.

In the infertile Tropics, herbivory and fire are complimentary and interdependent: if it does not get eaten, it burns; and unless it burns, subsequent regrowth cannot be eaten. This does not mean that all oxidation pathways have the same impact on the global atmosphere. For instance, in savanna fires, about 2% of the carbon is converted to methane, whereas if that same grass were eaten by a ruminant, about 15% would be released as methane, and if it decomposed in the soil, methane may even be consumed.

Fire

Fire is an oxidative pathway that short-circuits decomposition. Since organic matter can bypass the soil on its way to the atmosphere, the SOM pools in frequently burned ecosystems, such as savannas, are smaller than in infrequently burned vegetation, such as dry forest, on the same soils under the same climate (Zinke et al. 1986). Even in annually burned savannas, less than half of the NPP is burned, since most of the NPP is below ground, where it is protected from fire, and a portion of the aboveground NPP decays in-between fires.

A small portion of the fuel burned in each fire is converted to elemental carbon, which is either deposited with the ash or transported great distances in the smoke. Since elemental carbon is almost inert, this process in effect removes carbon from the atmosphere, via the biosphere, into the oceans or the soils. Thus although the short-term effect of fires is to release CO_2 to the atmosphere, and the medium-term effect is to reduce carbon sequestered in the soil and vegetation, the long-term effect is to pump CO_2 out of the atmosphere.

It is unclear whether the total amount of biomass burned in the world has changed over the past few centuries. The historical trend was towards prevention of wild fires, but now that the cost, futility, and ecological disadvantages of fire exclusion have became apparent in many areas, or the capacity to enforce regulations has disappeared, there is a trend towards more frequent burning.

Harvesting

Wood harvesting is practiced in many natural forests and woodlands. It has no impact on SOM unless it is accompanied by extensive disturbance of the soil surface, or if wood is harvested at a rate greater than it can be replaced by tree growth (Johnson 1992).

INDUSTRIAL AND URBAN LAND USES

Industrialization and urbanization are the dominant land-use changes in most developed countries, and in many developing countries as well. The burning of fossil fuels provides most of the energy required for industry and results in elevated levels of nitrogen and sulfur deposition in the surrounding areas. If these systems are nutrient-limited, this deposition results in increased primary production and ultimately increased SOM; provided that the capacity to absorb industrial emissions is not exceeded, which would lead to vegetation death.

A large part of the organic matter removed from croplands as harvest ends up in urban areas, where it is partly consumed and partly buried in landfills. As a consequence of gardening and nutrient, water, and organic matter importation, urban soils generally accumulate SOM (Lugo et al. 1986).

CONCLUSIONS

1. Relatively simple multiple-pool equilibrium models of SOM can account for the observed behavior of SOM in space and time, and following land-use change.
2. The current limitations to applying the models are (a) in understanding what the pools mean in chemical, physical, and biological terms, and therefore how to estimate their size from routine analyses; (b) in understanding the controls that soil mineralogy and structure exert on inter-conversion between pools; and (c) in obtaining long-term data for validation.
3. Crop agriculture typically results in a 20% to 50% decline in NLOM in the surface 30 cm of the soil over a period of 20–50 years; however, there are agricultural techniques which prevent or reverse this loss.
4. Land-use changes in nonarable areas, such as changes in the intensity of burning, grazing, and wood harvesting, can also have significant effects on carbon storage and release.
5. Liberation of NLOM following land-use change, due mainly to reduced organic inputs to the soil and to a lesser extent to tillage, is a significant source of atmospheric CO_2 loading. The current data on the rates and types of land-use change and the carbon stocks associated with different land uses have large margins of error.

REFERENCES

Bauer, A., and A.L. Black. 1992. Organic carbon effects on available water capacity on three soil textural groups. *Soil Sci. Soc. Am. J.* **56**:248–254.

Cambardella, C.A., and E.T. Elliott. 1992. Particulate soil organic matter changes across a grassland cultivation sequence. *Soil Sci. Soc. Am. J.* **56**:777–783.

Capriel, P., P. Harter, and D. Stephenson. 1992. Influence of management on the organic matter of the mineral soil. *Soil Sci.* **153(2)**:122–128.

Chan, K.Y., W.P. Roberts, and D.P. Heenan. 1992. Organic carbon and associated soil properties of a red earth after 10 years of rotation under different stubble and tillage practices. *Aust. J. Soil Res.* **30**:71–83.

Collins, H.P., P.E. Rasmussen, and C.L. Douglas. 1992. Crop-rotation and residue management effects on soil carbon and microbial dynamics. *Soil Sci. Soc. Am. J.* **56(3)**:783–788.

Dick, R.P. 1992 A review: Long-term effects of agricultural systems on soil biochemical and microbial parameters. *Agr. Eco. Environ.* **40**:25–36.

Goudriaan, J., and P. Ketner. 1984. A simulation study for the global carbon cycle, including Man's impact on the biosphere. *Clim. Change* **6**:167–192.

Havlin, J.L., D.E. Kissel, L.D. Maddux, M.M. Classen, and J.H. Long. 1990. Crop rotation and tillage effects on soil organic carbon and nitrogen. *Soil Sci. Soc. Am. J.* **54**:448–552.

Haynes, R.J., and P.H. Williams. 1992. Accumulation of soil organic matter and the forms, mineralization potential and plant-availability of accumulated organic sulfur: Effects of pasture improvement and intensive cultivation. *Soil Biol. Biochem.* **24(3)**:209–217.

Houghton, J.T., G.J. Jenkins, and J.J. Ephraums. 1990. Climate Change. The IPCC Scientific Assessment. Cambridge: Cambridge Univ. Press.

Houghton, R.A., R.D. Boone, J.R. Fruci, J.E. Hobbie, J.M. Melillo, C.A. Palm, B.J. Peterson, G.R. Shaver, and G.M. Woodwell. 1987. The flux of carbon from terrestrial ecosystems to the atmosphere in 1980 due to changes in land use: Geographic distribution of the global flux. *Tellus* **93B**:122–139.

Jenkinson, D.S., and J.H. Rayner. 1977. The turnover of soil organic matter in some of the Rothamsted classical experiments. *Soil Sci.* **123**:298–305.

Jenny, H. 1930. A study of the influence of climate apon the nitrogen and organic matter content of the soil. *Missouri Agr. Exp. Stn. Res. Bull.* **152**.

Johnson, D.W. 1992. Effects of forest management on soil carbon storage. *Water Air Soil Poll.* **64**:83–120.

Jones, M.J. 1973. The organic matter content of the savanna soils of West Africa *J. Soil Sci.* **24**:42–53.

Karlen, D.L., E.C. Berry, T.S. Colvin, and R.S. Kanwar. 1991. Twelve-year tillage and crop rotation effects on yields and soil chemical properties in northeast Iowa. *Comm. Soil Sci. Plant Anal.* **22(19 & 20)**:1985–2003.

Lugo, A.E., M.J. Sanchez, and S. Brown. 1986. Land use and organic carbon content of some subtropical soils. *Plant Soil* **96**:185–196.

Mazzarino, M.J., L. Szott, and M. Jimenez. 1993. Dynamics of soil total C and N, microbial biomass, and water-soluble C in tropical agroecosystems. *Soil Biol. Biochem.* **25(2)**:205–214.

Parton, W.J., J.W.B. Stewart, and C.V. Cole. 1988. Dynamics of C, N, P and S in grassland soils: A model. *Biogeochemistry* **5**:109–131.

Paul, E.A. 1984. Dynamics of organic matter in soils. *Plant Soil* **76**:275–285.

Paul, E.A., and J.A. van Veen. 1978. The use of tracers to determine the dynamic nature of organic matter. *Trans. 11th Intl. Cong. Soil Sci.* **3**:61–102.

Schlesinger, W.H. 1993. Response of the terrestrial biosphere to global climate change and human perturbation. *Vegetatio* **104/105**:295–305.

Scholes, R.J., R. Dalal, and S. Singer. 1993. Soil physics and fertility: The effects of water, temperature and texture. In: Biological Management of Tropical Soil Fertility, ed. P.L. Woomer and M.J. Swift. New York: Wiley-Sayce, in press.

Schulten, H.-R., and R. Hempfling. 1992. Influence of agricultural soil management on humus composition and dynamics: Classical and modern analytical techniques. *Plant Soil* **142**:259–271.

Schulten, H.-R., P. Leinweber, and G. Reuter. 1992. Initial formation of soil organic matter from grass residues in a long-term experiment. *Biol. Fert. Soils* **14(4)**:237–245.

Sparling, G.P. 1992. Ratio of microbial biomass carbon to soil organic carbon as a sensitive indicator of changes in soil organic matter. *Aust. J. Soil Res.* **30**:195–207.

Tiessen, H., I.H. Salcedo, and E.V.S.B. Sampaio. 1992. Nutrient and soil organic matter dynamics under shifting cultivation in semiarid northeastern Brazil. *Agr. Eco. Environ.* **38**:139–151.

Vitorello, V.A., C.C. Cerri, F. Andreux, C. Feller, and R.L. Victoria. 1989. Organic matter and natural carbon–13 distribution in forested and cultivated oxisols. *Soil Sci. Soc. Am. J.* **53**:773–778.

Wood, C.W., and J.H. Edwards. 1992. Agroecosystem management effects on soil carbon and nitrogen. *Agr. Eco. Environ.* **39**:123–138.

Wood, C.W., R.J. Mitchell, B.R. Zutter, and C.L. Lin. 1992. Loblolly pine plant community effects on soil carbon and nitrogen. *Soil Sci.* **154(5)**:410–419.

Zhang, H., M.L. Thompson, and J.A. Sandor. 1988. Compositional differences in organic matter among cultivated and uncultivated argiudols and hapludalfs derived from loess. *Soil Sci. Soc. Am. J.* **52**:216–222.

Zinke, P.J., A.G. Stangenbierger, W.M. Post, W.R. Emmanuel, and J.S. Olson. 1986. Worldwide organic carbon and nitrogen data. ONRL/CDIC–18. Oak Ridge, TN: ONRL Carbon Dioxide Information Center.

Standing, left to right:
Wolfgang Schäfer, Lex Bouwman, Gene Welch, Bill Sunda, Knute Nadelhoffer, Mary Scholes, Jerry Melillo
Seated, left to right:
Christian Sonntag, Bob Scholes, Ed Veldkamp, Peggy Delaney

14

Group Report: Effects of Climate Change and Human Perturbations on Interactions between Nonliving Organic Matter and Nutrients

K.J. NADELHOFFER, Rapporteur
A.F. BOUWMAN, M. DELANEY, J.M. MELILLO, W. SCHÄFER,
M.C. SCHOLES, R.J. SCHOLES, Ch. SONNTAG,
W.G. SUNDA, E. VELDKAMP, E.B. WELCH

INTRODUCTION

The dynamics of nonliving organic matter (NLOM) in terrestrial and aquatic ecosystems are closely linked to plant and microbial processes through exchanges of carbon (C) and growth-limiting nutrients (e.g., nitrogen [N], phosphorus [P], potassium [K], sulfur [S], iron [Fe], and trace elements) among soil, sediment, phytomass, detritus, aqueous, and atmospheric pools. Virtually all organic inputs to NLOM are ultimately derived from primary production by photosynthetic organisms. Mineral nutrients released during NLOM degradation are, in turn, critical regulators of primary production. Furthermore, properties conferred on soils and water columns by NLOM can influence growth environments of both primary producers and decomposers in ways not directly related to nutrient release. In soils and sediments, for example, NLOM influences properties such as porosity, ion exchange capacity, moisture retention, pH, and temperature. Also in soils, NLOM sequesters cationic nutrients (NH_4^+, K^+, and Mg^{2+}), preventing their loss, and acts as

Role of Nonliving Organic Matter in the Earth's Carbon Cycle
Edited by R.G. Zepp and Ch. Sonntag © 1995 John Wiley & Sons Ltd.

an important solubilizing agent for iron. In aquatic environments, NLOM reduces light penetration (especially for ultraviolet radiation), provides exchange surfaces for mineral elements and microorganisms, and enhances iron retention and supply to phytoplankton (see Sunda, this volume). In both terrestrial and aquatic ecosystems, NLOM binds toxic metals (e.g., Cu, Cd, Al), thereby decreasing their uptake by primary producers. Therefore, to assess the sensitivity of the global carbon cycle to changes in NLOM pools (the primary goal of this Dahlem Workshop), it is essential to consider how NLOM interacts with living components of ecosystems.

Our group concluded that developing or adapting "minimum models" capable of capturing essential interactions between NLOM and nutrient cycles in terrestrial and aquatic ecosystems will enhance our ability to predict the responses of C cycles to climate change and other perturbations. Because nutrient elements serve to constrain the dynamics of NLOM pools, minimum models should explicitly account for interactions between NLOM and key nutrient elements. By definition, minimum models are highly aggregated, with C and nutrients from small pools combined into large reservoirs. Such minimum models serve as useful conceptual frameworks for describing dynamics at large scales and should also be adaptable for use in simulation exercises.

MINIMUM MODELS TO DESCRIBE NLOM–ELEMENT INTERACTIONS

Our "minimum models" are intended to be used to predict how element interactions with NLOM might serve to regulate C sequestration in terrestrial and aquatic ecosystems. These models are specifically aimed at describing and predicting NLOM dynamics and element interactions over time scales of years or decades to centuries. Predictions at shorter or longer time scales would require different models. Shorter time scales, for example, would require explicit consideration of microbial biomass and short-lived classes of compounds, such as carbohydrates and proteins. Longer time scales might require additional NLOM components with turnover times ranging from centuries to millennia.

Terrestrial Model

Our minimum model for terrestrial ecosystems (Figure 14.1) has two reservoirs, each containing a rapid-cycling and a slow-cycling compartment. Compartments in the biomass reservoir include active and structural or storage tissues. Active tissues turnover at subannual to annual time scales, function to acquire resources (i.e., nutrient uptake, photosynthetic C fixation), and

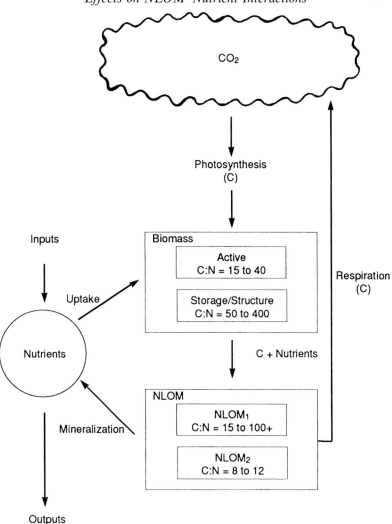

Figure 14.1 A minimum model for terrestrial ecosystems. Sizes of biomass and NLOM pools (boxes) vary with ecosystem type and disturbance history. Typical C:N mass ratios are shown for major compartments. Arrows show directions of net carbon and nutrient fluxes. See text for discussion.

have relatively low C:nutrient ratios. Examples include leaves and fine roots in vascular plants in which C:N mass ratios can range from 15 to 40. Implicitly included in the biomass reservoir, we assume that microbes consist entirely of active tissues with low C:N ratios (8 to 15). Structural and storage tissues, such as wood and rhizomes, have relatively high C:N ratios (up to 400) and turnover at decadal to century time scales.

The NLOM reservoir in terrestrial ecosystems consists of materials ranging from litter, whose source can be readily identified, to older, more amorphous organic matter. This range of materials can be grouped into two functional pools: $NLOM_1$ and $NLOM_2$. $NLOM_1$ has a shorter turnover time and consists of recently deposited plant detritus, soluble carbohydrates, amino compounds, and other materials subject to rapid microbial or chemical oxidation. C:nutrient ratios in $NLOM_1$ are wider than in $NLOM_2$ and reflect nutrient contents of inputs from biomass. The range of C:N mass ratios in $NLOM_1$ range from 15 to > 100, reflecting C:N ratios of plant inputs and narrowing of ratios due to decomposition. Relatively large supplies of readily degradable carbon compounds in $NLOM_1$ provide energy for microbial metabolism. As energy-rich C compounds are consumed and as microbes sequester inorganic nutrients from soil solution, C:nutrient ratios gradually decrease. $NLOM_2$ consists of materials that are more resistant to decomposition and mineralization. This resistance can result either from chemical characteristics, such as random or complex patterns of chemical bonding due to previous decomposition, or from physical encapsulation in aggregates composed of more resistant materials (see Alperin et al., this volume). C:nutrient ratios are lower in $NLOM_2$, with C:N ratios ranging from 8 to 12. Turnover times of $NLOM_1$ are subannual to < 10 years, whereas $NLOM_2$ turns over at decadal to century intervals. If, however, the $NLOM_2$ pool is subjected to physical disturbance, biochemically labile fractions that are physically protected, and therefore function as $NLOM_2$, can be released to the $NLOM_1$ pool.

The biomass reservoir acquires organic C via photosynthetic fixation of atmospheric CO_2. It acquires nutrients primarily from the soil and, to a lesser extent, from atmospheric sources via direct uptake of deposited nutrients and symbiotic N fixation. Detritus formation (e.g., litterfall), release of soluble organic compounds from living and dead tissues, and animal excretion facilitate the transfer of organic C and nutrients from the biomass to the soil reservoir. Microbial degradation of NLOM exports CO_2 to the atmosphere and releases mineralized nutrients to solution where they become available for uptake by plants or are lost from the system as a whole due to leaching or gaseous export (e.g., ammonia volatilization, denitrification). The NLOM reservoir also receives nutrient inputs from outside the system. Examples include wet and dry deposition of mineral nutrients, nitrogen fixation by free-living microbes, and weathering of nutrients from parent minerals.

Aquatic Model

Our minimum aquatic model (Figure 14.2) also consists of two reservoirs and can be applied to open ocean, coastal marine, or freshwater ecosystems.

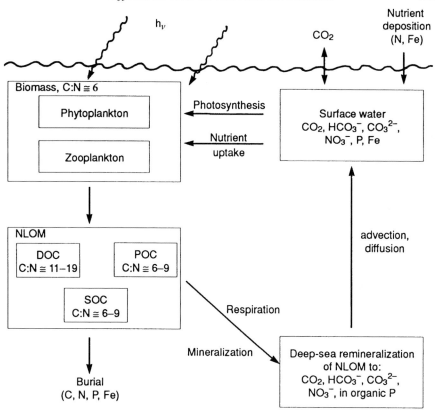

Figure 14.2　A minimum model for aquatic ecosystems. Biomass and NLOM pool sizes are variable but C:N mass ratios are relatively constant. In the open ocean, a biological pump transfers CO_2 to deep water against concentration gradients when NLOM diffuses or sinks to and is remineralized. The biological CO_2 pump does not function effectively in shallow coastal margins or fresh waters. See text. DOC = dissolved organic carbon; POC = particulate organic carbon; SOC = sedimentary organic carbon.

Phytoplankton and zooplankton comprise the biomass reservoir. As with the terrestrial model, heterotrophic microorganisms are implicit in the biomass reservoir and serve to mediate NLOM decomposition. In contrast to the terrestrial model, biomass turns over at short time scales, typically from less than one to several days for phytoplankton. Because structural or storage tissues are generally (though not always) unimportant, we have not included them. Nutrient concentrations in aquatic biomass are high. For example, plankton C:N mass ratios are close to 5.7. (Note that the equivalent molar ratio is 6.6. We use mass ratios throughout to allow for comparisons to mass fluxes.) Phytoplankton is the principal primary producer in the water column and is restricted to the euphotic zone. Zooplankton rapidly consume most of the phytoplankton. In certain environments, such as shallow lakes and some near-shore marine ecosystems, aquatic macrophytes could be added to the model since they are important primary producers. Other omissions are large planktivores and higher trophic level consumers whose total C and nutrient contents are generally small fractions of overall totals. Nevertheless, they can be important regulators of primary production through "top down" trophic effects.

The NLOM reservoir consists of dissolved organic carbon (DOC), particulate organic carbon (POC), and sedimentary organic carbon (SOC). DOC is operationally defined as material passing through a 0.2 or 0.4 μm filter and is composed of colloidal as well as truly dissolved materials. It is released by plankton and can also be produced during POC and SOC decomposition. POC consists of particulate matter derived from senescent plankton, fecal materials, and flocculated DOC aggregates. POC sinking rates are related to particle size, with larger particles sinking rapidly and smaller colloidal particles remaining suspended. SOC consists primarily of POC accumulated in sediments. Each of these pools has relatively labile and resistant fractions that are functionally similar to the $NLOM_1$ and $NLOM_2$ pools of the terrestrial model. However, we explicitly account for DOC, POC, and SOC in the aquatic model because these fractions can be readily identified in water columns and sediments, and because each of these pools is subject to different sets of dynamic processes. Unlike terrestrial ecosystems, C:nutrient ratios are relatively low in NLOM pools, with C:N mass ratios of 6 to 9 in POC pools, 8 to 10 in SOC, and 11 to 19 in DOC (Fuhs et al. 1972; Birch 1974; Bishop et al. 1977; Knauer and Martin 1981; Benner et al. 1992; Williams 1992).

Phytoplankton exist in CO_2-rich environments compared to terrestrial plants. This is particularly true for marine ecosystems, where total dissolved CO_2 (the sum of CO_2 + HCO_3^- + CO_3^{2-}) ranges from 1.9 to 2.4 mmol kg^{-1} in productive surface waters. These values are only 10% to 15% below those in the upwelling deep water sources, showing that elements other than C (such as N, P, or Fe) limit oceanic productivity (Broecker and Peng 1982). Because of high ambient dissolved inorganic C concentrations and a lack of moisture stress, mineral nutrient limitation is even more important in aquatic than in terrestrial ecosystems. Cycling of nutrients between biomass and NLOM in the water column is a key process serving to sustain productivity. When DOC and POC decompose in the water column, mineralized nutrients become available for uptake by phytoplankton. In general, decomposition of DOC and POC is highly efficient in the water column. For example, only about 1% of primary production in the open ocean and less than 10% of primary production in coastal waters is deposited as sinking POC at the sediment surface (Berger et al. 1989). Further decomposition within sediments results in < 0.05% and

< 1% of primary production finally accumulating as NLOM in open ocean and continental shelf sediments, respectively.

In the open ocean, NLOM plays a key role in the "biological CO_2 pump" by transporting carbon fixed in surface waters to deep waters (below 200 m), where it is then remineralized to CO_2 species. This "pump" utilizes NLOM (primarily POC) to transport CO_2 to deep waters against concentration gradients. Because the residence time of deep waters is generally about 1000 years, transport of CO_2 to deeper waters under present conditions does not necessarily match the upward transport of CO_2 to surface waters by advection and diffusion. As a result, the biological pump can isolate CO_2 from exchanging with the atmosphere on time scales of years to centuries.

Nutrient inputs to surface waters play a major role in controlling primary productivity in aquatic ecosystems. This is particularly true in upwelling oceanic regions, where inputs from nutrient-rich deep waters support high rates of productivity, and in continental margins where inputs from terrestrial sources and coastal upwelling can substantially increase productivity. Nutrient (especially P) exports from terrestrial ecosystems play a critical role in controlling productivity in fresh waters.

TERRESTRIAL ECOSYSTEMS

C storage and nutrient cycling rates in terrestrial ecosystems are strongly affected by feedbacks between soil microbial processes, NLOM dynamics, plant processes, and environmental factors such as temperature and moisture. The overall effects of temperature and moisture on NLOM dynamics are fairly well understood (cf. Paul and Clark 1989). Decomposition rates at constant moisture generally increase by factors of 1.5 to 2.5 for each 10°C increase between about 0 and 30°C. Microbial activity and decomposition rates are highest at about –0.01 MPa moisture tension and decrease as soils become saturated (≈ 0 MPa) or extremely dry (less than –5 to –10 MPa). Below we discuss interactions between plant, soil, and microbial processes using our minimum model for terrestrial ecosystems (Figure 14.1) as a contextual framework. First, we consider how NLOM composition, microbial processes, and environmental factors interact to control NLOM turnover and storage in soils. We then discuss how nutrients released from decomposing NLOM feed back to influence both primary production and subsequent inputs to NLOM pools. Finally, we show how element cycles can serve to constrain C balances and loss under changed environments using examples of C and nutrient budgets from forest and grassland ecosystems.

Effects of Substrate Quality, Microbial Processes, Chemical, and Physical Factors on NLOM Turnover and Storage

Developing an improved understanding of the effects of litter composition, microbial processes, and chemical/physical factors requires several lines of inquiry. First, we must identify how detritus composition and nutrients affect decomposition of materials comprising $NLOM_1$. We must also identify controls on the efficiency of transfer of materials from $NLOM_1$ to $NLOM_2$. Finally, we need to characterize factors controlling the decomposition of $NLOM_2$.

Effects of Detrital Composition and Nutrients on $NLOM_1$ Decomposition

Decomposition rates of plant litter (a major component of $NLOM_1$) typically decrease when chemical structural complexity increases and nutrient contents decrease. For example, various investigators have reported negative correlations between decomposition rates and initial lignin or polyphenolic concentrations during the first few years of litter decay (e.g., Fogel and Cromack 1977; Palm and Sanchez 1990; Berg 1986). Also, many studies report positive correlations between initial N content of litter and decomposition rate (Melillo and Gosz 1983). Initial lignin:N ratios of litter are often better predictors of decay rates than either lignin or N concentrations alone, with lignin:N ratios being inversely correlated with rates of mass loss for the first 2 to 3 years of litter decomposition (Melillo et al. 1982). The inverse relationship between lignin:N ratios and decomposition rates has been reported by many others over the past decade. It forms the basis for decomposition routines in various ecosystem simulation models. However, too few studies of key litter types with high initial concentrations of both lignin and N (e.g., fine roots) or extremely low lignin and N concentration (bolewood of many species) have been done to permit prediction of their decay rates.

   Although initial N concentrations in litter are positively correlated with rates of mass loss and nutrient release from decomposing plant litter, the effects of exogenous N on decomposition and nutrient mineralization are less well understood. A recent review by Fog (1988) highlights discrepancies among studies of the effects of N additions on litter decay rates, with large numbers of papers reporting either negative effects, no effect, or positive effects of N addition on decomposition. These contrasting results could be due to interactions among multiple limiting factors, with either C (energy availability) or nutrients assuming different levels of importance in limiting microbial activity under different conditions. For example, Fog's (1988) review noted a general trend of N additions inhibiting lignin and humus degradation but enhancing cellulose decomposition. He presented evidence suggesting that production of lignases and peroxidases by certain organisms may be repressed under high levels of exogenous N. Cellulase production, in contrast, seems to be either unaffected or stimulated by N additions. Also, effects of exogenous N may be less important when other elements limit microbial activity. For example, additions of phosphate have been shown to increase decomposition rates of oak leaves (Elwood et al. 1981) and N-rich alder wood (Melillo et al. 1984) where N was available in excess of microbial demands.

Effects of Detrital Composition and Nutrients on Transfer
Efficiencies of NLOM$_1$ into NLOM$_2$

Accumulation of NLOM in soils is a function of both the rate at which NLOM pools decompose and the efficiency by which materials are transferred from rapidly decomposing to slowly decomposing NLOM pools, with efficiency defined as the percentage of material entering a pool that is eventually transferred to a less reactive pool. It is important to understand what percentage of original plant debris is eventually transferred to NLOM$_2$ since this is the largest NLOM pool in most soils. Organic matter can be transferred from NLOM$_1$ to NLOM$_2$ by mechanisms including: (a) physical protection by association with fine-grained fractions of the soil material (clay and silt) or by incorporation into aggregates; and (b) polymerization of relatively simple compounds into more complex compounds with stable chemical bonds and random structures. The efficiency of physical protection by clay complexation is affected by several factors, including clay mineralogy and amount, location of detritus input to the soil (surface versus subsurface deposition), and soil faunal activity.

We have not yet developed a mechanistic understanding linking litter composition to the efficiency of transfer from NLOM$_1$ to NLOM$_2$. After 75% to 80% of initial litter mass is oxidized, the remaining 20% to 25% is resistant to further decay regardless of the initial chemistry of the material studied. This suggests that factors useful for predicting decomposition rates of materials in NLOM$_1$, such as lignin, phenolic, and N contents, are less useful predictors of decomposition rates of NLOM$_2$ pools. At present there is no clear evidence to suggest that mineral nutrient availability affects the final proportion of NLOM$_1$ that is eventually transferred to NLOM$_2$. Fog (1988), however, presents some evidence from N addition studies, which indicates that increased N in soils can lead to the formation of more labile nitrogenous compounds and less refractory "humus" (or NLOM$_2$).

The issue of transfer efficiency between NLOM$_1$ and NLOM$_2$ is critical to the development of a mechanistic understanding of the NLOM component of ecosystems. It must be considered a high-priority research area.

Controls on NLOM$_2$ Decomposition Rates

Resistance of NLOM$_2$ components to decomposition can result from either chemical characteristics or physical shielding. Chemical resistance to decomposition in NLOM$_2$ is conferred by complex, nonrepeating molecular structures. Cleaving bonds within such structures requires production of either general enzymes with low activities, a variety of specific enzymes, or combinations of general and specific enzymes. Production of these enzymes can be inhibited by high N availability (Fog 1988). This could result from end-product inhibition mechanisms with production of lignolytic enzymes, which cease when sufficient N is obtained. Therefore, high rates of net N mineralization from $NLOM_1$ might serve to decrease $NLOM_2$ decomposition rates. Enzyme activities within aggregates or clay-humus complexes may be less as a result of low O_2 concentrations (decreasing microbial activity) or binding of enzymes to charged clay surfaces.

Decay rates of $NLOM_2$, although slow by definition, can be stimulated by increased temperature (see Melillo et al., this volume). Also, additions of readily decomposable detritus can provide microbes with sources of metabolic energy for production of humus-degrading enzymes that would lead to more rapid degradation of $NLOM_2$ fractions and greater rates of nutrient mineralization. Disruption of aggregates and clay-humus complexes by physical processes, such as freeze/thaw and wet/dry cycles, faunal activity, and tillage, also increase rates of $NLOM_2$ decomposition. In addition to direct enzymatic degradation, nonspecific abiotic oxidation on the surfaces of iron and manganese oxides may also be of significance in the decomposition of $NLOM_2$. Of the two, Mn oxides are particularly effective in oxidatively degrading complex $NLOM_2$ molecules (Waite et al. 1988). Some of the oxidation products include low molecular weight organic compounds that are readily utilized by bacteria. This process, therefore, represents a potential mechanism for the transfer of $NLOM_2$ to labile $NLOM_1$ pools (Sunda and Kieber 1994). Because the formation of Mn oxides is mediated by bacteria, and because bacteria may use this mechanism to access C in $NLOM_2$ pools, the process cannot be considered strictly nonmicrobial or nonenzymatic.

Overall there is a great deal of evidence from studies of soil C and N kinetics (e.g., Deans et al. 1986) that NLOM exists in at least two general states as portrayed in our terrestrial model: one is a relatively small, labile C-rich pool derived mostly from recent plant litter inputs; the other is a much larger, less reactive and nutrient-rich pool composed largely of the end-products of the decomposition of both litter and microbes. However, no generally recognized method to identify or characterize the amounts and types of materials that make up these kinetic fractions yet exists. There is a pressing need for developing such methods (see Alperin et al., this volume).

Effects of Nutrient Availability on the Amounts and Quality of Inputs to NLOM

The availability of key nutrients, especially N and P, to primary producers is directly linked to the decomposition of $NLOM_1$ and $NLOM_2$ pools in many ecosystems. Elevated CO_2 concentrations will likely increase rates of C assimilation to the extent that nutrient demands of primary producers are met. Nutrient availability as well as CO_2 concentration, moisture availability, temperature, and fluxes of photosynthetically active radiation (PAR) control rates of net primary production and detrital inputs to $NLOM_1$. Thus, greater nutrient availability should lead to greater litter inputs to NLOM pools in soils (see review by Melillo and Gosz 1983). Changes in nutrient availability may not only change amounts of detrital inputs to NLOM, they will also alter the relative inputs of tissues from above- and belowground plant sources. Shifts in site-of-input have two potential consequences. First, materials entering $NLOM_1$ pools below ground are more likely than aboveground inputs to react with clays or to become shielded in soil aggregates and should thereby be less decomposable. Second, the chemical quality of belowground plant tissues differs considerably from that of aboveground tissues. Fine roots, in particular, have higher lignin concentrations than do aboveground tissues.

Soil nutrient availability can affect the chemical quality or ability of plant detritus to decompose as well as the total amounts of litter produced. Plants growing on less fertile sites generally have higher concentrations of phenolic defense compounds and higher lignin concentrations in litter. For example, phosphorus availability has been shown to affect the concentration of lignin in eucalyptus leaves in Australia (Loveless 1961). Also, higher nutrient availability in the soil is usually reflected in higher tissue nutrient concentrations in senesced tissues and litter (Vitousek and Sanford 1986).

Higher temperatures under global warming will probably lead to increased mineralization rates, greater productivity, and higher rates of litter input to NLOM pools. Also, elevated atmospheric CO_2 could increase plant C:N ratios. Lower nutrient contents and increased C-based defense compounds (e.g., condensed tannins) in foliage could lead to less herbivory and greater rates of NLOM inputs to soils.

If elevated CO_2, in combination with greater nutrient mineralization in warmer soils, increases both above- and belowground productivity, fluxes of litter into NLOM will increase. Also, higher CO_2 concentrations could increase the proportions of production allocated below ground (Norby et al. 1986, 1992), which could result in more NLOM storage. This could increase $NLOM_1$ pools and, over time, increase inputs to $NLOM_2$ pools, thereby increasing total NLOM storage. On the other hand, if warmer conditions lead to sufficiently greater decomposition rates, NLOM stocks could decrease despite greater inputs from biomass. Also, warmer conditions could lead to increased fire intensities or frequencies in some environments and thereby foster more rapid NLOM decomposition. However, increased fire frequencies would not necessarily lead to lower NLOM stocks, particularly if fire leads to increases in belowground (root) production, or if accumulations of charcoal lead to increased NLOM storage over long time scales.

Plant water-use efficiency will probably increase with atmospheric CO_2 concentrations (Tyree and Alexander 1993). The implications of this for nutrient-use efficiency are uncertain. Increased plant water-use efficiency will affect soil moisture and temperature regimes, thereby influencing the decomposition environment in soils. Uncoupling of N mineralization from plant uptake, such as could occur if higher winter temperatures stimulate mineralization when plant nutrient uptake rates are low, could lead to nutrient exports from terrestrial ecosystems via leaching and denitrification. A key question is how quickly plants will acclimate to changing climatic conditions, thereby damping the effects of increasing NLOM turnover on nutrient loss.

Consequences of Element Ratios in NLOM and Terrestrial Ecosystems for C Sequestration

Redistributions of key growth-limiting nutrients under current and altered climatic conditions have important implications for NLOM and C balances in terrestrial ecosystems. This is because C:nutrient ratios in NLOM pools and plant tissues in terrestrial ecosystems are constrained to characteristic ranges (Figure 14.1). Variations in the relative amounts of ecosystem components with different C:nutrient ratios will result in different patterns of C sequestration in NLOM and biomass pools in different ecosystem types. To illustrate this, we use our minimum terrestrial model as a framework for discussing how climate changes could affect C sequestration.

First, we assess the possible effects of a 5°C temperature increase on C and nutrient element distributions in a late successional temperate deciduous forest, based on simulations of Rastetter et al. (1991), who used a process-based biogeochemical model (GEM, a general ecosystem model) to examine changes in biogeochemical cycling resulting from a series of climatic perturbations. Under current climate conditions, the total C content of the modeled forest ecosystem is about 16.5 kg m^{-2} with about 8 kg m^{-2} in soil and 8.5 kg m^{-2} in forest biomass (Table 14.1). We estimate that about 1.3 of 8 kg C m^{-2} in soils resides in NLOM$_1$ (about 3 times annual leaf plus root detritus production) and that the C:N ratio in this pool is about 30. The remaining soil C is in NLOM$_2$, with C:N $\approx$ 12. About 0.5 kg of the 8.5 kg C m^{-2} in plant biomass exists as foliage and fine roots, with C:N $\approx$ 25, and about 8.0 kg C m^{-2} occurs in wood, with C:N $\approx$ 240. GEM predicted that after 50 years, under a 5°C warming, soil C would decrease by about 6% but that biomass C would increase by about 22% for an overall increase of 9% in total ecosystem C. The predicted increase in ecosystem C results from a net transfer of N from NLOM pools with low C:N ratios to plant pools (e.g., wood) with higher ratios.

It should be noted that GEM predicted a small decrease in C:N ratios in total soil (13.2 under ambient conditions versus 13.0 after 50 years of 5°C

Table 14.1 Possible carbon and nitrogen stocks of a temperate decidous forest under current climate and a 5°C warmer climate. Element stocks and C:N ratios of soil and plant subtotals and total ecosystem element stocks are from simulations by Rastetter et al. (1991), who used field data from an oak–maple forest in eastern North America (Harvard Forest, Petersham, Massachusetts) for current climate and a general ecosystem model (GEM) to predict changes in element stocks after 50 years under warmer conditions. We divided into $NLOM_1$ and $NLOM_2$ and plant biomass into active and storage pools based assuming that $NLOM_1$ was equivalent to three years' inputs of C and N from litterfall and root deposition. Active plant pools are estimates of biomass in foliage and fine roots during the growing season.

	Mean Annual Temperature					
	Current			+5°C		
	C:N	C $g m^{-2}$	N $g m^{-2}$	C:N	C $g m^{-2}$	N $g m^{-2}$
Soil						
$NLOM_1$	30	1300	31	27	1200	45
$NLOM_2$	12	6700	558	12	6300	525
Soil Subtotal	~13	8000	599	~13	7520	570
Plant						
Active pool (roots, leaves)	25	500	20	25	500	20
Storage pool (wood)	240	8000	33	240	9900	41
Plant Subtotal	~160	8500	53	~168	10400	62
Total Ecosystem		16500	652		17920	632

warming) and a slight increase in plant biomass C:N (from 160 to 168). We speculate that C:N ratios in bulk soil would narrow under warming due to larger decreases in C:N ratios of $NLOM_1$ relative to those of $NLOM_2$ (Table 14.1). Both pools would probably decrease in size, however, due to higher decomposition rates. Plant C:N ratios increased in the simulation because the proportion of wood in biomass increased under warming. If NLOM or plant C:N ratios actually decrease, then ecosystem C sequestration would decrease accordingly. However, simulating responses of this forest to a +5°C climate combined with a doubling of atmospheric CO_2 concentration showed higher C:N ratios in both plant and NLOM pools and about a 15% increase in total ecosystem C stocks (data not shown; see Rastetter et al. 1991). Finally, it is worth noting that under 5°C warming alone, about 20 g N m^{-2} was lost from the system. This was due to elevated losses (leaching and denitrification) associated with higher rates of net N mineralization from NLOM. If this N was taken up by biomass rather than being lost, C accumulation under warming would have been even greater. For example, accumulation of this lost N in wood (C:N = 240) would result in about 4.8 kg C m^{-2} being stored as biomass and a 35% increase in ecosystem C storage. This latter scenario represents an extreme end point as some of the retained N would likely enter pools with lower C:N ratios than wood. However, it serves to illustrate the sensitivity of C stocks to variations in nutrient distribution among biomass and NLOM pools with widely different C:nutrient ratios.

Although a warmer, CO_2-enriched environment could increase C storage in forest ecosystems, where C can be stored in woody tissues with high C:nutrient ratios, ecosystems dominated by plant growth forms with lower C:N ratios will likely respond differently. For example, Rastetter et al. (1991) also used GEM to simulate responses of a tundra ecosystem on Alaska's North Slope to temperature and CO_2 increases. GEM predicted that increases in net primary production in tundra ecosystems (with almost no woody biomass) will likely result from transfers of N from NLOM pools to plants. However, the simulations predict that relative increases in tundra C stocks would likely be small compared to those in forest ecosystems: GEM predicted that tundra ecosystem C would increase by $< 2\%$ under $+ 5°C$ warming and by $< 4\%$ under warming and doubled CO_2 concentrations. Also, Schimel et al. (1990) used CENTURY, a process-based biogeochemical model with one plant biomass pool and three soil pools, to simulate responses of grasslands in the central United States to $4°C$ mean annual temperature increases. Their simulations showed net primary productivity increases of about 7% and 20% in the southern and northern Great Plains sites after 50 years. The increase in the southern site resulted from redistribution of N from NLOM to plant biomass while the increase in the northern site resulted from both redistribution of N from soil and greater moisture availability. (Climate models providing input data predicted increases of $\sim 8\%$ in precipitation in the northern site.) However, soil organic C in stocks decreased by slightly greater amounts at both sites over this 50-year period, resulting in slight decreases in ecosystem C stocks.

AQUATIC ECOSYSTEMS

Interactions among primary producers, herbivorous zooplankton, microbial decomposers, and chemical, physical, and photolytic processes control C and nutrient cycles and NLOM accumulation in marine and freshwater ecosystems. Nutrient limitation is a critical determinant of primary production in marine and freshwater ecosystems because neither inorganic C nor water is limiting in aquatic environments. Primary producers rely heavily on nutrients mineralized from dissolved, particulate, and sedimentary NLOM (Figure 14.2). Because decomposition rates and efficiencies in water columns and sediments are often high, extremely small proportions of C fixed by primary producers accumulate as NLOM in sediments.

Below we discuss key interactions among primary producers, decomposers, nutrients, and the physical environment as they affect sedimentation and

NLOM burial in marine and freshwater systems using our minimum aquatic model (Figure 14.2). We then identify possible effects of climate changes on these interactions and on potential changes in NLOM storage in aquatic ecosystems.

Controls on C Inputs and Sequestration in Aquatic Environments

The Biological CO_2 Pump

The biological CO_2 pump, a key process unique to the open ocean, utilizes the downward flux of NLOM through the water column to transport CO_2 to deep water against concentration gradients. The pump has important implications for global C cycles at decadal to century (and longer) time scales. It consists of the fixation of CO_2 in surface water by phytoplankton, the sinking of particulate organic matter (most of which is NLOM), and remineralization of NLOM to inorganic carbon at depth. The upper 100–200 m of the ocean are relatively well mixed, and CO_2 in these waters is approximately in equilibrium with the atmosphere on time scales of years (Vanderborght and Wollast 1994). Below about 700 m, however, dissolved CO_2 species are effectively cut off from atmospheric exchange on time scales of about 1000 years (the mean residence time of deep seawater).

The pump is important because it significantly increases the capacity of the ocean to store inorganic carbon. Because of the pump there is a 12% higher inorganic carbon concentration in seawater below 500 m (90% of the ocean) than there is in surface seawater equilibrated with the atmosphere (Vanderborght and Wollast 1994). As about 38×10^{18} g of inorganic carbon resides in the ocean, this 12% increase amounts to about 4.7×10^{18} g or approximately 7 times the atmospheric CO_2 pool (0.66×10^{18} g), approximately 5 times the amount of carbon in land plant tissue and approximately 3 times the amount in soils (based on data from Olson et al. 1985). Therefore, any changes in rates of CO_2 transport to deep waters via the biological pump will have important implications for the global carbon budget.

Carbon Burial in Marine Sediments

Primary productivity in the euphotic zone exerts a major control on C inputs to the deep sea and to sediments. Strong positive correlations exist between rates of primary production and organic C accumulation in sediments of coastal margins and the open ocean (Emerson and Hedges 1988; Pedersen and Calvert 1990). High rates of primary production tend to favor high rates of C delivery to deep waters and sediments, not only because of higher formation rates of biogenic particles in surface waters, but also because particles that form under nutrient rich conditions, such as fecal pellets and cellular aggregates, tend to be large and sink rapidly (Longhurst 1992). Because they spend less time in transit than smaller particles, larger particles are less likely to decompose before reaching either the deep sea or the sediment surface (Fowler and Knauer 1986).

The relationship between productivity and biogenic particle size reflects the fact that low productivity generally occurs as a result of nutrient limitation. At low nutrient availability, growth of small plankton is favored because smaller cells can sustain higher nutrient uptake rates per unit of cell volume (Pasciak and Gavis 1974). Small cells, whose growth is favored under low nutrient conditions, lead to the formation of small, slowly sinking biogenic particles that are more likely than larger particles to decompose before reaching deep waters or sediments.

Rates of NLOM accumulation in marine sediments correlate strongly with primary production and bulk sedimentation rates rather than with bottom water chemistry or oxygen availablity. Conventional "geologic wisdom" might suggest that bottom water oxygen content would play a critical role in organic matter preservation. However, recent geochemical evidence suggests that oxygen levels, even at anoxic extremes, do not play a major role in controlling organic carbon burial (Pedersen and Calvert 1990; see also Reeburgh, this volume).

Once particulate NLOM settles to depths greater than about 700 m in marine systems, it is effectively cut off from exchange with the atmosphere on a time scale of about 1000 years, regardless of whether it is remineralized in the deep sea (which most is), remineralized in the sediments (the fate of most sedimentary NLOM), or buried as NLOM in the sediments. Sediment burial of NLOM, however, becomes important to global carbon cycling and to atmospheric CO_2 levels on geologic time scales.

According to Hedges (1992), more than 80% of marine carbon burial in sediments occurs in deltaic-shelf environments and at least some portion of this is $NLOM_2$ of terrestrial origin. However, unlike in deep-sea sediments, CO_2 that is regenerated in coastal sediments can diffuse to overlying surface waters and ultimately diffuse back into the atmosphere, at annual to decadal time scales. Time scales of CO_2 release from sediments should vary, however, depending upon the depth in the sediment at which the remineralization occurs, bioturbation rates, and CO_2 and HCO_3^- diffusion rates in sediments.

Carbon Sequestration in Freshwater Ecosystems

In contrast to marine (especially open ocean) environments, where extremely small percentages of primary production are transferred to sediments as NLOM, freshwater lakes often transfer large fractions of phytomass to sediment NLOM pools. Whereas sediment accumulation rates (and contents) in the open ocean are extremely low, rates of NLOM accumulation in sediments can be considerably higher in lacustrian environments with shallow water columns. Accumulation rates increase with eutrophication (high levels of inorganic N and P), and organic C concentrations in sediments of eutrophic lakes can be as high as 15% (I. Ahlgren and E. Welch, pers. comm.). Relatively large proportions of this C consist of the remains of dead plankton and function as labile $NLOM_1$ under aerobic conditions.

Rates of NLOM decomposition in freshwater sediments are probably slow due to anoxic conditions that usually prevail during the productive season. At these times, oxidation of lake sediments is strongly limited by diffusion of molecular oxygen (Wetzel 1975, p. 221). Shallow water (through which NLOM particles are quickly transported), anoxia and low concentrations of alternative electron acceptors (such as SO_4^{2-} in marine systems) often combine to account for high rates of NLOM accumulation in lake sediments. Yet although NLOM deposition and accumulation rates in lakes can be large compared to rates in marine ecosystems, any net global flux of C into NLOM of freshwater ecosystems is minor due to the relatively small areas of these ecosystems.

Effects of Nutrient Availability on Marine Productivity and Carbon Cycling

Historically, marine productivity was thought to be limited largely by N and occasionally P availability (Thomas et al. 1971; Codispoti 1989). For example, the highest rates of primary productivity occur in coastal margins with high rates of N inputs from terrestrial sources and in upwelling regions where nitrate-rich deep water is advected to the surface. However, there is increasing evidence that iron (Fe) as well as N regulates primary productivity in open ocean regions (Martin et al. 1991). Unlike N, which is added to the euphotic zone primarily via upwelling of nutrient-rich subsurface waters, Fe is delivered to the surface ocean largely via aeolean deposition of continental mineral aerosols (Duce and Tindale 1991). As a consequence, the tendency of marine ecosystems to be either Fe or N limited will be controlled by the relative rates of upwelling and deposition of continental dust. Because dust deposition decreases dramatically with distance from continental source regions, surface waters in remote oceanic upwelling systems tend to have extremely low ratios of Fe to N availability. These regions, represent sizable areas of the ocean, including the Southern Ocean, equatorial Pacific, and subarctic Pacific. Small Fe additions to these waters (1–10 nmol l^{-1}) in shipboard bottle experiments markedly stimulate phytoplankton growth, resulting in depletion of soluble pools of available N and phosphorus (Martin et al. 1991). Additions of Fe preferentially stimulate the growth of larger cells, leading to the formation of larger biogenic particles (cells, cell agregates, and zooplankton fecal pellets), which settle out of the euphotic zone at higher rates than smaller particles. Thus, higher Fe concentrations should not only increase productivity but also increase the fraction of fixed carbon and nutrients that are lost via settling to deeper waters and sediments.

Limitation of ocean primary production by Fe has important implications for the transport of CO_2 to deep waters via the biological CO_2 pump (above). The efficiency of the pump is regulated by primary productivity which, in turn, is largely controlled by nutrient supplies in surface water. Major nutrients (N and P) and Fe both appear to be important in limiting primary productivity in different regions of the ocean. A key difference between macronutrients versus Fe in the open ocean is that N and P are supplied primarily from vertical advection and mixing of subsurface water. Transport of these recycled nutrients to the surface is associated with transport of regenerated CO_2, which is then free to exchange back into the atmosphere. In contrast, atmospheric deposition of continental Fe occurs without accompanying increases in CO_2 concentrations. Input of continental iron is important because after organically bound Fe is remineralized in seawater, much of it is scavenged from solution and sedimented as Fe oxides or as Fe adsorbed to particles. Therefore, in Fe-limited systems, net rates of CO_2 transfer to deep waters and sediments should be closely linked to rates at which continental Fe is transported to the ocean surface.

Consequences of Element Ratios in NLOM and Aquatic Ecosystems for C Sequestration

Ranges of C:nutrient ratios are considerably more restricted in aquatic than in terrestrial ecosystems. C:N mass ratios in marine ecosystems are approximately 5.7 in phytoplankton, 6–9 in sedimented NLOM, and about 11–19 in dissolved organic matter (Knauer and Martin 1981; Williams 1992). As in marine ecosystems, the Redfield C:N ratio of 5.7:1 is often used to represent freshwater phytoplankton. C:N ratios of NLOM in lake water columns, however, often range from 6–9 (Fuhs et al. 1972; Birch 1974). C:N ratios in freshwater sediments tend to be more constant and similar to ratios in marine sediments, ranging from 8–10 (I. Ahlgren, pers. comm.). However, in forested watersheds, where $NLOM_1$ derived from forest litter accumulates on lake bottoms, C:N ratios can approach 20 (Bauer 1971).

Because C:nutrient ratios are generally low and do not differ greatly among components of aquatic ecosystems (Figure 14.2), it is unlikely that shifts in the relative amounts of different components within marine or freshwater ecosystems would have major effects on C storage in NLOM pools. Accumulation of large amounts of C in biomass would be unlikely due to the absence of structural tissues with high C:nutrient ratios. Increased NLOM accumulation in aquatic ecosystems would require chronically high inputs of growth-limiting nutrients and the subsequent burial of these nutrients and associated C as NLOM in sediments or as suspended NLOM in deep ocean waters. The combination of low C:nutrient ratios in sediment NLOM (C:N mass ≤ 10) and high decomposition efficiencies in aquatic ecosystems would serve to minimize NLOM accumulation in sediments of most aquatic systems, even if nutrient inputs increased.

Under current climatic conditions, NLOM burial in marine sediments occurs primarily along coastal margins. However, the annual total amount of C buried in near-shore and open ocean environments is a minor component of the global C budget and is not an appreciable sink for atmospheric CO_2 (Berner 1992). Our group agreed that neither warmer conditions nor elevated atmospheric/aquatic CO_2 concentrations would likely change C:nutrient ratios in aquatic ecosystems or NLOM burial sufficiently to increase C accumulation substantially in marine ecosystems. Nutrient deposition or inputs to coastal areas and freshwater ecosystems could increase sediment NLOM accumulations locally, but not enough to serve as significant sinks for CO_2 at a global scale.

Considering sedimentary NLOM accumulation alone as a possible C sink ignores the potential of oceanic NLOM to transfer C between the atmosphere and deep sea. Two factors unique to open ocean environments could serve to change C storage under climate change: the biological CO_2 pump and possible Fe limitation of primary productivity. At present, large amounts of major nutrients (N and P) remain unutilized in surface waters of major oceanic upwelling systems, such as the Southern Ocean. The reasons for this are not clear, but it is becoming increasingly evident that Fe plays a central role in limiting major nutrient (and CO_2) drawdown in these regions (Martin et al. 1991; results of a recent mesoscale Fe addition experiment in the Equatorial Pacific). Computer models have shown that complete utilization of N and P in surface waters of the Southern Ocean would substantially mitigate atmospheric increases in CO_2 concentrations (Sarmiento and Orr 1991). If low Fe availability does, in fact, lead to an inefficient use of N (or P) in otherwise nutrient-rich ocean regions, then changes in Fe delivery to these regions due to climate change could increase primary productivity and increase the rates of C delivery to deep ocean waters by stimulating the biological CO_2 pump (see above). Increasing the delivery of NLOM to deep waters and ocean sediments, even if NLOM is remineralized to CO_2, could serve to impact atmospheric CO_2 concentrations by transferring greater amounts of carbon to deep ocean waters, where turnover times range from 4000 to 6000 years (Druffel et al. 1992).

ATTRIBUTES OF TERRESTRIAL AND
AQUATIC ECOSYSTEMS

From a global perspective, C:nutrient ratios of living tissues can be viewed as being constrained by their metabolic functions. For example, leaf mesophyll tissues and phytoplankton, both of which function to fix CO_2 into organic C, have low C:N ratios due to high concentrations of metabolic proteins and low concentrations of structural compounds. Supportive and storage tissues in terrestrial plants have high C:nutrient ratios because of the presence of large amounts of structural compounds, such as cellulose and lignin or stored carbohydrates. Although C:nutrient ratios within tissue types are not firmly fixed and ratios can vary somewhat among species or in response to changes in resource availability, living tissues must remain within certain bounds to perform their required functions. As a result, differences in C:nutrient ratios among organisms and species result largely from variations in tissue composition.

NLOM is not under the same physiological constraints as living tissues. As such, there is no apparent a priori reason to suspect that C:nutrient ratios in NLOM could not vary widely. Nevertheless, $NLOM_2$, the more stable and larger of the two pools in our models (Figures 14.1 and 14.2), converges to relatively constant C:N mass ratios in a wide range of ecosystem types. For example, in both marine and terrestrial ecosystems, where C:N ratios of materials entering NLOM differ considerably (ca. 6–9 in marine versus 20–200+ in terrestrial ecosystems, depending on tissue type), the C:N ratios in NLOM remaining after removal of more labile material ($NLOM_1$) is about 10:1 in soils and sediments and approximately 15 in marine DOC.

It is not entirely clear why C:nutrient ratios in $NLOM_2$ should converge to a relatively constant value in different ecosystems. This convergence of C:N ratios suggests a necessary "stoichiometry" in the process of $NLOM_2$ formation. The 10:1 C:N mass ratio could represent the "break-even" point for bacteria in terms of the energy cost of nutrient extraction from low quality (low energy yielding) compounds. Alternatively, it could represent a necessary composition for the structure of refractory compounds comprising $NLOM_2$.

Hedges (1992) has suggested a plausible mechanism for decreases in soil organic matter C:N ratios with age. He noted that organic molecules containing amino groups would be more positively charged than molecules without such groups. Therefore, compounds with more amino groups (and N) might be more tightly adsorbed to negatively charged mineral surfaces (e.g., clays) and thereby more physically protected from degradation. Organic compounds containing a high content of negatively charged carboxyl groups and less amino N (e.g., fulvic acids), on the other hand, would be more likely to be retained in pore waters, where they would be subject to microbial degradation or leached from soils. Such a mechanism would not act in water columns, where most organic matter is dissolved or in colloidal suspension, and is not protected by sorption onto minerals. In aquatic systems, preferential uptake of amino acids by bacteria and deamination reactions catalyzed by enzymes on algal surfaces would tend to increase C:N ratios. The net result would be a convergence of C:N ratios in the two systems.

The convergence of $NLOM_2$ C:N ratios to relatively constant values in widely different ecosystem types suggests that C:N ratios of a large fraction of the Earth's NLOM pools will remain stable in a warmer, CO_2-enriched environment. However, C:nutrient ratios and C contents of entire ecosystems could change in response to changes in climate and atmospheric composition. Changes in distributions of growth-limiting nutrient elements among ecosystem pools with widely different C:nutrient ratios could alter both C:nutrient ratios of ecosystems and C sequestration. In forest ecosystems, for example, redistributions of N from $NLOM_2$ to woody biomass could increase C storage and C:N ratios. Any increases in C:N ratios of biomass, litter, and $NLOM_1$ that accompany increases in atmospheric CO_2 concentrations could also increase ecosystem C storage. Finally, changes in efficiency of transfer of $NLOM_1$ to $NLOM_2$ resulting from changes in climate or litter quality could change the relative amounts of C stored in NLOM pools with high and low C:nutrient ratios. Higher $NLOM_1$ decompostion efficiencies would tend to decrease C storage in $NLOM_1$ while lower efficiencies would increase C storage. Climatic changes will likely affect both overall C stocks and C:nutrient ratios quite differently in terrestrial and aquatic ecosystems.

Any changes in C storage within terrestrial ecosystems that occur under a changed climate will likely result from shifts in distributions of growth-limiting elements among pools with widely varying C:nutrient ratios. Changes in C storage in aquatic, particularly ocean, ecosystems are more likely to result from changes in functioning of the CO_2 pump resulting from changes in the delivery of limiting nutrients (N, P, or Fe) to surface waters. If changes in climate or land use transfer greater amounts of nutrients from continents to the oceans, the CO_2 pump could serve as a more efficient mechanism for transfering C (via NLOM) to deep waters and sediments. There is no equivalent CO_2 pump in the terrestrial ecosystems because when CO_2 is regenerated in soils, diffusion back to the atmosphere is not restricted.

In summary, a fundamental difference between terrestrial and marine ecosystems is that C storage potentials of terrestrial ecosystems depend strongly upon distributions of nutrients among biomass and NLOM pools with widely different C:nutrient ratios. Storage of C in marine environments depends primarily upon the biological CO_2 pump which, in turn, is regulated by nutrient delivery to surface waters.

HUMAN EFFECTS

How might human activities change the amounts and distributions of nutrient elements in ecosystems and thereby affect C storage? In addition to increasing atmospheric concentrations of CO_2 and other "greenhouse" gases as a result of industrial activity and biomass burning, humans have directly altered global patterns of C and nutrient cycling. We expect that industrial and agricultural activities will continue to have major impacts on ecosystems and biogeochemical cycles well into the next century. These activities will continue to have profound effects on interactions between nutrient elements and NLOM pools in ecosystems. It is important to consider temporal and spatial scales when assessing the effects of human activities. Also, although C contents of individual ecosystems can change dramatically due to changes in resource use or nutrient cycles, the summary effects at global scales are more difficult to assess.

Nutrient Redistribution

The increase in industrial N fixation (conversion of N_2 into biologically available N forms such as NH_4^+ and NO_3^-) during the 20th Century has been particularly dramatic. Annual rates of N fertilizer production are now approximately equal to annual rates of biological N fixation on a global scale (J. Galloway, pers. comm.). In addition, deposition of combustion-derived N oxides on the Earth's surface are increasing the amounts of N available to primary producers in ecosystems (Aber et al. 1989). Increasing population and industrial activity during the coming decades will probably result in greater inputs from both fertilizer use and NO_x deposition. As with N, other important nutrients such as P, K, and Ca have been added to agricultural ecosystems, thereby increasing the amounts of these elements available to primary producers. Additions of these growth-limiting nutrients to ecosystems should eventually serve to sequester some C. As yet, however, no strong evidence suggests that this is taking place. Nevertheless, unless nutrient losses equal or exceed inputs or C:nutrient ratios of ecosystem components decrease, C gains by ecosystems should accompany nutrient inputs.

Land-use Change

Changes in land-use practices typically lead to changes in both C:nutrient ratios and C storage in NLOM and ecosystems. Conversion of forests to croplands, for example, replaces high C:nutrient biomass components with annual cultivars characterized by low C:nutrient ratios and low C stocks. Also, nutrient cycles become more open (greater inputs and outputs relative to internal cycling), litter inputs to NLOM decrease, C:nutrient ratios in NLOM narrow, and overall soil C stocks decline due to mineralization and erosion. A review by Davidson and Ackerman (1993) concluded that 20% to 40% of soil C in upper profiles is typically lost following cultivation and that most C loss occurs within five years of conversion to cultivation. This suggests that decreases in $NLOM_1$ account for most of the initial C loss. Decreases in NLOM following forest conversion generally serve to decrease fertility at decadal scales by lowering soil nutrient mineralization rates and ion retention capacities. However, specific tillage practices can be used to ameliorate some of these effects (see Scholes and Scholes, this volume).

Conversion of "natural" ecosystems to grazed pastures can have complex effects on ecosystem C budgets and NLOM dynamics depending upon the type of ecosystem converted and the intensity of grazing that follows land conversion. As with conversion to cropland, conversion of forests into pastures removes high C:N biomass, changes the quality and amount of plant inputs to NLOM pools, and alters nutrient cycles (Scholes and Scholes, this volume). Although NLOM stocks decline for several years following conversion to pasture, the magnitude of the decline can be affected by soil texture and characteristics (productivity, nutrient use) of pasture grasses (Veldkamp 1994). Also, although conversion of forests to pasture generally decreases C storage overall and in NLOM, overgrazing of grasslands can result in the eventual dominance of woody plant species that are less palatable than grasses. This could lead to slower nutrient cycles, greater biomass C stocks and higher C:N ratios in both plants and NLOM. In some pastures, the presence of N-fixing species can serve to maintain NLOM stocks by increasing N inputs to soils (Veldkamp 1993).

Reforestation of degraded or abandoned croplands can also lead to C increases in biomass and NLOM pools. Forested lands in eastern North America and Europe, for example, are likely to continue increasing in area as marginal agricultural areas are abandoned (Scholes and Scholes, this volume). As with fire suppression (below), reversion of croplands to woodlands and forests will increase C contents in biomass and NLOM. Rates at which NLOM and biomass C pools recover, however, will be greatly dependent upon rates at which growth-limiting nutrients (e.g., N and P) can be supplied by soils and mineralizable NLOM. If pools of readily mineralizable nutrients in NLOM are depleted due to past land-use practices, then C recovery will be slowed. N-limited ecosystems with high rates of atmospheric N inputs or with active N-fixing species sources will likely recover C stocks more rapidly.

Fire Regimes

Frequent burning occurs over large regions of the Earth, particularly in grasslands and savannas. Fires typically decrease woody biomass and increase forage quality. Although the short- and intermediate-term result is a decrease in ecosystem C:nutrient ratios and NLOM stocks, preservation of small amounts of organic C as charcoal during individual burns can increase NLOM amounts at long time scales (millenia). In contrast, fire suppression in fire-prone ecosystems such as boreal forests, chaparral, and savannas can increase the C contents of an ecosystem and NLOM pools at time scales of years to centuries. In the absence of fire, production is increasingly concentrated in woody plants with higher C:nutrient ratios than nonwoody vegetation. As aboveground biomass and C pools increase, NLOM pools also increase due to greater litter inputs and accumulation of larger $NLOM_1$ pools at the soil surface. In savannas, for example, long-term fire exclusion can double C sequestration, from $5-11 \, kg \, C \, m^{-2}$ (Jones et al. 1990; Scholes and Hall 1994). Given the large land area occupied by savannas ($16 \times 10^6 \, km^2$) fire suppression in this ecosystem type could result in an important medium-term C sink.

Aquatic Ecosystems

Intensive agriculture, industrialization, population growth, and urbanization can all lead to increased inputs of NLOM and nutrients to aquatic ecosystems. Additions of growth-limiting nutrients to aquatic systems typically lead to increases in primary productivity with P and N being the principal limiting nutrients in freshwater and marine ecosystems, respectively. Although eutrophication of freshwater ecosystems has important effects on water quality and species composition of lakes, streams, and rivers, we focus here on marine ecosystems because of their more important role in global C and nutrient cycling. Because primary productivity and NLOM dynamics in continental margins and open oceans differ markedly, we discuss these regions separately.

Continental Margins

Coastal waters often receive large sediment and nutrient loadings derived from agricultural, industrial, and developmental activities, which result in greatly increased algal production. Many coastal areas are experiencing significant increased rates of eutrophication, for example, the Skagerak and Kattegat areas adjacent to the North Sea (Rosenberg et al. 1990) and the Mississippi River Plume area (Rabalais et al. 1991). The extent to which coastal margins are sequestering increased NLOM in response to eutrophication is not known, however, because most studies so far have emphasized eutrophication effects on productivity and fisheries. The role of near-shore ecosystems as NLOM sinks needs to be explored more fully. It is also important to distinguish the relative amounts of NLOM derived from inputs of sediments from terrestrial sources versus amounts derived from primary production in the water column. Although NLOM accumulation rates in certain near-shore areas can be quite high and although accumulation rates are higher on continental margins than in the open ocean as a whole, the net flux of C into sediment NLOM in continental margins appears to be small relative to the rate at which CO_2 is accumulating in the atmosphere (Berner 1992).

Oceans

Burial of NLOM in open ocean sediments is small relative to burial on continental margins (Berner 1992). However, about 2 Gt C yr^{-1} (ca. one-third of fossil-fuel emissions) is currently being stored in surface and deeper ocean waters (Siegenthaler and Sarmiento 1993). Land-use practices and the limitation of oceanic productivity by Fe have implications for the global carbon cycle. Because Fe appears to limit primary productivity in many open ocean ecosystems (above), increases in Fe inputs to the ocean due to greater aeolean deposition of continental dust would tend to increase oceanic productivity and export more fixed C and nutrients to the deep waters and bottom sediments. Aeolean transport of Fe to the ocean should be highest under dry, windy conditions, such as those thought to have occurred during the Last Glacial Maximum. If global warming resulted in a wetter climate and lower overall wind speeds, as some climate models predict, then this might result in lower dust inputs into the ocean, lower productivity in oceanic upwelling systems, and lower rates of carbon sequestration in the deep ocean and sediments. Such an effect, however, could be partially offset by increased cultivation, degradation of grazing lands, and desertification, all of which would increase wind erosion of soils and result in increased deposition of Fe-containing dust particles in the oceans. Dust deposition in open ocean environments has the potential to interact with changes in wind and currents and nutrient upwelling and thereby influence primary productivity and NLOM burial.

KEY QUESTIONS AND EXPERIMENTS

Because C sequestration in terrestrial ecosystems is constrained by C:nutrient ratios of ecosystem components, there are three means by which C contents of terrestrial ecosystems can change in response to climatic or direct human perturbation:

- redistribution of nutrients among pools with different C:nutrient ratios,
- shifts in C:nutrient ratios within ecosystem compartments, and
- changes in the balance between nutrient inputs and outputs.

Sequestration of C by marine ecosystems is constrained more by rates of delivery of key limiting nutrients to surface waters and by interactions between nutrient inputs, primary productivity, water circulation, and the biological CO_2 pump.

Key research questions, based on the above terrestrial and marine processes, that can be addressed using minimum models (Figures 14.1 and 14.2) as frameworks are:

- How will different types of disturbance (e.g., warming, land conversion, nutrient deposition) change nutrient distributions among ecosystem components?
- How flexible are C:nutrient ratios within individual components under current and possible future environments?
- How will changes in nutrient supplies to plants affect above- and below-ground transfers of biomass to $NLOM_1$, decomposition rates of $NLOM_1$ and $NLOM_2$, and $NLOM_1$ to $NLOM_2$ transfer efficiencies?
- How will the quality of inputs to $NLOM_1$ change in a warmer, CO_2-enriched environment?
- How might mechanisms and pathways of $NLOM_1$ and $NLOM_2$ formation respond to climate change?
- Will nutrient cycles become more open (outputs > inputs) or more closed (inputs > outputs) under possible future environments?
- What are the roles of different plant species and growth forms in regulating C:nutrient ratios in biomass and NLOM pools?
- What physical factors influence the stabilization of NLOM pools in terrestrial and aquatic environments?
- How will nutrient transfers between terrestrial and marine ecosystems influence C and NLOM dynamics? For example, how will changes in climate or land use affect the potential Fe feedback loop to oceanic productivity?
- What is the relative importance of iron as opposed to major nutrients (N and P) in controlling POC export to the deep sea?
- What are the effects of deliberate (or inadvertent) manipulations such as reforestation and nutrient deposition on C and NLOM pools at regional and global scales?
- What are the mechanistic relationships among concentrations of limiting nutrients (N or Fe), algal productivity, trophic structure, and export of fixed C to the deep sea?

Experiments designed to test these questions would be best conducted in ecosystems or regions with high potentials for changes in C:nutrient ratios and in rates of total C sequestration. Boreal forest and tundra ecosystems in high latitudes, aggrading forests in middle latitudes and forest–savanna–grassland transition areas in low latitudes are critical areas where the effects of nutrient–NLOM interactions on C sequestration should be studied.

Large-scale experiments simulating climate change scenarios using intact plant-soil or aquatic systems are critical for improving our understanding of how nutrient and C cycles will interact in changing environments. Manipulations of whole ecosystems will be required to resolve uncertainties about how feedbacks and nutrient exchanges between NLOM and live biomass pools might change under different climate and land-use patterns. Examples include multiyear experiments designed to test the effects of changes in CO_2 concentration, temperature, and nutrient inputs on NLOM dynamics, primary productivity, and detritus quantity and quality.

Experiments at smaller scales than whole ecosystems would also help resolve key questions about NLOM–nutrient interactions. For example, long-term manipulations of the amounts of detrital inputs to NLOM pools in soils or sediments would allow us to estimate the potentials of sediments and different soil types for accumulating NLOM at varing levels of primary productivity. Simultaneous measurements of C and nutrient mineralization potentials of soils and sediments under controlled conditions (different temperatures and moistures) would yield valuable information about how turnover times and C:nutrient ratios of NLOM pools will respond to climate changes. We emphasize that simulation models should be implemented in designing and interpreting whole system, meso-scale, and microcosm experiments.

REFERENCES

Aber, J.D., K.J. Nadelhoffer, J.M. Melillo, and P. Steudler. 1989. Nitrogen saturation in northern forest ecosystems. *BioScience* **39**:378–386.

Bauer, D.H. 1971. Carbon and nitrogen in the sediments of selected lakes in the Lake Washington drainage. M.Sci. thesis, Univ. of Washington, Seattle.

Benner, R., J.D. Pakulski, M. McCarthy, J.I. Hedges, and P.G. Hatcher. 1992. Bulk chemical characteristics of dissolved organic matter in the ocean. *Science* **255**: 1561–1564.

Berg, B. 1986. Nutrient release from litter and humus in coniferous forest soils: A mini review. *Scan. J. For. Res.* **1**:359–369.

Berger, W.H., V.S. Smetacek, and G. Weber, eds. 1989. Ocean productivity and paleo-productivity: An overview. In: Productivity of the Ocean: Past and Present, pp. 1–34. Dahlem Workshop Report LS 44. Chichester: Wiley.

Berner, R.A. 1992. Comments on the role of marine sediment burial as a repository for anthropogenic CO_2 . *Glob. Biogeochem. Cyc.* **6**:1–2.

Birch, P.B. 1974. Sedimentation in lakes of the Lake Washington drainage basin. M.Sci. thesis, Univ. of Washington, Seattle.

Bishop, J.K.B., J.M. Edmond, D.R. Ketten, M.P. Bacon, and W.B. Silker. 1977. The chemistry, biology, and vertical flux of particulate matter from the upper 400 m of the equatorial Atlantic Ocean. *Deep-Sea Res.* **24**:511–548.

Broecker, W.S., and T.-H. Peng. 1982. Tracers in the Sea. Lamont-Doherty Geological Observatory, Columbia Univ. Palisades. New York: Eldigio Press.

Codispoti, L.A. 1989. Phosphorus vs. nitrogen limitation of new and export production. In: Productivity of the Ocean: Past and Present, ed. W.H. Berger, V.S. Smetacek, and G. Weber, pp. 1–34. Dahlem Workshop Report LS 44. Chichester: Wiley.

Davidson, E.A, and I.L. Ackerman. 1993. Changes in soil carbon inventories following cultivation of previously untilled soil. *Biogeochemistry* **20**:161–193.

Deans, J.R., J.A.E. Molina, and C.E. Clapp. 1986. Models for predicting potentially mineralizable nitrogen and decomposition rate constants. *Soil Sci. Soc. Am. J.* **50**:323–326.

Druffel, E.R.M., P.M. Williams, J.E. Bauer, and J.R. Ertel. 1992. Cycling of dissolved and particulate organic matter in the open ocean. *J. Geophys. Res.* **97**:15,639–15,659.

Duce, R.A., and N.W. Tindale. 1991. Atmospheric transport of iron and its deposition in the ocean. *Limnol. Oceanogr.* **36**:1715–1726.

Elwood, J.W., J.D. Newbold, A.F. Trimble, and R.W. Stark. 1981. The limiting role of phosphorus in a woodland stream ecosystem: Effects of P enrichment on leaf decomposition and primary producers. *Ecology* **62**:146–158.

Emerson, W., and J.I. Hedges. 1988. Processes controlling the organic carbon content of open ocean sediments. *Paleoceanography* **3**:621–634.

Fog, K. 1988. The effect of added nitrogen on the rate of decomposition of organic matter. *Biol. Rev.* **63**:433–462.

Fogel, R., and K. Cromack, Jr. 1977. Effect of habitat and substrate quality on Douglas fir litter decomposition in western Oregon. *Can. J. Bot.* **55**:1332–1340.

Fowler, S.W., and Knauer, G.A. 1986. Role of large particles in the transport of elements and organic compounds through the oceanic water column. *Prog. Oceanogr.* **16**:147–194.

Fuhs, G.W., S.D. Demmerle, E. Canelli, and M. Chew. 1972. Characterization of phosphorus limited plankton. In: Nutrients and Eutrophication: The Limiting Nutrient Controversy, ed. G.E. Likens. Special Symposium. *Limnol. Oceanogr.* **1**:113–133.

Hedges, J.I. 1992. Global biogeochemical cycles: Progress and problems. *Mar. Chem.* **39**:67–93.

Jones, C.L., N.L. Smithers, M.C. Scholes, and R.J. Scholes. 1990. The effect of fire frequency on the organic componets of a basaltic soil in the Kruger National Park. *S. Afr. J. Plant Soil* **7(4)**:236–238.

Knauer, G.A., and J.H. Martin. 1981. Primary production and carbon-nitrogen fluxes in the upper 1,500 m of the northeast Pacific. *Limnol. Oceanogr.* **26**:181–186.

Longhurst, A.R. 1992. The role of the marine biosphere in the global carbon cycle. *Limnol. Oceanogr.* **36**:1507–1526.

Loveless, A.R. 1961. A nutritional interpretation of Sclerophylly based on differences in the chemical composition of Sclerophyllous and mesophytic leaves. *Ann. Bot. N.S.* **25**:168–184.

Martin, J.H., R.M. Gordon, and S. Fitzwater. 1991. The case for iron. *Limnol. Oceanogr.* **36**: 1793–1802.

Melillo, J.M., J.D. Aber, and J.F. Muratore. 1982. Nitrogen and lignin control of hardwood leaf litter decomposition dynamics. *Ecology* **63**:621–626.

Melillo, J.M., and J.R. Gosz. 1983. Interactions of biogeochemical cycles in forest ecosystem. In: The Major Biogeochemical Cycles and Their Interactions, ed. B. Bolin and R.B. Cook, pp. 177–222. SCOPE 21.

Melillo, J.M., R.J. Naiman, J.D. Aber, and A.E. Linkins. 1984. Factors controlling mass loss and nitrogen dynamics of plant litter decaying in northern streams. *Bull. Mar. Sci.* **35**:341–356.

Norby, R.J., C.A. Gunderson, S.D. Wullschieger, E.G. O'Neill, and M.K. McCracken. 1992. Productivity and compensatory responses of yellow-poplar trees in elevated CO_2. *Nature* **357**:322–324.

Norby, R.J., E.G. O'Neill, and R.J. Luxmore. 1986. Effects of atmospheric CO_2 enrichment on the growth and mineral nutrition of *Quercus alba* seedlings in nutrient-poor soil. *Plant Physiol.* **82**:83–89.

Olson, J.S., R.M. Garrels, R.A. Berner, T.V. Armentano, M.I. Dyer, and D.H. Taalon. 1985. The natural carbon cycle. In: Atmospheric Carbon Dioxide and the Global Carbon Cycle, ed. J.R. Trablaka, pp. 175–213. Washington, D.C.: U.S. Dept. of Energy.

Palm, C.A., and P.A. Sanchez. 1990. Decomposition and nutrient release patterns of the leaves of three tropical legumes. *Biotropica* **22(4)**:330–338.

Pasciak, W.J., and J. Gavis. 1974. Transport limitation of nutrient uptake in phytoplankton. *Limnol. Oceanogr.* **19**:881–888.

Paul, E.A., and F.E. Clark. 1989. Soil Microbiology and Biochemistry. San Diego: Academic.

Pedersen, T.F., and S.E. Calvert. 1990. Anoxia vs. productivity: What controls the formation of organic-carbon-rich sediments and sedimentary rocks? *Bull. AAPG* **74**:454–466.

Rabalais, N.N., R.E. Turner, W.J. Wiseman, Jr., and D.F. Boesch. 1991. A brief summary of hypoxia on the northern Gulf of Mexico continental shelf: 1985–1988. In: Modern and Ancient Continental Shelf Anoxia, ed. R.V. Tyson and T.H. Peason. London: Geol. Soc. Spec. Publ. No. 58.

Rastetter, E.B., M.G. Ryan, G.R. Shaver, J.M. Melillo, K.J. Nadelhoffer, J.E. Hobbie, and J.D. Aber. 1991. A general biogeochemical model describing the responses of the C and N cycles in terrestrial ecosystems to changes in CO_2, climate, and N deposition. *Tree Physiol.* **9**:101–126.

Rosenberg, R., R. Elmgren, S. Fleischer, P. Jonsson, G. Persson, and H. Dahlin. 1990. Introduction: Marine eutrophication in Sweden. *Ambio* **3**:102–108.

Sarmiento, J.L., and J.C. Orr. 1991. Three-dimensional simulations of the impact of Southern Ocean nutrient depletion on atmospheric CO_2 and ocean chemistry. *Limnol. Oceanogr.* **36**:1928–1950.

Schimel, D.S., W.J. Parton, T.G.F. Kittel, D.S. Ojima, and C.V. Cole. 1990. Grassland biogeochemistry: Links to atmospheric processes. *Clim. Change* **17**:13–25.

Scholes, R.J., and D.O. Hall. 1994 The carbon budget of tropical savannas, woodlands and grasslands. In: Global Change: Effects on Coniferous Forests and Grasslands, ed. J.M. Melillo et al. New York: Wiley, in press.

Siegenthaler, U., and J.L. Sarmiento. 1993. Atmospheric carbon dioxide and the ocean. *Nature* **365**:119–125.

Sunda, W.G., and D.J. Kieber. 1994. Oxidation of humic substances by manganese oxides yields low molecular weight organic substrates. *Nature* **367**:62–64.

Thomas, W.H., D.H. Renger, and A.N. Dodson. 1971. Near-surface organic nitrogen in the Eastern Tropical Pacific Ocean. *Deep-Sea Res.* **18**:65–71.

Tyree, M.T., and J.D. Alexander. 1993. Plant water relations and the effects of elevated carbon dioxide: A review and suggestions for future research. *Vegetatio* **104/105**:47–62.

Vanderborght, P., and R. Wollast. 1994. Aquatic carbonate systems: Chemical processes in natural waters and global cycles. In: Chemistry of Aquatic Systems: Local and Global Perspectives, ed. W. Stumm and G. Bidoglio. Ispra, Italy: Commission of the European Communities.

256   *K.J. Nadelhoffer et al.*

Veldkamp, E. 1993. Soil organic carbon dynamics in pastures established after defor-
estation in the humid tropics of Costa Rica. Ph.D. thesis. Wageningen Agricultural
Univ., Wageningen, The Netherlands, 117 pp.
Veldkamp, E. 1994. Organic carbon turnover in three tropical soils under pasture after
deforestation. *Soil Sci. Soc. Am. J.*, in press.
Vitousek, P.M., and R.L. Sanford. 1986. Nutrient cycling in moist tropical forest. *Ann.
Rev. Ecol. Syst.* **17**:137–167.
Waite, T.D., I.C. Wrigley, and R. Szymczak. 1988. Photoassisted dissolution of colloi-
dal manganese oxide in the presence of fulvic acid. *Environ. Sci. Tech.* **22**:778–785.
Wetzel, R.G. 1975. Limnology. Philadelphia: W.B. Saunders.
Williams, P.M. 1992. Measurement of dissolved organic carbon and nitrogen in natural
waters. *Oceanogr.* **5**:107–116.

15

Processes that Control Storage of Nonliving Organic Matter in Aquatic Environments

W.S. REEBURGH

Earth System Science, University of California, Irvine,
California 92717–3100, U.S.A.

ABSTRACT

This article addresses processes important in the cycling of nonliving organic matter (NLOM) by focusing on storage of organic matter in sediments. It considers differences in the oxidation capacity of freshwater and marine sediments, and shows that the oxidizing capacity in marine sediments is at least 100-fold larger. The controversy surrounding the role of productivity and carbon supply versus the availability of oxygen in bottom waters as controls on organic matter storage draws information from carbon-rich rocks and contemporary sediments. Anoxic bottom water conditions and lack of bioturbation are indicated as factors preserving organic matter in carbon-rich sedimentary rocks; however, studies in contemporary environments point to productivity and sedimentation rate as the controlling factors. Contemporary studies cover a wide range of sedimentation rates and experimental conditions, but questions dealing with observation time scales and postdepositional reworking complicate a straightforward comparison with sedimentary rocks. Other processes that may also be important in carbon storage, such as stabilization and protection by sorption on mineral particles and the concept of a sediment microbial loop, have emerged during the debate and measurements suggested by them are outlined.

INTRODUCTION

Several recent books and reviews have contributed to our understanding of the processes that control cycling of organic matter in aquatic environments. Berger et al. (1989) synthesized the current state of knowledge on ocean

Role of Nonliving Organic Matter in the Earth's Carbon Cycle
Edited by R.G. Zepp and Ch. Sonntag © 1995 John Wiley & Sons Ltd.

productivity and loss of organic matter to sediments, and Wollast et al. (1993) reviewed interactions in the C, N, S, and P cycles and their effect on global change. Canfield (1993) summarized organic matter oxidation in sediments, Henrichs (1992) reviewed organic matter diagenesis, and Lee and Wakeham (1988, 1992) considered water column processes affecting organic matter. Hedges (1992) reviewed the role of active reservoirs in the global carbon cycle and enumerated some key questions, Holmen (1992) emphasized recent trends in the global carbon cycle, and Smith and Hollibaugh (1993) reviewed organic matter metabolism in the coastal ocean.

In this chapter I consider the processes that control cycling of organic matter by focusing on the preservation of organic matter in marine sediments. I chose this approach for several reasons: Sedimentation and storage occur at the sediment–water interface, a convenient boundary in the carbon cycle to take stock of fluxes. There is a substantial amount of data in this area as a result of the controversy surrounding controls on organic matter storage. Resolution of this controversy requires an explanation consistent with the sedimentary record, another area with a growing data base. There are pitfalls in this "inverse" approach. Organic matter decomposition is very efficient, so that the quantity of organic matter stored is the small difference between two large terms, *gross deposition* and *oxidation*, and may be sensitive to errors or small changes in either term. The range of conditions encountered in nature is large, and it is likely that several processes controlling organic matter storage are important to varying degrees over this range.

Recently, as they sought explanations for sedimentary carbon maxima in Pleistocene glacial horizons, paleoceanographers and sedimentologists have examined the processes controlling accumulation and storage of organic matter in marine sediments. These maxima and other carbon-rich sediments are generally considered to have accumulated under anoxic or low bottom water oxygen conditions (Demaison and Moore 1980; Arthur et al. 1984; Ingall et al. 1993).

Studies on contemporary systems include reviews of field data compilations from a range of environments (Henrichs and Reeburgh 1987; Canfield 1989; Betts and Holland 1991), interpretation of new data in selected high sedimentation rate environments with low-oxygen overlying water (Calvert and Pedersen 1992; Pedersen and Calvert 1990; Reimers et al. 1992; Jahnke 1990), and experimental studies of the metabolism of specific compounds under oxic and anoxic conditions (Lee 1992). These contemporary studies generally conclude that productivity and flux of organic matter to the seafloor are the principal controls on formation of organic-rich sediments.

In the following discussion I summarize the debate on organic matter storage, discuss possible explanations that have emerged during the debate, and suggest measurements that may help reconcile the conflicting observations.

CARBON CYCLE IN AQUATIC SYSTEMS

Carbon is stored in a range of reservoirs that turn over and exchange with other carbon reservoirs over a range of time scales. It is crucial to place the processes and reservoirs important at various time scales in perspective. Sundquist (1985) and Walker (1993) outlined a hierarchy of interacting carbon reservoirs that controls atmospheric carbon dioxide on all time scales. These reservoirs consist of small, rapidly exchanging reservoirs (atmosphere, shallow sea, biota), which are coupled by slower transfers to the large deep-sea and sedimentary rock reservoirs. For example, at the hundred million year time scale (Berner 1992), the chemical weathering of Ca–Mg silicate rocks is the principal process for removal of CO_2 from the atmosphere. The weathering of kerogen in shales uplifted on continents imposes a control on the carbon cycle and atmospheric oxygen on time scales of several million years, as burial of organic carbon in modern sediments generates a continuous source of oxygen that must have a compensating sink (Hedges 1992). Studies focusing on the past million years consider changes in carbon storage on continents and continental shelves as a result of glaciation and deglaciation (Harden et al. 1992), while studies considering the effects of anthropogenic alterations with a time scale of centuries (Smith and Hollibaugh 1993) focus on metabolism in the coastal ocean. Studies in coastal sediments, which usually focus on understanding a particular environment, consider time scales of a few months to decades (Henrichs and Reeburgh 1987).

Below, I briefly describe the aquatic portion of the global carbon cycle; this description stems largely from Hedges (1992) and Emerson and Hedges (1988). Remineralization of organic matter is a very efficient process on a global basis: only ~0.1% (0.1×10^{15} gC yr^{-1}) of global primary production (terrestrial: 60×10^{15} gC yr^{-1}; marine: 50×10^{15} gC yr^{-1}) escapes oxidation and is ultimately buried in marine sediments. Continental reservoirs for carbon storage are small and near steady state, so that burial in ocean sediments is the ultimate fate of biosynthesized organic matter that escapes remineralization at time scales of several million years. Export to the sea is the only route for long-term preservation of significant amounts of terrestrial organic matter at this time scale. Most of the particulate fraction of this exported terrestrial organic matter, estimated to be 0.2×10^{15} gC yr^{-1}, is deposited close to shore. The dissolved fraction of exported terrestrial organic carbon, estimated to 0.2×10^{15} gC yr^{-1}, enters the large seawater DOC pool (0.7×10^{18} gC) and is apparently oxidized. Although primary productivity is higher in the coastal ocean, about 80% of the total ocean primary production occurs in the open ocean, which accounts for about 90% of the ocean area. Much of the ocean primary production is recycled in the upper 100 m of the ocean; the recycling process in the upper 100 m is more efficient in the open ocean (80–90%) than in coastal waters (50%). The portion of the particulate flux from the upper 100 m of the ocean (7×10^{15} gC yr^{-1}) reaching the sediments is a function of water depth. Only $\sim$1% of the primary production reaches depths of 4000 m; this decrease is accompanied by selective removal of more reactive biochemicals.

Differences in primary production, surface recycling efficiency, and oxidation of particulate organic carbon (POC) settling through varying water column depths result in deposition of a major fraction of the organic matter that reaches the seafloor in near-shore sediments. More than 80% of the preserved organic carbon is deposited in shelf and continental margin sediments. Further reworking and oxidation of deposited organic matter by diagenetic reactions occurs in shelf and pelagic sediments. The POC deposited in continental margin sediments has a marine origin as indicated by δ^{13}C measurements and consists of two components (Anderson et al. 1994; Jahnke 1990): one component is labile and is remineralized in the surface (a few millimeters of the sediments) on time scales of less than a year to several years; the other component is refractory and is derived from older material. Uniform mixed layer organic carbon distributions suggested decomposition time scales greater than mixed layer residence times; recent ^{14}C measurements show little accumulation of bomb ^{14}C and suggest a remineralization half-life for this refractory material of greater than 1000 years. The fraction of organic carbon that is oxidized on a 1000-yr time scale is higher in continental margin sediments (30–50%) than in pelagic sediments (20–40%) (Emerson et al. 1987). Despite the efficiency of organic matter remineralization, the resulting low storage rate in sediments has led to accumulation of a huge (1.5×10^{22} gC) reservoir of organic matter in sedimentary rocks over geologic time, and this has played a central role in modulating atmospheric oxygen concentrations.

OXIDATION OF ORGANIC MATTER

Oxidation Sequence

Degradation of complex organic matter is microbially mediated and generally proceeds through depolymerization reactions, which may involve digestive enzymes or exoenzymes, fermentation reactions, which produce simple compounds, and terminal metabolism of the simple compounds. Aerobic oxidizers can oxidize a wide range of substrates to CO_2, and are responsible for oxidation of over 90% of the organic matter in sediments. Anaerobic oxidizers have more specific substrate requirements, and can oxidize a restricted number of molecules, often incompletely. Aerobic and anaerobic organic matter degradation rates may be limited by these initial reactions. Anaerobic processes are important for their role in determining the amount of carbon preserved in sediments (Henrichs and Reeburgh 1987). One model for organic matter degradation (Froelich et al. 1979) considers that organic matter oxidation proceeds using the available electron acceptor with the greatest free energy yield according to the sequence: O_2 > NO_3^- > Mn(IV) > Fe(III) > SO_4^{2-} > CO_2 (Table 15.1). Thus, oxidation of organic matter proceeds through a series of reactions consuming the most energetically favorable oxidant in sequence, beginning with oxygen reduction and followed by denitrification, metal oxide (Mn^{+4}, Fe^{+3}) reduction, sulfate reduction, and finally methanogenesis. This results in separation of specific microbial processes into zones of age or depth in sediments.

This sequence is generally followed in most environments; thus it serves as a reasonable framework. Yet, Canfield (1993) has demonstrated overlap of processes, i.e., anoxic processes occurring in oxic conditions, in some environments, so this sequence may be an oversimplification of the complex microbial interactions that can occur in sediments. The explanation seems to lie in competition for reduced substrates rather than free energy yield alone (Lovely and Klug 1986). Canfield (1993) emphasized the importance of focusing on rates of microbial processes rather than chemical distributions, as we have a poor understanding of how microbially reduced species interact with adjacent oxidized species as well as the overall importance of these reactions. Direct rate measurements of denitrification, sulfate reduction, and methane oxidation can be made with stable and radioisotope tracers (Reeburgh 1983); no direct rate measurements are available for oxygen, Fe oxide, or Mn oxide reduction. The rates of the latter reactions can be estimated from gradients; however, this approach may lead to underestimates for the Fe oxide and Mn oxide reduction rates (Canfield 1993). Canfield, Jørgensen et al. (1993) performed incubations on amended and unamended anoxic coastal sediments sealed in gas impermeable plastic bags to estimate metal oxide reduction rates as well as rates of interaction between oxidized and reduced species of Mn, Fe, and S. The amended experiments involved additions of MoO_4^{2-} to inhibit sulfate reduction and slow interactions between Fe and Mn oxides and sulfide, and addition of Ferrozine to complex Fe^{2+}, reducing the possibility of Mn reduction by Fe^{2+}. Canfield, Thamdrup, and Hansen et al. (1993) concluded that most of the oxygen flux into these sediments was used to oxidize reduced species and that in some sediments the importance of metal oxide reduction in organic matter oxidation has been underestimated.

Oxidation Capacity

Composition differences between sea- and fresh waters result in large oxidation capacity differences. The above oxidation sequence is combined with typical electron acceptor concentrations and reaction stoichiometries to estimate closed sediment system oxidation capacity in Table 15.1. This oxidation capacity estimate considered a hypothetical oxygen-saturated marine sediment and estimated the amount of a hypothetical organic compound, CH_2O, that could

Table 15.1 Sequence and capacity of organic matter oxidation processes important in freshwater and marine sediments.

Reaction	p(e)	Eh (mV)	Energy Yield KJ/mole CH_2O	Typical Oxidant Conc. (mM)	Hypothetical Sed. Conc. mmol/l_{sed}	Oxidizing Capacity mmol/l_{sed}	Freshwater Wetland	Characteristic Depth Scales (cm) Marine — Low OM flux	High OM flux
Oxygen reduction $[CH_2O + O_2 \rightarrow CO_2 + H_2O]$	12.1–12.5	720–740	−475	0–0.09	0–0.85	0–0.8	±5 mm of water table	0–1	
Denitrification $[5CH_2O + 4NO_3^- \rightarrow 2N_2 + 4HCO_3^- + CO_2 + 3H_2O]$	~12	710	−448	0–0.04	0–0.037	0–0.03	?	15	
Mn-oxide reduction $[CH_2O + 3CO_2 + H_2O + MnO_2 \rightarrow 2Mn^{2+} + 4HCO_3^-]$	8.0	470	−349	≥1.5% (0.27 mmol/g)	43.2	21.6(?)	?	15–30	≥1 m ~1–5 cm
Fe-oxide reduction $[CH_2O + 7CO_2 + 4Fe(OH)_3 \rightarrow 4Fe^{3+} + 8HCO_3^- + 3H_2O]$	1.0	60	−114	≥4.0% (0.7 mmol/g)	112	28(?)	?	≥20	
Sulfate reduction $[2CH_2O + SO_4^{2-} \rightarrow H_2S + 2HCO_3^-]$	−3.8	−200	−77	0–30	0–28.2	56.4	—	—	≥10
Methane production $[CH_2O + 2H_2O \rightarrow 2CO_2 + 4H_2, \&$ $4H_2 + CO_2 \rightarrow CH_4 + 2H_2O]$ or $[CH_3COOH \rightarrow CH_4 + CO_2]$	−4.2	−250	−58				major process	—	≥25

Adapted from Reeburgh (1983) and Canfield (1993). Hypothetical sediment has an oxygen-saturated seawater content of 85% and a density of 1.1 g cc^{-3}.

be oxidized under closed system conditions. This approach yields a minimum estimate, as oxygen and sulfate is supplied to sediments from the overlying water, and depending on organic matter loading, oxygen demand, and bioturbation and irrigation, can be exhausted in a matter of centimeters to meters. Nitrate is rarely present in concentrations exceeding tens of micromoles, and the role of Mn and Fe oxides remains unclear. The concentration of sulfate is the major difference between freshwaters and marine systems. Seawater sulfate concentrations are 20–30 mmol l^{-1}, while sulfate concentrations rarely exceed 100–200 μmol l^{-1} in fresh waters. A closed-system estimate of oxidation capacity (Reeburgh 1983) shows that a marine sediment has an oxidation capacity at least 60-fold larger than an oxygen-saturated sediment containing no sulfate. The actual oxidizing capacity of a natural sediment will exceed this estimate in non-euxenic environments, because sulfide and reduced metal oxides can be reoxidized by oxygen from overlying waters added by diffusion or by irrigation processes. Thus, there appears to be unutilized oxidizing capacity in many recent marine sediments, even euxinic sediments. Dissolved oxidants are exhausted only in areas of very high organic matter settling flux, such as the Peru and Namibian margins.

ORGANIC MATTER STORAGE

Freshwater Aquatic Environments

Peatlands are end-member environments that illustrate the important differences in organic carbon degradation in terrestrial freshwater environments with low oxidation capacity and marine sediments with high oxidation capacity. Peatlands cover about 3% of the Earth's land surface and have been estimated to contain 180×10^{15} gC. Seventy-five percent of the global peatland area is concentrated in two large areas: the Hudson Bay Lowlands (3.2×10^5 km^2) in Canada and the Vasuugan peatland (5.4×10^5 km^2) in Siberia. Organic matter produced in terrestrial wetland environments contains structural polymers such as lignin and cellulose, and has a relatively high C:N ratio (20–200), in contrast to marine organic matter, which consists of more labile structural polymers and has a low C:N ratio (7). Peatlands consist of two layers: the acrotelm, an oxic 10–50 cm thick surface layer where primary production by mosses and sedges and rapid decay occur, and the catotelm, which receives about 10% of the primary productivity, is usually anoxic, and has much lower decay rates. The catotelm forms the largest part of the peat mass, is waterlogged, and has low hydraulic conductivity. Since the oxidation capacity of fresh waters is low, the dominant degradation reactions in peatland catotelms are fermentation and methanogenesis, which do not result in net oxidation of carbon (Capone and Kiene 1988). Clymo (1984) emphasized that although degradation rates in the catotelm are low, they are not zero, and that most peatlands reach a thickness, usually 5–10 m, where the net rate of accumulation is zero.

Marine Sediments

Intrinsic Differences in Oxic and Anoxic Rates?

Several recent studies suggest that there are no large differences in the rates of oxic and anoxic processes in degrading fresh organic matter. Laboratory microcosm experiments (reviewed in Henrichs and Reeburgh 1987) involving decomposition of added algae under oxic and anoxic conditions give similar rates of decomposition and amounts of refractory material in ~200-day experiments. Lee (1992) studied the metabolism of individual radiolabeled compounds in the oxic and anoxic portions of several stratified water columns. Her work emphasized measurements of turnover times and pool sizes of simple molecules and avoided complications due to interactions with sediments. These sediment-free systems also have the advantage that carbon incorporation as well as respiration can be measured. Lee (1992) observed no major differences in the intrinsic rates of organic matter decomposition under oxic and anoxic conditions.

Conclusions drawn from the behavior of fresh algae and simple molecules are instructive, but their applicability to interpretations of long-term organic matter storage in nature may be limited. The organic matter used in these rate determinations is representative of the labile, rapidly reacting organic matter fraction, so less information on the resistant, slowly reacting organic matter pool is obtained. Also, experiments should be designed to ensure that the amount of organic matter added does not exceed natural inputs.

Role of Bottom Water Oxygen in Organic Matter Preservation

Emerson (1985) and Emerson and Hedges (1988) used a simple model involving sedimentation and bioturbation to examine carbon storage in deep-sea sediments. All carbon was assumed to be of plankton origin and was homogeneously degradable, i.e., one decay constant was used in the model. Oxygen was the only electron acceptor considered; all other electron acceptors were neglected. Emerson (1985) supported this assumption with correlations that showed low organic carbon contents in the presence of high bottom water oxygen concentrations. Emerson and Hedges (1988) point out that rate constants for organic matter degradation are a function of carbon age, i.e., the rates are dependent upon the observation time scales. The model results showed that bioturbation had no effect on carbon storage at high sedimentation rates but that it enhanced carbon storage at low sedimentation rates. The sedimentation rate only affected the average carbon content of sediments at rates > 10 cm kyr^{-1}. Emerson suggested that it might be possible to understand the effect of bottom water oxygen with studies in California Borderland basins, where there is a fairly constant rain rate of organic carbon and a wide range of bottom water oxygen contents. Jahnke (1990) observed no correlation between burial efficiency and bottom water oxygen in sediments from the Borderland Basins of California, and Reimers et al. (1992) reported no indication that fluxes or remineralization rates in Central California slope and rise sediments were influenced by conditions in the oxygen minimum zone. Sarnthein et al. (1987) also concluded that the effect of bottom water oxygen on carbon storage was small. Isolines of carbon accumulation rate on a particulate carbon flux versus bottom water oxygen concentration plot showed sediment accumulation rate increases only at very low oxygen concentrations.

Sediments of the Oman margin of the Arabian Sea, a site of high monsoon upwelling driven production and sedimentation and one of the most intense oxygen minima, have been studied by Pedersen et al. (1992). There they found no correlation between bottom water oxygen and the abundance of sedimentary organic carbon, which was mostly of marine origin. The organic matter progressively degrades, as shown by the C_{org}:N ratio and the I:C_{org} ratio, accompanied by reworking and winnowing as shown by correlations with Cr:Al and Zr:Al ratios, which are proxies for sediment texture. Thus the spatially variable high C_{org} sediments reflect a high settling flux and postdepositional reworking.

Calvert et al. (1992) compared data from Guaymas Basin surface sediments contacting oxygenated bottom water and those contacting the oxygen minimum and concluded that bottom water oxygen levels exerted no major influence on the concentration and composition of organic matter in the sediments.

Factors Affecting Burial Efficiency

Henrichs and Reeburgh (1987) examined the relationship between burial efficiency (the ratio of carbon buried long-term to the sediment input rate, i.e., the fraction that survives remineralization; see Blackburn 1991) and sedimentation rate in a range of environments. Burial efficiency ranged from 0.6% to 16% in pelagic sediments to 8% to 28% in slope sediments and 8% to 79% in estuarine environments. Data from environments with low or zero bottom water oxygen and thin sediment oxic layers did not cluster at the high sedimentation rate:high burial efficiency end of the data set, but were distributed along the sedimentation rate axis. The data provided no striking evidence for the importance of low oxygen conditions in preserving organic matter. Sarnthein et al. (1987) considered open ocean sediments and found a strong correlation between organic carbon accumulation rate and the carbon-free sedimentation rate.

Canfield (1989) considered a similar data set and estimated the importance of sulfate reduction and oxygen reduction at given sedimentation rates with plots of depth-integrated sulfate reduction and oxygen consumption versus sediment burial rate. Data from these plots were combined to produce a plot of integrated carbon oxidation rate versus sedimentation rate. This plot showed that for sedimentation rates $> 0.1\, g\,cm^{-2}\,yr^{-1}$, sulfate reduction and oxygen reduction are equally important in oxidizing organic carbon. For sedimentation rates below $10^{-3}\, g\,cm^{-2}\,yr^{-1}$, oxic respiration remineralized 100- to 1000-fold more organic carbon than sulfate reduction. Canfield's plot of burial efficiency versus sedimentation rate using data from euxinic and semi-euxinic environments suggested high burial efficiency at low sedimentation rates for Black Sea data not included in the Henrichs and Reeburgh (1987) analysis. Canfield concluded that anoxia could enhance carbon preservation and pointed out that small changes in depth integrated organic carbon decomposition rates could have large effects on burial efficiency. Calvert et al. (1991) pointed out that the sedimentation rates for the Black Sea data points used by Canfield were in error and overestimated carbon accumulation rates. Their work indicated that the modern anoxic Black Sea is not a site of anomalously high carbon accumulation.

Betts and Holland (1991) determined the burial efficiency of organic carbon in a data set involving 69 sites and estimated errors in the various burial efficiency estimates. As with the previous work of Henrichs and Reeburgh (1987) and Canfield (1989), they found that sedimentation rate exerts a dominant influence on burial efficiency. Although scatter in the burial efficiency data was large, Betts and Holland concluded that bottom water oxygen has, at most, a minor effect on burial efficiency of organic matter in marine sediments.

Calvert and Pederson (1992) have emphasized that increased primary productivity resulting from increased wind-driven upwelling is the likely cause of organic-rich Pleistocene sediments. Several indices of oxic conditions in Panama Basin sediments, such as size and abundance of benthic foraminifera, size and number of meiofaunal fecal pellets, no changes in the $I:C_{org}$ ratio, and the lack of Mo enrichment, are taken as evidence that the waters overlying these sediments have never been anoxic.

Sediment Microbial Loop

Lee (1992) suggested that extending the concept of the "microbial loop" in water column environments to sediments might offer an explanation for enhanced carbon storage in anoxic sediments. The microbial loop in water column environments involves bacterial uptake of dissolved organic matter (DOM). The bacteria grow and retain the organic carbon as cellular material. Grazers (protozoans) consume the bacteria, efficiently remineralize the nutrients, and release DOM to the water column for another cycle. Lee (1992) suggested that the efficiency of grazing by protozoans in sediments may be limited by sediment compaction and oxygen concentrations in the sediments. The reduction of protozoan numbers and efficiency by compaction and anoxic conditions could result in enhanced storage of bacterial biomass and bacterial products. A number of observations are consistent with a sediment microbial loop (Lee 1992), but direct observations and experimental manipulations are needed.

Sorption by Particles

Recent determinations of organic carbon concentration and specific surface area on a range of marine sediments from a variety of environments (Mayer 1994b) show that in most cases organic matter is present at monolayer equivalent (ME) concentrations on mineral surfaces. The organic matter appears to be associated with the high density or mineral phase. The topography of these natural mineral grains is such that ~80% of the surface area is inside pores of < 8 nm width (Mayer 1994a). Mayer hypothesizes that organic matter is stabilized and protected by its incorporation into pores and slits in the mineral surfaces too small to allow functioning of hydrolytic enzymes involved in the initial stages of degradation. This hypothesis offers an explanation for the "refractory background" level of organic carbon observed in shelf sediments and also provides a plausible mechanism for the carbon burial:sedimentation rate relationship discussed earlier. Exceptions to this relationship are areas with high riverine sediment input (deltas), which have lower than possible ME concentrations, and areas with organic pollution and low oxygen water columns, which have greater than possible ME concentrations. The lack of ME levels of organic carbon in high sedimentation rate deltaic sediments suggests that a finite time is necessary to accumulate organic carbon concentrations to ME levels, as in soils. Interactions between adsorption and slower processes like condensation reactions may also be important in the slow accumulation to ME concentration levels.

SUMMARY

The preceding discussion presents a number of puzzling and apparently conflicting observations. The studies on carbon-rich marine shales point to anoxic conditions and lack of bioturbation as important in formation of laminated organic-rich shales (Ingall et al. 1993), and differences between the organic C:P ratio in laminated and bioturbated shales offer additional support. Studies on contemporary sediments overlain by anoxic and low oxygen waters, however, show no differences in storage that cannot be explained by higher productivity and organic matter deposition when compared to nearby sediments in contact with oxic bottom waters. Field studies in a range of environments show that the major control on sediment burial efficiency appears to be primary production and the supply of organic matter. Differences in the intrinsic rates of oxic and anoxic degradation reactions on fresh organic matter appear to be small; however, small changes in sediment degradation rates can result in large changes in burial efficiency.

The controversy surrounding carbon storage has polarized into two camps, and although information is being added to the debate, it consists of similar information from different environments. Can these conflicting observations be reconciled? What additional information is needed? Since the controversy has drawn on data from a range of water depths, the following points should be carefully considered when drawing comparisons:

- the 10^4- to 10^6-fold range of sedimentation and organic matter supply rates,
- the range of reactivity of organic matter,
- questions relating to observation time scales, time since deposition, and organic matter age.

Emerson and Hedges (1988) observed that decomposition rates are a function of carbon age and observation time scale; this suggests that rate measurements from short-term field and microcosm experiments on fresh organic matter may be inapplicable to explanations of long-term storage. Van Capellen and Canfield (1993) point out that the surface sediments studied in high deposition rate areas have been subjected to diagenetic processing for only a decade or more and that the organic matter remaining in sediments after extensive microbial degradation is another end member that should be studied. They argue that the degree of preservation of organic matter in the surface 1–5 cm of sediments only set limits on how long diagenetic processing must take place before effects can be distinguished. The refractory organic matter concentrated by reworking and redeposition is another important component of this carbon age-observation time scale range. The work of Anderson et al. (1994) shows that about half of the organic matter deposited in the Middle Atlantic Bight (U.S. East coast) margin is refractory and originated from older deposits. The amount deposited and the nature and age of this refractory sediment organic matter is very important to a general explanation of controls on organic matter storage.

INFORMATION NEEDED TO FOCUS THE DEBATE

1. What is the role of sediment texture and mineralogy in organic matter transport and storage?

Recent work (Mayer 1994a, b; Keil et al. 1994) indicates that organic material is present as monolayer equivalent coatings on sediments and that organic matter loadings increase linearly as a function of sediment specific surface area. This hypothesis offers an explanation for the "refractory background" level of organic carbon observed in shelf sediments and provides a plausible mechanism for the carbon burial:sedimentation rate relationship. It should be considered one of the strongest possibilities for resolving the carbon storage controversy. Separation by SPLITT (split flow, thin cell, lateral flow) fractionation (Giddings 1985) permits sorting sediment samples into size/density fractions under conditions that mimic natural winnowing processes with little loss of organic material. Quantities of particular sediment fractions large enough to permit more extensive analyses can be isolated with this technique. Fractions collected with this technique would provide a starting point for long-term reactivity and kinetic (Canfield 1994) experiments on intact sediment-organic associations under oxic and anoxic conditions. Collection of large quantities of selected fractions would permit more complete studies of the composition, structure, and possible origin of sorbed organic matter, including compound specific isotope ratio measurements (Hayes et al. 1990; Freeman and Hayes 1992).

2. What is the composition of the "resistant when anaerobic" and recalcitrant fractions of sediment organic matter? What is the age and origin of this material?

Many studies have suggested that a fraction of sediment organic matter is refractory under anoxic but not oxic conditions, while others have demonstrated a fraction that is recalcitrant, even under oxic conditions. Henrichs (1992) cautions that a small fraction of this "resistant when anaerobic" material cannot explain the entire range of burial efficiencies, as a large fraction of the incoming organic matter is buried in the most rapidly depositing sediments. More information on the composition of the organic matter that is deposited and survives burial below the oxygenated sediment layers is needed to understand its susceptibility to microbial remineralization. Henrichs (1992) points out that complete characterization of the composition of sediment organic matter is not a realistic goal, but that identification of differences in structure between the remineralized and preserved fraction of organic matter buried at high and low preservation sites would be valuable as a starting point for understanding reactivity as a control on storage. These studies could be conducted by either examining down-core differences in sediments from selected sites or by fractionating quantities of sediments, as in Question 1 above, to determine age and reactivity as a function of grain size and mineralogy. Organic matter ^{14}C measurements from all of the contemporary environments discussed in this chapter, particularly the environments with the highest sedimentation rates where bomb ^{14}C is detectable, are needed to determine how much of the deposited material has been recycled.

3. What is the role of metal oxides in organic matter remineralization?

Metal oxides are capable of serving as electron acceptors in biologically mediated organic matter remineralization (Lovely and Phillips 1983; Sørensen 1982; Wilson et al. 1986), and have high oxidizing potential and capacity. Canfield et al. (1993a, b) showed that metal oxides in coastal sediments are more important as electron acceptors than previously believed. Sediment environments where metal oxides are important in organic matter remineralization should be identified and estimates of their global importance should be made. Metal oxide surfaces are important organic matter adsorption sites; does organic matter adsorption affect oxidizing capacity?

4. What is the role of the "microbial loop" in storage of sediment organic matter?

The role of the microbial loop promises to be more difficult to observe in sediments than in water column environments, but it could be clarified by several approaches. Studies of the distribution of bacterial biomarkers (Lee 1992) with depth may give some indication of the depth at which grazing becomes unimportant. It may be possible to develop hybridization probes for protozoan grazers and assess their numbers and depth distributions directly.

5. What is the role of animals and bioturbation/bioirrigation processes in the preservation of sediment organic carbon, reoxidation of reduced metal and sulfur species, and the effectiveness of storage of microbial organic matter?

Lee (1992) and Emerson and Hedges (1988) pointed out that animals deserve attention as an explanation for the differences in organic carbon preservation in oxic and anoxic systems. Animals play an important role in remineralization of organic matter by fragmentation and digestion of organic matter or by direct uptake or further degradation of polymers by intra- or extra-cellular enzymes. These processes provide new surfaces and substrates that stimulate further microbial degradation. Protozoan grazers can help maintain microbial populations in a state of high growth rate, and their absence may be important in storage of microbial organic matter. Sediment irrigators and bioturbators appear to enhance degradation of POC and PON (Kristensen and Blackburn 1987) and may be important in the reoxidation of metal oxides and reduced sulfur compounds (Stone and Morgan 1987). Emerson's (1985) model showed that bioturbation played an important role in enhancing storage of carbon in pelagic sediments, but that it was of less importance in high deposition rate coastal sediments. The work of Ingall et al. (1993) shows striking differences in the organic matter content between bioturbated and laminated (unbioturbated) shales.

ACKNOWLEDGEMENTS

This manuscript benefited from constructive reviews by Cindy Lee and Susan Henrichs. Don Canfield and Larry Mayer supplied preprints of papers in press, and Clare Reimers, Susan Trumbore, and David Shaw commented on versions of the manuscript. This work was supported by grants from NSF, NASA, and EPA.

REFERENCES

Anderson, R.F., G.T. Rowe, P. Kemp, S. Trumbore, and P. Biscaye. 1994. Carbon budget for the mid-slope depocenter of the middle Atlantic Bight: A test of the shelf export hypothesis. *Deep-Sea Res., Part II, Special Topics,* in press.

Arthur, M.A., W.E. Dean, and D.A. Stow.1984. Models for the deposition of Mesozoic-Cenozoic fine grained organic-carbon-rich sediment in the deep sea. Geol. Soc. London Spec. Publ. 26, pp. 527–562.

Berger, W.H., V.S. Smetacek, and G. Wefer, eds. 1989. Productivity of the Ocean: Present and Past. Dahlem Workshop Report LS 44. Chichester: Wiley.

Berner, R.A. 1992. Weathering, plants, and the long-term carbon cycle. *Geochim. Cosmochim. Acta* **56**:3225–3231.

Betts, J.N., and H.D. Holland. 1991. The oxygen content of ocean bottom waters, the burial efficiency of organic carbon and the regulation of atmospheric oxygen. *Glob. Planet. Change* **97**:5–18.

Blackburn, T.H. 1991. Accumulation and regeneration: Processes at the benthic boundary layer. In: Ocean Margin Processes in Global Change, ed. R.F.C. Mantoura, J.-M. Martin, and R. Wollast, pp. 181–195. Dahlem Workshop Report PC 9. Chichester: Wiley.

Calvert, S.E., R.M. Buston, and T.F. Pederson. 1992. Lack of evidence for enhanced preservation of sedimentary organic matter in the oxygen minimum of the Gulf of California. *Geology* **20**:757–760.

Calvert, S.E., R.E. Karlin, L.J. Toolin, D.J. Donahue, J.R. Southon, and J.S. Vogel. 1991. Low organic carbon accumulation rates in Black Sea sediments. *Nature* **350**:692–695.

Calvert, S.E., and T.F. Pedersen. 1992. Organic carbon accumulation and preservation in marine sediments: How important is anoxia? In: Organic Matter: Productivity, Accumulation and Preservation in Recent and Ancient Sediments, ed. J. Whelan and J. Farrington, pp. 231–263. New York: Columbia Univ. Press.

Canfield, D.E. 1989. Sulfate reduction and oxic respiration in marine sediments: Implications for organic carbon preservation in euxinic environments. *Deep-Sea Res.* **36**:121–138.

Canfield, D.E. 1993. Organic matter oxidation in sediments. In: Interactions of C, N, P, and S Biogeochemical Cycles and Global Change, ed. R. Wollast, F.T. Mackenzie, and L. Chou, pp. 333–363. New York: Springer.

Canfield, D.E. 1994. Factors influencing organic carbon preservation in marine sediments. *Chem. Geol.* **114**:315–329.

Canfield, D.E., B.B. Jørgensen, H. Fossing, R. Glud, J. Gundersen, J.W. Hansen, L.P. Nielsen, and P.O.J. Hall. 1993. Pathways of organic carbon oxidation in three continental margin sediments. *Mar. Geol.* **113**:27–40.

Canfield, D.E., B. Thamdrup, and J.W. Hansen. 1993. The anaerobic degradation of organic matter in Danish coastal sediments: Fe reduction, Mn reduction, and sulfate reduction. *Geochim. Cosmochim. Acta* **56**:3867–3883.

Capone, D.G., and R.P. Kiene. 1988. Comparison of microbial dynamics in marine and freshwater sediments: Contrasts in anaerobic carbon metabolism. *Limnol. Oceanogr.* **33**:725–749.

Clymo, R.S. 1984. The limits to peat bog growth. *Phil. Trans. R. Soc. Lond. B* **303**:605–654.

Demaison, G.J., and G.T. Moore. 1980. Anoxic environments and oil source bed genesis. *Org. Geochem.* **2**:9–31.

Emerson, S.E. 1985. Organic Carbon Preservation in Marine Sediments. In: The Carbon Cycle and Atmospheric CO_2: Natural Variations Archaean to Present, ed. E.T. Sundquist and W.S. Broecker, pp. 78–87. Geophys. Monogr. 32. Washington, D.C.: Am. Geophysical Union.

Emerson, S.E., and J.I. Hedges. 1988. Processes controlling the organic carbon content of open ocean sediments. *Paleoceanography* **3**:621–634.

Emerson, S., C. Stump, P.M. Grootes, M. Stuiver, G.W Farwell, and F.H. Schmidt. 1987. Organic carbon in surface deep-sea sediment: C-14 concentration. *Nature* **329**:51–54.

Freeman, K.H., and J.M. Hayes. 1992. Fractionation of carbon isotopes by phytoplankton and estimates of ancient CO_2 levels. *Glob. Biogeochem. Cyc.* **6**:185–198.

Freeman, K.H., J.M. Hayes, J.-M. Trendel, and P. Albrecht. 1990. Evidence from carbon isotope measurements for diverse origins of sedimentary hydrocarbons. *Nature* **343**:254–256.

Froelich, P.N., G.P. Klinkhammer, M.L. Bender, N.A. Leudtke, G.R. Heath, D. Cullen, P. Dauphin, B. Hartman, and V. Maynard. 1979. Early oxidation of organic matter in pelagic sediments of the eastern equatorial Atlantic: Suboxic diagenesis. *Geochim. Cosmochim. Acta* **43**:1075–1090.

Giddings, J.C. 1985. A system based on split-flow lateral transport (SPLITT) thin cells for rapid and continuous particle fractionation. *Sep. Sci. Technol.* **20**:749–768.

Harden, J.W., E.T. Sundquist, R.F. Stallard, and R.K. Mark. 1992. Dynamics of soil carbon during deglaciation of the Laurentide ice sheet. *Science* **258**:1921–1924.

Hayes, J.M., K.H. Freeman, B.N. Popp, and C. Hoham. 1990. Compound-specific isotope analyses: A novel tool for reconstruction of ancient biogeochemical processes. *Org. Geochem.* **16**:1115–1128.

Hedges, J.I. 1992. Global biogeochemical cycles: Progress and problems. *Mar. Chem.* **39**:67–93.

Henrichs, S.M., and W.S. Reeburgh. 1987. Anaerobic mineralization of marine sediment organic matter: Rates and the role of anaerobic processes in the oceanic carbon economy. *Geomicrobiol. J.* **5**:191–237.

Henrichs, S.M. 1992. The early diagenesis of organic matter in marine sediments: Progress and perplexity. *Mar. Chem.* **39**:119–149.

Holmen, K. 1992. The global carbon cycle. In: Global Biogeochemical Cycles, ed. S.S. Butcher, R.J. Charlson, G.H. Orians, and G.V. Wolfe, pp. 239–262. San Diego: Academic.

Ingall, E.D., R.M. Bustin, and P. Van Cappellen. 1993. Influence of water column anoxia on the burial and preservation of carbon and phosphorus in marine shales. *Geochim. Cosmochim. Acta* **57**:303–316.

Jahnke, R.A. 1990. Early diagenesis and recycling of biogenic debris at the seafloor, Santa Monica Basin, California. *J. Mar. Res.* **48**:413–436.

Keil, R.G., C.B. Fuh, C.A. Giddings, and J. Hedges. 1994. Mineralogical and textural controls on the organic composition of coastal marine sediments: Hydrodynamic separation using SPLITT-fractionation. *Geochim. Cosmochim. Acta* **58**:879–893.

Kristensen, E., and T.H. Blackburn. 1987. The fate of organic carbon and nitrogen in experimental marine sediment systems: Influence of bioturbation and anoxia. *J. Mar. Res.* **45**:231–257.

Lee, C. 1992. Controls on organic carbon preservation: The use of stratified water bodies to compare intrinsic rates of decomposition in oxic and anoxic systems. *Geochim. Cosmochim. Acta* **56**:3323–3335.

Lee, C., and S.G. Wakeham. 1988. Organic matter in seawater: Biogeochemical Processes. In: Chemical Oceanography, ed. J.P. Riley, vol. 9, pp. 1–51. New York: Academic.

Lee, C., and S.G. Wakeham. 1992. Organic matter in the water column. Future research challenges. *Mar. Chem.* **39**:95–118.

Lovely, D.R., and M.J. Klug. 1986. Model for distribution of sulfate reduction and methanogenesis in freshwater sediments. *Geochim. Cosmochim. Acta* **50**:11–18.

Lovely, D.R., and E.J. Phillips. 1983. Organic matter remineralization with reduction of ferric iron in anaerobic sediments. *Appl. Environ. Microbiol.* **51**:683–689.

Mayer, L.M. 1994a. Relationships between mineral surfaces and organic carbon concentrations in soils and sediments. *Chem. Geol.* **114**:347–363.

Mayer, L.M. 1994b. Surface area control of organic carbon accumulation in continental shelf sediments. *Geochim. Cosmochim. Acta* **58**:1271–1284.

Pederson, T.F., and S.E. Calvert. 1990. Anoxia versus productivity: What controls the formation of organic rich sediments and sedimentary rocks? *Am. Assn. Pet. Geol.* **74**:454–466.

Pederson, T.F., G.B. Shimmield, and N.B. Price. 1992. Lack of enhanced preservation of organic matter in sediments under the oxygen minimum of the Oman region. *Geochim. Cosmochim. Acta* **56**:546–551.

Reeburgh, W.S. 1983. Rates of biogeochemical processes in anoxic sediments. *Ann. Rev. Earth Planet. Sci.* **11**:269–298.

Reimers, C.E., R.A. Jahnke, and D.C. McCorkle. 1992. Carbon fluxes and burial rates over the continental slope and rise off central California with implications for the global carbon cycle. *Glob. Biogeochem. Cyc.* **6**:199–224.

Sarnthein, M., K. Winn, and R. Zhan. 1987. Paleoproductivity of the ocean upwelling and the effect of atmospheric CO_2 and climate change during deglaciation times. In: Abrupt Climate Change: Evidence and Implications, ed. W.H. Berger and L.D. Labeyrie, pp. 311–337. Dordrecht: Reidel.

Smith, S.V., and J.T. Hollibaugh. 1993. Coastal metabolism and the oceanic organic carbon balance. *Rev. Geophys.* **31**:75–89.

Sørensen, J. 1982. Reduction of ferric iron in anaerobic marine sediment and interaction with reduction of nitrate and sulfate. *Appl. Environ. Microbiol.* **43**:319–324.

Stone, A.T., and J.J. Morgan. 1987. Reductive dissolution of metal oxides. In: Aquatic Surface Chemistry: Chemical Processes at the Particle–Water Interface, ed. W. Stumm, pp. 221–252. New York: Wiley.

Sundquist, E.T. 1985. Geological perspectives on carbon dioxide and the carbon cycle. In: The Carbon Cycle and Atmospheric CO_2: Natural Variations Archaean to Present, ed. E.T. Sundquist and W.S. Broecker, pp. 5–59. Geophys. Monogr. 32. Washington, D.C.: Am. Geophysical Union.

Van Cappellen, P., and D.E. Canfield. 1993. Comment on "Lack of evidence for enhanced preservation of sedimentary organic matter in the oxygen minimum of the Gulf of California." *Geology* **21**:570–571.

Walker, J.C.J. 1993. Biogeochemical cycles of carbon on a heirarcy of time scales. In: Biogeochemistry of Global Change: Radiatively Active Trace Gases, ed. R.S. Oremland, pp. 3–28. New York: Chapman and Hall.

Wilson, T.R.S., J. Thomson, D.J. Hydes, S. Colley, F. Culkin, and J. Sarensen. 1986. Oxidation fronts in pelagic sediments: Diagenetic formation of metal-rich layers. *Science* **232**:972–975.

Wollast, R., F.T. Mackenzie, and L. Chou. 1993. Interactions of C, N, P, and S. Biogeochemical Cycles and Global Change. Heidelberg: Springer.

16

Formation of Refractory Organic Matter from Biological Precursors

C. LARGEAU

Laboratoire de Chimie Bioorganique et Organique Physique,
UA CNRS D 1381, École Nationale Supérieure de Chimie de Paris,
11, rue Pierre et Marie Curie, 75231 Paris cedex 05, France

ABSTRACT

In this article, the consequences of the recent discovery of a new family of biomacro-molecules on nonliving organic matter (NLOM) cycling and global carbon cycle are presented. Such constituents are characterized by a very high resistance to microbial degradations and can escape mineralization, as shown by comparison with fossil materials; thus they provide a very efficient carbon sink. The occurrence, abundance, and main features of these refractory compounds (so far identified in higher plants, micro-algae, and bacteria) are first described. Quantitative information, presently restricted to microalgae and bacteria, indicates that such resistant biomacromolecules shall account, under most depositional conditions, for a major part of the carbon sequestered in sediments via burial of bacteria- and algae-derived organic matter. Various environmental changes, including human perturbations, should induce large variations in the production of such compounds and, hence, in the efficiency of carbon sequestration by NLOM. Moreover, feedbacks may be involved in some of the above variations. Finally, the possible contribution of resistant biomacromolecules to dissolved organic carbon in the ocean and the resulting existence of an additional biospheric carbon sink are also considered.

INTRODUCTION

In this chapter I examine the consequences to nonliving organic matter (NLOM) cycling and the global carbon cycle, which can be anticipated from

Role of Nonliving Organic Matter in the Earth's Carbon Cycle
Edited by R.G. Zepp and Ch. Sonntag © 1995 John Wiley & Sons Ltd.

the recent discovery of a widespread group of biomacromolecules and insoluble in organic solvents, which are characterized by an unusually high resistance to chemical and microbial degradations (reviewed in Largeau, Derenne, Clairay et al. 1990; de Leeuw et al. 1991; de Leeuw and Largeau 1993). Indeed, these macromolecular compounds can survive, unaffected, drastic acid and basic hydrolysis and the intensive microbial attacks that take place upon deposition in aquatic environments. They are also likely to withstand the intense biodegradations associated with incorporation in soils. Accordingly, such biomacromolecules will escape mineralization into CO_2 during NLOM cycling and will thus act as highly efficient carbon sinks.

This new type of biomacromolecule commonly occurs and has been identified in the cell walls or protective envelopes of a number of organisms, including higher plants (cutans and suberans), microalgae (algaenans), cyanobacteria, and bacteria (bacterans). All these nonhydrolyzable and diagenetically resistant[1] macromolecular materials are based on a network of long, saturated hydrocarbon chains (hence the suffix "an," which is intended to stress their highly aliphatic nature; Tegelaar et al. 1989). The basic chemical structure of the above macromolecules is therefore sharply different from that of lignins.

I begin by discussing the main features of such refractory biomacromolecules and their contribution to NLOM and will illustrate these using examples from microalgae and bacteria. Thereafter I consider the consequences to be expected from their occurrence, in connection with the extent of NLOM mineralization and the role of the biosphere in CO_2 mass balance and carbon sequestration.

NONHYDROLYZABLE, HIGHLY ALIPHATIC BIOMACROMOLECULES: OCCURRENCE IN EXTANT ORGANISMS, TYPICAL FEATURES, AND GEOCHEMICAL IMPORTANCE

Botryococcus braunii Algaenan

B. braunii is a colonial green microalga that is widely distributed throughout the world in freshwater, brackish, and saline lakes. Its colonies show a characteristic organization (Figure 16.1), and the basal part of cells is deeply embedded in thick (1–3 μm) outer walls. As illustrated in Figure 16.1c, the skeleton of these basal walls is composed of algaenan. This nonhydrolyzable macromolecular material is a major constituent accounting for 10% to 33% of

---

[1]Early diagenesis: organic matter transformations occurring in sediment upper layers. Such transformations mainly originate from extensive microbial degradations and no thermal processes are implicated.

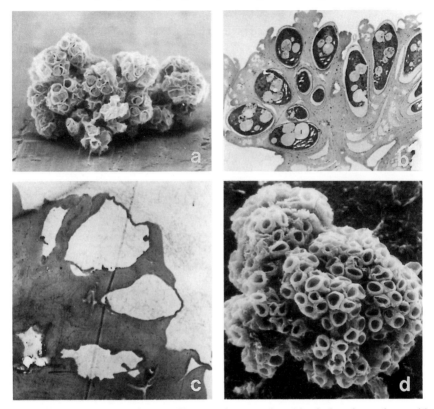

Figure 16.1 *Botryococcus braunii.* Untreated extant alga: (a) whole colony observed by scanning electron microscopy (SEM; ×400); (b) portion of colony examined by transmission electron microscopy (TEM; ×1500); the cells are deeply embedded in thick basal outer walls building up colony matrix. Treated biomass observed by TEM (×2800) after drastic hydrolysis: (c) *B. braunii* cells appear as "ghosts," due to a total elimination of their constituents, whereas the algaenan-composed outer walls are fully retained. Fossil *Botryococcus* examined by SEM (×800): (d) colony from a 4×10^4 years old lacustrine sediment; cell contents were entirely removed by microbial degradation, in sharp contrast the algaenan-composed outer walls completely escaped mineralization and were thus selectively preserved.

B. braunii dry biomass, depending on growth conditions and/or considered strain. Further studies (Metzger et al. 1991) indicate that (a) large amounts of lipids, including long chain hydrocarbons and high molecular weight compounds not previously observed in living organisms (Figure 16.2), are stored in *B. braunii* outer walls, (b) *B. braunii* algaenans originate from the polymerization of this type of lipids, and (c) such macromolecules are end products of *B.*

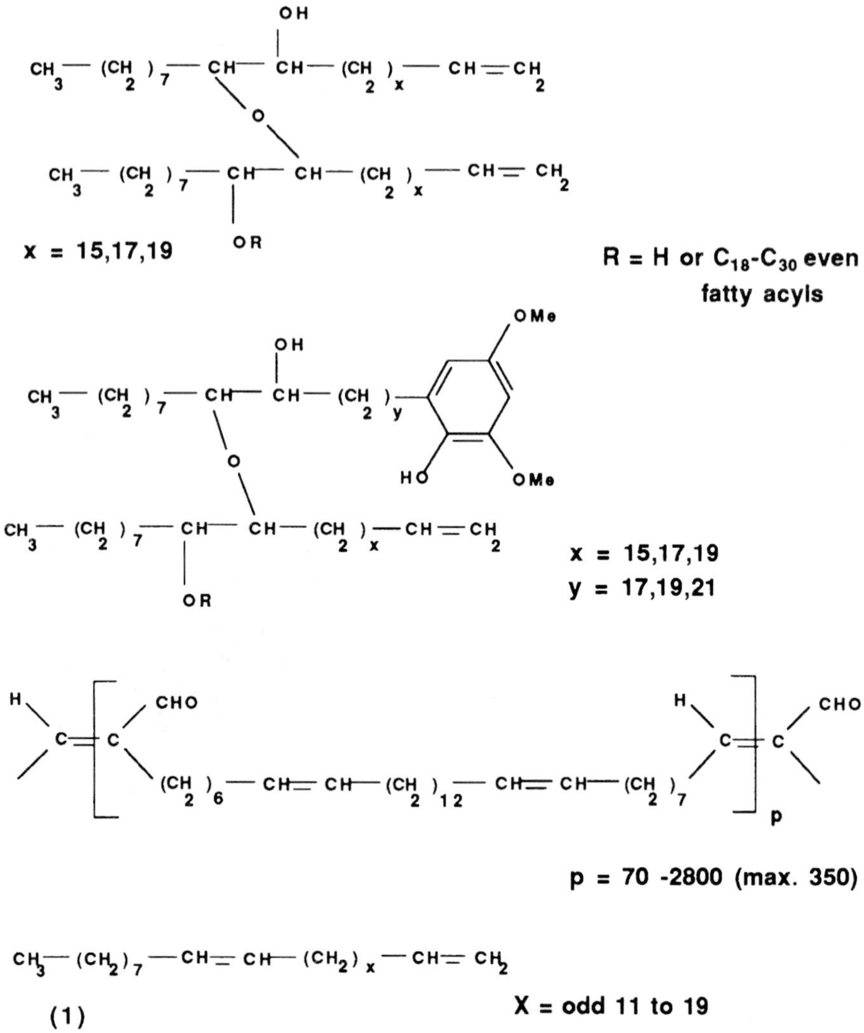

C. *Largeau*

Figure 16.2 Some of the high molecular weight lipids identified in *B. braunii* strains producing *n*-alkadienes (1) as hydrocarbons. Polymerization of such lipids finally results in the formation of the algaenan (i.e., the insoluble, hydrolysis-resistant macromolecules material) building up outer walls in these microalgae.

braunii metabolism that cannot be degraded by the alga and used for its respiration. Finally, comparative studies between the algaenans isolated from extant *B. braunii* and the fossil organic matter of Torbanites[2] reveal very close

---

[2]Torbanites: organic-rich deposits formed by accumulations of fossil colonies of *Botryococcus*.

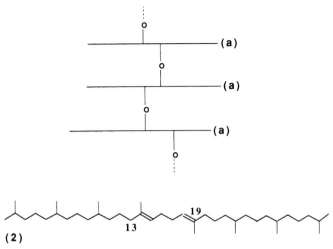

Figure 16.3 Basic bulk chemical structure of the algaenan of *B. braunii* strains producing lycopadiene (2) as hydrocarbon. (a): C_{40} chains with a lycopane-type skeleton, the central carbons of these chains (C_{14} to C_{19}) are implicated in ether bridge formation.

chemical structures (Largeau et al. 1986; Derenne et al. 1988). *B. braunii* fossilization is therefore associated with a complete preservation of the algaenan-composed outer walls, whereas all the other constituents are heavily mineralized due to the extensive biodegradations taking place in the water column and sediment upper layers (Figure 16.1d). The ability of *B. braunii* algaenans to survive drastic chemical hydrolysis is thus paralleled by a very high resistance to biodegradation.[3] Such features originate from the steric protection provided by the network of hydrocarbon chains to potentially weak bonds, thus hindering attacks by enzymes and chemical reagents. These chains can be linked by ether bridges (Figure 16.3), and they sometimes bear fatty acid esters. The type of bonds described above are known to be easily cleaved upon hydrolysis, when they occur in low molecular weight compounds. However, once protected within the macromolecular network of *B. braunii* algaenans, they can no longer be cleaved by hydrolytic enzymes or chemical reagents; hence the unusual diagenetic stability of such materials and their high efficiency as carbon sinks.

---

[3]Including biodegradations under highly oxic conditions, as shown by the study of Coorongites (rubbery materials formed, following *B. braunii* blooms, when large amounts of algal biomass are deposited on shorelines) (Dubreuil et al. 1989). A high resistance of *B. braunii* algaenan to photodegradations is also indicated by the chemical composition of Coorongites.

These parallel studies on *B. braunii* algaenans and Torbanites provided the first direct evidence of fossil organic matter formation via a new pathway: the "selective preservation" of macromolecules occurring, as such, in the initial biomass (Largeau et al. 1986). As I stress below, this pathway is sharply different, in regard to the extent of mineralization, when compared to the classical "degradation-recondensation" process.

All of the above studies on algaenans were carried out from vegetative cells. In addition, diagenetically resistant macromolecular constituents are also likely to occur in the walls of specialized cells involved in microalga reproduction or in the survival of numerous species under adverse conditions (i.e., spores and cysts). In fact, close morphological similarities with ubiquitous microfossils point to the presence of algaenans in such walls.

Algaenan of *Scenedesmus quadricauda* and Related Species

The systematic examination of extant microalgae revealed the common occurrence of algaenan-composed outer walls (Derenne et al. 1992). However, important morphological differences were generally noted relative to *B. braunii*. Thin resistant outer walls (ca. 10–20 nm thick), entirely surrounding each cell (Figure 16.4), were thus observed in a number of species.[4] In such microalgae the shape of the cells is no longer retained following algaenan isolation, via hydrolytic degradation of all the other cell constituents. In fact, the outer walls then appear highly distorted and associated into small bundles (Figure 16.4c). In all of these species, algaenans only account for approximately 1% to 5% of dry biomass of living cells. The extent of algaenan production from such microalgae is, however, substantially higher than suggested by the above values. In fact, following cell divisions, the mother cell walls burst and are freed along with the daughter cells. On the contrary, the mother basal outer walls of *B. braunii* remain tightly associated with daughter cells and thus build colony matrix. Accordingly, actively growing populations of microalgae with thin outer walls shall produce large amounts of free, algaenan-composed, outer walls.

This type of algaenans also originates from polymerization of high molecular weight lipids and comprises a network of long hydrocarbon chains. They exhibit, however, specific chemical features, especially the systematic presence of amide functions (Derenne et al. 1991) whereas such groups do not occur in *B. braunii* algaenans. In classical biomacromolecules, like proteins, the highly sensitive nature of amide groups to chemical hydrolysis and biodegradations has been well documented. In sharp contrast, the amide functions occurring in

---

[4]The presence of algaenan-composed outer walls has been so far demonstrated, via chemical tests, in 21 species belonging to 11 genera and chiefly corresponding to freshwater species. Furthermore, based on electron microscopy observations, this type of thin resistant outer wall is likely to occur in a number of other microalgae.

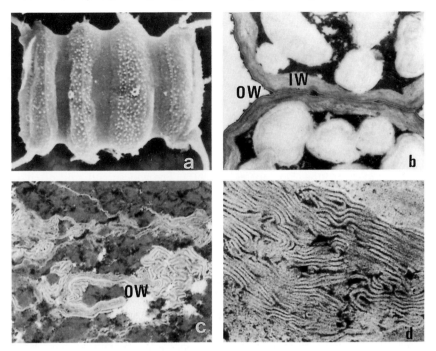

Figure 16.4 *Scenedesmus quadricauda.* Untreated extant alga: (a) four-celled colony observed by SEM (×4000); (b) portion of colony examined by TEM (×12,000); the cells are entirely surrounded, in addition to a thick polysaccharidic inner wall (IW), by a very thin outer wall (OW). Treated biomass observed by TEM (×20,000) after drastic hydrolysis: (c) cell contents and polysaccharidic walls were totally eliminated, algaenan-composed outer walls are retained but appear now highly distorted and associated into bundles. (d) Ultralaminae detected by TEM (×25,000) in a Kimmeridgian source rock (ca. 150 × 10⁶ years old); this fossil material corresponds to algaenan-composed thin outer walls that escaped mineralization whereas all the other constituents of these microalgae were heavily degraded.

the above algaenans are characterized by a very high resistance, owing to the steric protection provided by the macromolecular network.

As demonstrated by comparison with fossil materials (Derenne et al. 1991), these algaenans will escape mineralization and will thus act as carbon sinks during NLOM deposition. Indeed, transmission electron microscopy (TEM) examinations at high magnifications (Raynaud et al. 1988; Largeau, Derenne, Casadevall et al. 1990) recently revealed the presence of well-defined lamellar structures, termed ultralaminae, in samples so far considered as amorphous from light microscopy observations (Figure 16.4d). These ultralaminae originate from the selective preservation of algaenan-composed thin outer walls, as

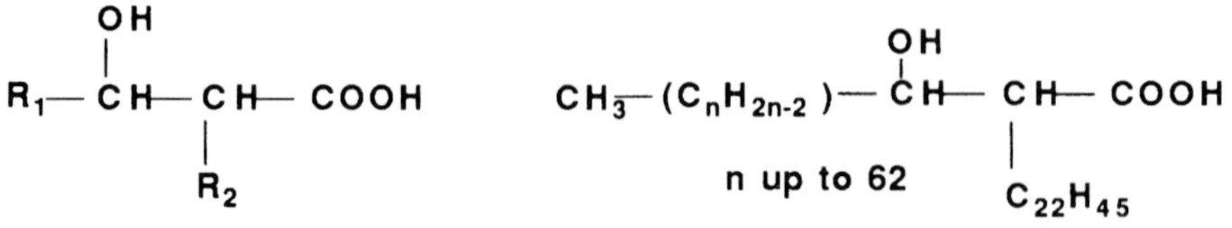

$$R_1 - \overset{\overset{\displaystyle OH}{|}}{C}H - CH - COOH$$
$$\underset{\displaystyle R_2}{|}$$

$$CH_3 - (C_nH_{2n-2}) - \overset{\overset{\displaystyle OH}{|}}{C}H - CH - COOH$$
$$n \text{ up to } 62 \qquad \underset{\displaystyle C_{22}H_{45}}{|}$$

Figure 16.5   General structure of mycolic acids and example of a typical mycolic acid of *Mycobacterium smegmatis.*

revealed by close morphological and chemical similarities, including the presence of amide groups (Derenne et al. 1991). The widespread, and often abundant, occurrence of ultralaminae in fossil organic matter thus reflects (a) the ability of this type of algaenans to survive mineralization and sequester large amounts of carbon, and (b) their presence in a number of ubiquitous species.

Bacterans

Nonhydrolyzable, highly aliphatic biomacromolecules were recently shown to occur also in bacteria belonging to different families, including both Archaebacteria (methanogens and halophilic species) and Eubacteria (ubiquitous aerobes, five out of six cyanobacteria tested). Further studies on three of the above species, namely *Corynebacterium glutamicum, Nocardia opaca,* and *Mycobacterium smegmatis,* indicated bacteran levels around 3% of total dry biomass (Le Berre et al. 1991).

   In sharp contrast to algaenans, bacterans do not build the skeleton of cell walls alone. In fact, all bacterium walls are known to be mainly based on a peptidoglycan skeleton, i.e., on macromolecules easily cleaved upon hydrolysis. As a result, such walls are no longer recognizable following drastic acid and basic treatments, and the bacterans then isolated appear as amorphous materials, even at high TEM magnifications. These three bacterans were shown to be derived from the polymerization of some typical lipids of *C. glutamicum, N. opaca,* and *M. smegmatis* comprising long hydrocarbon chains, like mycolic acid esters (Figure 16.5). Comparison with amorphous fossil materials indicated close similarities in chemical structure. Accordingly, such bacterans are capable of escaping mineralization and are selectively preserved during fossilization, thus contributing to carbon sequestration.

Cutans and Suberans

It has long been recognized that higher plant cuticles[5] comprise soluble waxy compounds and an insoluble matrix composed of polyesters termed cutins.

---

[5]Cuticle: extracellular layer covering the aerial parts of higher plants, e.g., leaves, where no secondary growth occurs.

Recent studies (Nip et al. 1986; Tegelaar et al. 1991), however, revealed the widespread presence of another group of cuticular constituents, corresponding to nonhydrolyzable, highly aliphatic macromolecules, termed cutans. Such resistant compounds were detected in leaf cuticles of a large number of species (15 out of 19 tested) and represent various families of higher plants, including angiosperms and gymnosperms. Comparison with morphologically preserved fossil leaf cuticles, isolated from sedimentary rocks, revealed the same chemical structure as in cutans from extant counterparts. Cutans were thus shown to be selectively preserved during leaf deposition in aquatic environments and, hence, to provide efficient carbon sinks.

Parallel examination of the periderm layer in several extant angiosperms indicated that suberin polyesters are not the only macromolecular constituents occurring in outer bark tissues. Indeed, the presence of nonhydrolyzable, highly aliphatic compounds, termed suberans, was observed in the six angiosperm species tested. Comparison with fossil materials confirmed that such macromolecules can also escape mineralization and are selectively preserved upon deposition in aquatic environments.

So far, no study has reported on with the fate of cutans and suberans in soils. However, recent examinations of complex lipids in soils (Amblès et al. 1991) suggest the presence of highly aliphatic polymers based on long hydrocarbon chains. Such compounds could thus reflect the preservation of cutans and/or suberans even under the conditions (strongly oxic and promoting biodegradations) prevailing in soils. Due to the presence of various functional groups, suberans and cutans should be strongly associated with clay surfaces, hence further increase in their resistance. Indeed, all of the clay fractions so far examined by solid-state ^{13}C NMR exhibit an important contribution of aliphatic structures (J.M. Oades, pers. comm.).

Finally, it should be noted that highly aliphatic, diagenetically resistant biomacromolecules can also occur in specialized cells involved in plant reproduction: seed coats of water plants (J.W. De Leeuw, pers. comm.) as well as outer walls (exines) of spores and pollen grains (reviewed in De Leeuw and Largeau 1993).

INFLUENCE OF NONHYDROLYZABLE, HIGHLY ALIPHATIC BIOMACROMOLECULES ON BIOSPHERE CONTRIBUTION TO THE GLOBAL CARBON CYCLE

The direct role of the biosphere in the Earth's carbon cycle can be assessed from a single parameter. This key value corresponds to the percentage of the CO_2, initially fixed as organic carbon by photosynthetic organisms (biospheric CO_2 sink), that is not ultimately recycled either via respiration of these organisms themselves or via transformation and biodegradation by heterotrophic species (biospheric CO_2 source). Such a carbon sink at a geological time scale corresponds, in fact, to the organic matter ultimately incorporated into sediments and to the "CO_2 potential," which will only be released upon fossil-fuel burning.

Until the last few years, sedimentary organic matter was considered to have originated primarily from a degradation-recondensation process (Tissot and Welte 1984). According to this mechanism, the constituents of the source organisms (e.g., classical macromolecules like polysaccharides and proteins) are heavily biodegraded during deposition. Only a tiny fraction escapes mineralization, owing to random condensation reactions of a few degradation products. These reactions ultimately result in the formation of complex, resistant, geomacromolecules that are buried in sediments. As discussed above, however, recent observations demonstrate that a second pathway, sharply different from the classical degradation-recondensation process, is also involved in the formation of the sedimentary organic matter: the selective preservation of the diagenetically resistant biomacromolecules occurring in a number of organisms. The latter constituents are incorporated as such in sediments and escape mineralization under any depositional condition.

The importance of the CO_2 sink provided by the biosphere, via the formation of fossil organic matter, is therefore controlled by two factors (Largeau and Derenne 1993): (a) the amount of nonhydrolyzable, highly aliphatic, macromolecular compounds occurring in the source organisms, and (b) the depositional conditions and resulting level of mineralization of the other constituents. If x represents the average content (%) of diagenetically resistant biomacromolecules in the source organisms and y the amount (%) of their "nonresistant" constituents that finally escape mineralization, then x% of the organic matter will be selectively preserved and directly incorporated into sediments. In sharp contrast, the $(100 - x)$% corresponding to the biodegradable constituents will be incorporated through the degradation-recondensation pathway, thus yielding a sedimentary organic fraction corresponding to $\frac{y}{100}(100 - x)$% of the initial biomass. As a result, the total amount of carbon, z, finally sequestered in sediments would account for $x + \frac{y}{100}(100 - x)$% of this biomass.

Quantitative studies on diagenetically resistant biomacromolecules have so far been only carried out on microalgae and bacteria. In the following discussion, which will estimate the ranges to be expected for x and y values, I consider aquatic environments with low land-plant contributions. The source of biomass corresponds primarily to microalgae and bacteria.

According to Tissot and Welte (1984), in their classical description of the degradation-recondensation pathway, less than 0.1% of the nonresistant biomolecules would escape mineralization on a global scale and be incorporated into sediments. Incorporation yields of up to 4% could occur in environments like the Black Sea, where microbial degradations are limited by oxygen deficiency. However, further studies led one to question the occurrence of such relatively high levels of preservation of biodegradable constituents. Indeed, as recently stressed (Blackburn 1991), there is no unambiguous evidence that organic carbon oxidation by heterotrophic organisms is inherently slower in anoxic conditions than in oxic ones. Mineralization in the water column can be reduced by fast sedimentation rates (e.g., via flocculation of algae in high productivity areas) and/or a relatively low depth, resulting in large amounts of detrital matter reaching the sediment–water interface. This organic material will, however, undergo extensive biodegradations, either in the upper layers of the sediment or during subsequent resuspension and transport in the case of shallow areas (Wollast 1991). Accordingly, the extent of mineralization of the nonresistant constituents does not seem to be tightly controlled by depositional conditions unless very unusual environments are considered.[6] Values from 0% to 4% can be therefore used for y and most environments should plot below 1%.

Regarding x, as stressed above, the presence of resistant macromolecules has been observed in numerous genera of microalgae, commonly occurring in the mixed populations found in natural environments (Derenne et al. 1992). For a given genus, for example *Chlorella* or *Scenedesmus*, species comprising thin resistant outer walls and species completely devoided of algaenan were both identified; nevertheless numerous ubiquitous species were shown to correspond to the former group. From all of these results it can be assumed that most natural populations of microalgae contain an average level of algaenan corresponding, at least, to 1% of the total biomass. Regarding bacteria, as already stressed, recent studies point to a widespread occurrence of bacterans, and levels of approximately 3% of total biomass were observed (Le Berre et al. 1991). Values around 1% should therefore correspond, on a global scale, to the minimal level of diagenetically resistant macromolecular constituents likely to occur in most algal and bacterial populations.

Values from 0% to 1% and around 1% can therefore be anticipated for y and x, respectively; hence total sequestration yields

$$z = x + \frac{y}{100}\,(100 - x),$$

which corresponds to approximately 1% to 2% of the initial biomass. Such total preservation yields are consistent with those generally reported for

---

[6]Analysis of sediments from the Orca Basin, on the Gulf of Mexico continental slope, revealed that seaweeds, *Sargassum sp.* encountered few diagenetic alterations, relative to living tissues, for hundreds of years (Kennicutt et al. 1992). This unique environment is characterized by a very weak bacterial activity due to the co-occurrence of anoxic and hypersaline conditions, combined with a low temperature and with the formation of antibacterial exudates by these seaweeds.

organic matter burial in sediments (Wollast 1991; Butcher et al. 1992). Taken together, the above values indicate that the level of diagenetically resistant macromolecular products in the source organisms (x) should play a major role in the efficiency of the biosphere as a carbon sink (as measured by z).[7] Indeed, when $y = 0.5$, an increase of only 1% to 1.5% in the average content of resistant biomacromolecules in the deposited biomass will result in a one-third increase in the amount of organic carbon ultimately sequestered within sediments. For lower values of y (e.g., 0.1%, as observed in a number of environments), a still higher influence of x will, of course, take place. Accordingly, relatively low changes in the abundance of resistant biomacro-molecular constituents in the source organisms will have important outcomes on biosphere contribution to global carbon cycle. All of the above features are summarized in Figure 16.6 by a schematic diagram illustrating the role of the biosphere in carbon sequestration.

Environmental Control on the Production of Nonhydrolyzable, Highly Aliphatic Biomacromolecules

As stressed above, weak variations in the abundance of diagenetically resistant biomacromolecules strongly influence the efficiency of the biosphere to act as a carbon sink. The production of such macromolecular compounds by a natural microbial population, e.g., microalgae, is controlled by the nature, relative abundance, and physiological state of the contributing species. In fact, such a mixed population will comprise both algaenan-containing and algaenan-devoid species. Furthermore, considerable differences in algaenan levels can occur in the former group (e.g., less than 1% in some *Chlorella* against above 30% in some *B. braunii* strains). In addition, for a given algaenan-producing species, the level of resistant macromolecular material can exhibit marked changes with physiological stage (as shown in the case of *B. braunii* by Plain et al. [1993] where outer wall thickness and algaenan content strongly vary with growth conditions). It should be noted that the aerial alga *Phycopeltis epiphyton* exhibits a multilayered outer wall comprising several thin, algaenan-composed layers. Such a stacking may protect the cells against desiccation. Indeed, all other species so far shown to comprise thin resistant outer walls live under normal aquatic conditions and only show a single outer layer. However, no precise information is presently available about the way variations in natural environments may influence the productivity of resistant macromolecular

---

[7]In addition, resistant biomacromolecules could play an indirect role in promoting carbon sequestration through interactions with biodegradable compounds. As already stressed the preservation of the latter occurs via polycondensations. The resistant macromolecules could act as a matrix for such reactions. Owing to the resulting increase in condensation rates and to incorporation into inert macromolecules, a larger fraction of the potentially biodegradable compounds could escape mineralization.

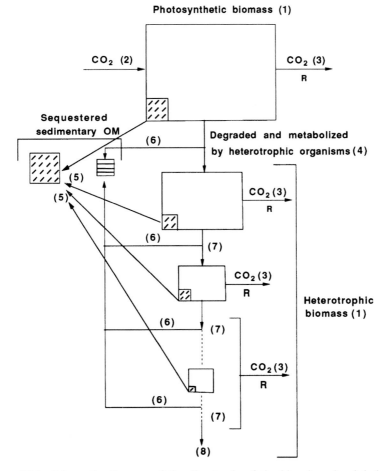

Figure 16.6 Schematic diagram of the direct role of the biosphere in global carbon cycle. (1) Mixed populations comprising species with diagenetically resistant macromolecular constituents. (2) Photosynthetic CO_2 sink. (3) CO_2 recycled by respiration, (4) including zooplankton grazing and bacterial degradations of nonresistant constituents in the water column, and in sediment upper layers. (5) Direct incorporation of unaltered, diagenetically resistant biomacromolecules in sediments (selective preservation pathway). (6) Incorporation of a tiny fraction of the degradation products of nonresistant biomolecules in sedimentary organic matter via polycondensation reactions (degradation-recondensation pathway). (7) Degradation of the nonresistant biomolecules of heterotrophic biomass. As the organic matter (OM) is progressively buried in sediments, successive populations of heterotrophs are implicated in these biodegradations. During such a process, as along any food web, total biomass sharply decreases from each heterotrophic population to the next one. (8) Final population of heterotrophic organisms. This population should correspond to a negligible biomass (unless high sedimentation rates occur, thus resulting in a fast burial into deep zones of sediments where no biological activity takes place any more). R: respiration; ▨: diagenetically resistant constituents; □biodegradable constituents; ▤: highly altered sedimentary OM originating from the degradation-recondensation pathway.

material in microalgal populations. Thus, conspicuous blooms of *B. braunii* are sometimes observed in nature; however, the cause(s) of such prolific growth of this algaenan-rich species must still be elucidated.

Anthropogenic perturbations could also influence algaenan production and, hence, CO_2 sink efficiency. In this connection, studies concerned with detergent influence on various microalgae belonging to the same genus, *Chlorella* or *Scenedesmus*, led to interesting observations (Biedlingmaier et al. 1987). It was noted that algaenan-containing species can survive much higher concentrations of detergents when compared to algaenan-free ones. Some pollutants, therefore, may increase the contribution of algaenan-producing species in natural populations. Climatic perturbations may influence the production of diagenetically resistant material through variations in microalgal populations (contributing species and their relative abundances) but also through changes in metabolic pathways. Thus the increase in water temperature associated with a decrease in the total amount of carbon fixed as lipidic constituents has been well documented. As already discussed, algaenans are known to be derived from lipid condensation. Accordingly, one may speculate that temperature increase, due to buildup in atmospheric CO_2, could be associated with decrease in the average level of algaenan in a given population, hence a positive feedback via the decrease of the efficiency of the derived NLOM as a carbon sink.

Changes in bacterial populations, when associated with variations in bacteran levels, will also strongly influence the efficiency of carbon sequestration as bacterially derived fossil organic matter. Practically no information is available at present, regarding the possible influence of environmental parameters on such changes.

Possible Contribution of Nonhydrolyzable, Highly Aliphatic Biomacromolecules to Ocean Dissolved Organic Carbon

Dissolved organic carbon (DOC) is the primary form of organic matter in the ocean and nearly corresponds to the pool of atmospheric carbon. DOC is thus generally considered to be approximately 30 times more abundant than ocean particulate organic carbon (POC), including both living and dead organisms (Butcher et al. 1992).[8] Accordingly, the elucidation of DOC amount, residence time, budget, nature, and source(s) has become a priority in relation to climatic changes and the greenhouse effect. Nevertheless, little is known about the origins and precise chemical structure of DOC. In fact, the presently available methods for DOC isolation, i.e., adsorption and ultrafiltration, led to recovery

---

[8]Observations carried out in the last few years, using a new analytical method, suggest that the actual amount of DOC could be 2–3 times higher than previously considered and that DOC/POC ratios of approximately 10^2 could indeed occur in the ocean (Suzuki and Tanoue 1991).

yields below 30% (Benner et al. 1992), so that even bulk chemical information on total DOC is missing. Moreover, comparison of the results derived from the two methods mentioned above clearly indicates that the chemical features of isolated DOC are procedure dependent. Polysaccharide units have been detected in DOC. Such units seem to be relatively abundant in surface waters but their contribution sharply decreases with depth, probably due to intense mineralization (Benner et al. 1992). DOC would also comprise a "refractory humic material," with an apparent age above 3×10^3 years (Williams and Druffel 1987). Moreover, the presence of highly aliphatic moieties has been noted in DOC (Hedges et al. 1992). A substantial part of ocean DOC could therefore correspond to colloidal particles[9] composed of nonhydrolyzable macromolecules. Such a DOC fraction would then represent an important carbon sink, in addition to the organic matter buried in sediments. Testing the above assumption and assessing the percentage of total DOC possibly derived from such diagenetically resistant biomacromolecules, therefore, appears to be a major goal for future studies.

MAIN CONCLUSIONS AND OPEN PROBLEMS

The recent discovery of a new family of biomacromolecules that occur widespread in several groups of living organisms requires us to reconsider, at least in part, the role of NLOM in global carbon cycle. The above constituents are characterized by a very high resistance to chemical and microbial degradations. Comparison with fossil materials demonstrated that such macromolecules are thus diagenetically resistant and completely escape mineralization, at least in aquatic environments. Accordingly, they act as very efficient carbon sinks. The average level of this type of biomacromolecules relative to total biomass and the degree of mineralization of all the other constituents (i.e., the nonresistant biomolecules) indicate a major contribution of the former to carbon sequestration via organic matter burial in sediments. It appears that relatively low variations in the content of these diagenetically resistant compounds in the source organisms can result in important changes in the amount of organic matter finally buried within sediments. This factor therefore plays an important role in the control of NLOM efficiency as a carbon sink and, hence, in biosphere contribution to global carbon cycle; on the contrary, the role of depositional conditions would be much less important than considered so far.

A number of studies is now required for a better assessment of the consequences of the above observations. Further information on the following points is, in particular, required:

---

[9]DOC comprises truly dissolved material (particle size below 0.001 μm) and colloidal particles (0.001–0.1 μm) (Suzuki and Tanoue 1991).

1. Occurrence and level of nonhydrolyzable, highly aliphatic macromolecules in living organisms (additional studies on higher plants and on microalgae and bacteria in marine and lacustrine environments; screening for new groups, such as yeasts and fungi; examination of soil microorganisms). Chemosynthetic organisms, i.e., species that derive energy from the oxidation of inorganic products but which use CO_2 as a carbon source, should also be examined. For the new biomacromolecular resistant constituents so identified, correlations with fossil materials from sedimentary rocks will have to be found in order to ascertain that we are actually dealing with biomacromolecules capable of escaping mineralization, at least in aquatic environments, and hence playing a major role in carbon sequestration. When isolated from terrestrial organisms, the fate of such macromolecules during incorporation in soils will have also to be examined so as to test the lack of mineralization even under strongly oxic conditions. In the case of microalgae, the lack of degradation of these biomacromolecules, following zooplankton grazing, will have also to be ascertained.

2. Total production of diagenetically resistant biomacromolecules in typical environments, especially in high productivity zones like continental margins and restricted seas.

3. Influence of environmental parameters (e.g., temperature) and anthropogenic perturbations (e.g., pollutants) on the above production and resulting variations in NLOM efficiency as carbon sink. Possible occurrence of feedback processes linking the production of diagenetically resistant biomacromolecules and global environmental changes.

REFERENCES

Amblès, A., J.C. Jacquesy, P. Jambu, J. Joffre, and R. Maggi-Churin. 1991. Polar lipid fraction in soil: A kerogen-like matter. *Org. Geochem.* **17**:341–349.

Benner, R., B.J. Eadie, and P.G. Hatcher. 1992. Bulk chemical characteristics of dissolved organic matter in the ocean. *Science* **255**:1561–1564.

Biedlingmaier, S., G. Wanner, and A. Schmidt. 1987. A correlation between detergent tolerance and cell wall structure in green algae. *Z. Naturforsch. C.* **42**:245–250.

Blackburn, T.H. 1991. Accumulation and regeneration: Processes at the benthic boundary interface. In: Ocean Margin Processes in Global Change, ed. R.F.C. Mantoura, J.M. Martin, and R. Wollast, pp. 181–209. Dahlem Workshop Report PC 9. Chichester: Wiley.

Butcher, S.S. 1992. Global Biogeochemical Cycles. New York: Academic.

De Leeuw, J.W., and C. Largeau. 1993. A review of macromolecular organic compounds that comprise living organisms and their role in kerogen, coal and petroleum formation. In: Organic Geochemistry, ed. M.H. Engels and S.A. Macko, pp. 23–72. New York: Plenum.

De Leeuw, J.W., P.F. Van Bergen, B.G.K. Van Aarsen, J.P. Gatellier, J.S. Sinninghe-Damsté, and M.E. Collinson. 1991. Resistant biomacromolecules as major contributors to kerogen. *Phil. Trans R. Soc. Lond. B* 333:329–337.

Derenne, S., C. Largeau, C. Berkaloff, B. Rousseau, C. Wilhelm, and P.G. Hatcher. 1992. Nonhydrolyzable macromolecular constituents from outer walls of *Chlorella fusca* and *Nanochlorum eucaryotum*. *Phytochemistry* 31:1923–1929.

Derenne, S., C. Largeau, E. Casadevall, C. Berkaloff, and B. Rousseau. 1991. Chemical evidence of kerogen formation in source rocks and oil shales via selective preservation of thin resistant outer walls of microalgae: Origin of ultralaminae. *Geochim. Cosmochim. Acta* 55:1041–1050.

Derenne, S., C. Largeau, E. Casadevall, and J. Connan. 1988. Comparison of Torbanites of various origins and evolutionary stages. Bacterial contribution to their formation. Cause of the lack of botryococcane in bitumens. *Org. Geochem.* 12:43–59.

Dubreuil, C., S. Derenne, C. Largeau, C. Berkaloff, and B. Rousseau. 1989. Mechanism of formation and chemical structure of Coorongite. I. Relationships with the resistant biopolymer of *Botryococcus braunii*. *Org. Geochem.* 14:543–553.

Hedges, J.I., P.G. Hatcher, J.R. Ertel, and K. Meyers-Schulte. 1992. A comparison of dissolved humic substances from seawater with Amazon River counterparts by ^{13}C NMR spectrometry. *Geochim. Cosmochim. Acta* 56:1753–1757.

Kennicutt, M.C., S.A. Macko, H.R. Harvey, and R.R. Bidigare. 1992. Preservation of *Sargassum* under anoxic conditions: Molecular and isotopic evidence. In: Organic Matter: Productivity, Accumulation and Preservation in Recent and Ancient Sediments, ed. J.K. Whelan and J.R. Farrington, pp. 123–141. New York: Columbia Univ. Press.

Largeau, C., and S. Derenne. 1993. Relative efficiency of the selective preservation and degradation-recondensation pathways in kerogen formation. Source and environment influence on their contributions to type I and II kerogens. *Org. Geochem.* 20:611–615.

Largeau, C., S. Derenne, E. Casadevall, C. Berkaloff, M. Corolleur, B. Lugardon, J.F. Raynaud, and J. Connan. 1990. Occurrence and origin of "ultralaminar" structures in "amorphous" kerogens of various source rocks and oil shales. *Org. Geochem.* 16:889–895.

Largeau, C., S. Derenne, E. Casadevall, A. Kadouri, and N. Sellier. 1986. Pyrolysis of immature Torbanite and of the resistant biopolymer (PRB *A*) isolated from extant alga *Botryococcus braunii*. Mechanism of formation and structure of Torbanite. *Org. Geochem.* 10:1023–1032.

Largeau, C., S. Derenne, C. Clairay, E. Casadevall, J.F. Raynaud, B. Lugardon, C. Berkaloff, M. Corolleur, and B. Rousseau. 1990. Characterization of various kerogens by Scanning Electron Microscopy (SEM) and Transmission Electron Microscopy (TEM): Morphological relationships with resistant outer walls in extant microorganisms. *Meded. Rijks Geol. Dienst.* 45:91–101.

Le Berre, F., S. Derenne, C. Largeau, J. Connan, and C. Berkaloff. 1991. Occurrence of nonhydrolyzable, macromolecular, wall constituents in bacteria. Geochemical implications. In: Organic Geochemistry. Advances and Applications in Energy and the Natural Environment, ed. D.A.C. Manning, pp. 428–431. Manchester: Manchester Univ. Press.

Metzger, P., C. Largeau, and E. Casadevall. 1991. Progress in the Chemistry of Organic Natural Products, vol. 57, ed. W. Herz, G.W. Kirby, W. Steglich, and C. Tamm, pp. 1–70. New York: Springer.

Nip, M., E.W. Tegelaar, H. Brinkhuis, J.W. De Leeuw, P.A. Schenck, and P.J. Holloway. 1986. Analysis of modern and fossil plant cuticles by Curie point PY–

GC and Curie point PY–GC–MS: Recognition of a new, highly aliphatic and resistant biopolymer. *Org. Geochem.* **10**:769–778.

Plain, N., C. Largeau, S. Derenne, and A. Couté. 1993. Variabilité morphologique de *Botryococcus braunii* (Chlorococcales, Chlorophyta) corrélation avec les conditions de croissance et la teneur en lipides. *Phycologia* **32**:259–265.

Raynaud, J.F., B. Lugardon, and G. Lacrampe-Couloume. 1988. Observation de membranes fossiles dans la matière organique "amorphe" de roches mères de pétrole. *C.R. Acad. Sci. Paris* **307–II**:1703–1709.

Suzuki, Y., and E. Tanoue. 1991. Dissolved organic carbon enigma: Implication for ocean margins. In: Ocean Margin Processes in Global Change, ed. R.F.C. Mantoura, J.M. Martin, and R. Wollast, pp. 197–209. Dahlem Workshop Report PC 9. Chichester: Wiley.

Tegelaar, E.W., J.W. De Leeuw, S. Derenne, and C. Largeau. 1989. A reappraisal of kerogen formation. *Geochim. Cosmochim. Acta* **53**:3103–3106.

Tegelaar, E.W., J.H.F. Kerp, H. Visscher, P.A. Schenck, and J.W. De Leeuw. 1991. Bias of the paleobotanical record as a consequence of variations in the chemical composition of higher vascular plant cuticles. *Paleobiology* **17**:133–144.

Tissot, B.P., and D.H. Welte. 1984. Petroleum Formation and Occurrence. Berlin: Springer.

Williams, P.M., and E.R.M. Druffel. 1987. Dissolved organic matter in the ocean: Comments on a controversy. *Oceanogr. Mag.* **1**:14–17.

Wollast, R. 1991. The coastal organic carbon cycle: Fluxes, sources and sinks. In: Ocean Margin Processes in Global Change, ed. R.F.C. Mantoura, J.M. Martin, and R. Wollast, pp. 365–381. Dahlem Workshop Report PC 9. Chichester: Wiley.

17

An Overview of Processes Affecting the Cycling of Organic Carbon in Soils

J.M. OADES
Cooperative Research Centre for Soil and Land Management,
Department of Soil Science, Waite Agricultural Research Institute,
The University of Adelaide, Glen Osmond, South Australia 5064

ABSTRACT

In this chapter, various factors influencing the additions, dynamics, and loss of carbon from soils are considered. The inputs to soil from plants, the major primary producers, are in aboveground material, deposited as litter and as rhizodeposition below ground. The form, distribution, and chemistry of the inputs all impact on the rate of decomposition of the added photosynthate and the incorporation of the decomposition by-products into the soil.

Decomposition is initiated by soil fauna; however, the major decay process is completed by the soil flora. The process is controlled by the temperature and the water regimes.

The concept of position of organic materials within the soil matrix is introduced. The protection of biopolymers in the soil is stressed in terms of general protection of organic debris by encapsulation in inorganic materials and by interaction with clay surfaces and exchangeable ions such as calcium.

Finally, the integration of this information into simulation models of carbon flow is introduced, followed by a listing of the gaps in knowledge.

INTRODUCTION

Organic carbon plays a vital role in the formation and properties of soils that retain about 1.7×10^{18} g C. This represents a major pool in the global carbon cycle and is about twice as much C as is present in the atmosphere. The

Role of Nonliving Organic Matter in the Earth's Carbon Cycle
Edited by R.G. Zepp and Ch. Sonntag © 1995 John Wiley & Sons Ltd.

concentration of carbon in soils ranges from less than 1% to almost 40% in peats. The major factors controlling the carbon contents of soils are the temperature and water regimes, because they control the amount of carbon fixed by photosynthesis and the rate of decomposition and mineralization of organic carbon compounds. Generally, annual inputs of carbon to soils represent about 3% to 5% of the soil carbon, and a similar amount of carbon is released to the atmosphere as CO_2 by respiration processes. Turnover times of carbon in soils thus appear to be on the order of a few decades.

Turnover times are only a rough guide of the speed with which carbon cycles through the soil, since we know that some carbon compounds added to soils are retained for only a day or so while others remain in soils for thousands of years. Apart from the temperature and water regimes, numerous other factors influence the ability of the fauna and flora (the decomposers) to utilize organic materials as a source of food and energy, respiring CO_2 to the atmosphere in the process. Below, I consider these factors under two major headings: the nature of inputs to the soil and the influence of the soil matrix on the decomposition of a range of biopolymers and the living soil microbial biomass.

CARBON INPUTS TO SOILS

The input of carbon to soils is one of the most important factors controlling the carbon cycle and ultimate amount of organic carbon retained in soils. Amounts of input are most important; however, physical form, distribution in the soil profile, position within the soil matrix and the chemistry of inputs all influence the rates and pathways of decomposition.

Aboveground Inputs

The amount of carbon added to soil in natural ecosystems is controlled by primary production, and various estimates have been made for major terrestrial ecosystems (Bolin 1983). These data are being improved continuously for grasslands, forests, and agroecosystems. We can measure litterfall in forests quantitatively (Raich and Nadelhoffer 1989) and assess carbon turnover in various litter layers above the mineral soil. It is more difficult, however, to determine the actual addition of carbon, via litterfall and litter decomposition, to the mineral portion of the soil profile. In grassland systems it is difficult to assess additions of litter to soils because of problems created by large herbivores and detrivores. Aboveground additions of photosynthate are relatively easy to measure, but the paucity of good quantitative data in agroecosystems is a concern. In noncultivated situations, most of the litter resides on the surface and considerable decomposition can occur before incorporation of carbon compounds into the soil matrix, either as particulate material mixed by faunal activity in neutral to alkaline soils or as various soluble biopolymers in acidic conditions (such as Spodosols). In neutral to alkaline soils, the comminution of plant debris by fauna increases the rate of decomposition of plant biopolymers, but simultaneously provides an intimate mixing of organic materials, microbial biomass, and metabolic products with clays in the inorganic matrix.

The constraints on further decay due to organomineral associations may offset the initial rapid decay stimulated by the soil fauna compared with acid soils where decomposition is dominated by fungi. In agroecosystems, except in humid conditions or waterlogged soils, a cover of trash (stubble, straw) on the soil surface may delay decomposition compared with incorporation by tillage. This is a particular problem in semiarid regions used for cereal production, where harvest is followed by a hot, dry summer. The cereal straw remains undecomposed through the summer period because it is dry and not in intimate contact with soil materials, including soil organisms. The surface litter causes physical problems during the establishment of the next crop. However, a protective surface cover of debris that is promoted by minimum tillage will, over a period of years, stimulate soil fauna. The fauna may increase the rate of decomposition of the debris by incorporating the products of decomposition into the soil matrix, as in neutral to alkaline soils.

Rhizodeposition

The transfer of photosynthate below ground by roots in all ecosystems is probably the most important and quantitatively the least understood of all the carbon cycling processes in soil. It is technically very difficult to measure rhizodeposition in soils, even under laboratory conditions. The growing root pumps carbon into soils as root cells, root hairs, exudates, mycorrhizas, and respired CO_2. Root systems are extremely adaptable and while one part may die because of adverse conditions, compensatory growth will occur elsewhere. Thus growth and death of root systems occur simultaneously. During the growing season there is root growth and respiration, root death and turnover, and turnover of all the carbon compounds that have been derived from the root to create the rhizosphere. The process is complex and dynamic.

Currently, we are able to apply a range of techniques to assess rhizodeposition by different plants. These techniques include the sequential harvesting of plants grown under controlled environments or in the field, pulse labeling of plants with $^{14}CO_2$ to trace the labeled photosynthate below ground, and the use of ^{13}C abundance measurements during the growth of maize in soils formed under C_3 plants (Johansson 1992; Keith et al. 1986; Balesdent and Balabane 1992). For agricultural crops, such as cereals and leys, the indications are that during the growth of roots to maturity, 20% to 50% of the carbon pumped below ground leaves the roots and is subject to decomposition (Paustian et al. 1990; Hansson and Steen 1984). For natural grasslands, ratios of underground biomass to green shoot biomass range from 2 to 13 for temperate regions to 0.5 for tropical grasslands (Coupland 1979). Rhizodeposition in forest ecosystems has been estimated by using C and N budgets (Nadelhoffer and Raich 1992). Root:shoot ratios for forests are thought to be considerably less than for grasslands although estimates of fine root biomass appear to increase with time as workers investigate root systems in more detail. Fine roots may undergo more than 100% annual turnover in spite of the fact that fine roots have high "lignin" (acid insoluble fiber) contents.

Form and Distribution

The physical size and shape of litter and roots is important for decomposition. Large plant fragments, such as branches and twigs, will decompose slowly because until the plant tissues become physically accessible to the soil flora, decay will be slow. It is assumed that similar principles apply to roots. One might speculate that organic debris in large pieces may initially decay slowly, but subsequently more rapidly once the decomposers are established on their substrate. Fine roots will present a significant surface to the decomposers initially, but because the roots are well distributed in the soil matrix, subsequent decomposition may be slowed by the organo-mineral associations formed.

The depth to which organic matter is pumped into the soil is also important. There is evidence from ^{14}C dating of soils that mean residence times increase with depth (Oades 1989). Most of the data on carbon cycling in soils is based on what happens in the top 20 cm. Turnover times for organic matter at depth are not available and there are many soils where more than half the carbon content is in deeper layers.

Chemistry

The chemistry of plant materials controls palatability to organisms. Routine analytical procedures have been developed for litter to determine contents of cellulose, lignin, and water soluble components as well as carbon and nitrogen contents. Cellulose and lignin comprise the structural components of plants and are more resistant to decomposition than other carbohydrates and nitrogenous components. Natural ecosystems have developed suites of organisms and associated enzymes capable of decomposing all naturally occurring organic compounds. Otherwise soils and sediments would become a repository for the stable biomolecules which would dominate nonliving organic matter on land and in the oceans. Decomposition rates, or turnover times, have been determined using ^{14}C-labeled compounds for a wide range of chemicals (Paul and Van Veen 1978; Trumbore and Druffel, this volume). The turnover of simple monomers such as glucose and amino acids is measured in terms of hours or days. Polymers have longer turnover times, e.g., weeks, but even complex heteropolymers turnover within months. The most resistant polymers are those containing aromatic rings, such as in lignin, and a range of poly-methylenic molecules, such as lipids and waxes. However, in the absence of interactions with the inorganic matrix, even these compounds have turnover times of only about one year. Recent studies using ^{13}C CPMAS NMR (cross polarization, magic angle spinning nuclear magnetic resonance) techniques have shown that in soils, lignin is altered relatively quickly. Oxygenated functional groups on the aromatic ring disappear during early stages of decomposition, and it is not clear how the transformation of phenolic aromatic lignin moieties to carboxylated aromatic rings occurs in nature. When the annual production of lignin is considered, this transformation is quantitatively a very important mechanism.

Fats and waxes are probably the most resistant group of chemicals in terrestrial systems, and we now know, from application of solid-state NMR to a range of soils, that polymethylenic materials dominate organic materials in many soils. They originate in cutins and suberins, but a synthesis of polymethylene materials by soil organisms also occurs. Yet we still must investigate the possibility that highly resistant biomacromolecules are produced in soils and survive geologic time. Based on the studies of Largeau et al., as described in this volume (see chapter 16), this possibility may exist.

Finally, a neglected carbon source in soils is charcoal. Fire has been a natural phenomenon, aided in recent times by human activities. Fire can create a range of complex aromatic chemicals collectively known as charcoal. Charcoal is resistant to microbial degradation and is widespread in terrestrial systems. Charcoal may represent a significant pool in certain soils, e.g., under grasslands, and it may well be the "inert" pool sought by the C cycle simulation modelers.

No standard combination of chemical groupings exists to assess biological availability. Lignin:nitrogen, lignin:cellulose, and structural:solubles ratios have been shown to be useful in predicting decomposition rates of forest litter (Melillo et al. 1982). Similar approaches have been used in agroecosystems and lignin-nitrogen ratios have been shown to be linearly related to values for lignin plus cellulose (Paustian et al. 1992). For some time we have known that adding nitrogen to surface litters with high C:N ratios stimulates decomposition. Across ecosystems, however, the impact of additional nitrogen on carbon dynamics requires further investigation.

Recently, Baldock et al. (1992) proposed that the ratio of carbohydrate to polymethylene components could be used as a measure of the bioavailability of organic materials in aerobic environments. This approach requires further testing.

J.M. Oades
THE DECOMPOSERS

We generally accept that most of the cycling of organic matter in soils is biologically mediated. Soil fauna may play a major role in comminuting and mixing plant materials into the soil matrix but utilize little more than one-tenth of the available energy. Most of the energy is utilized by the soil microbial biomass, which consists of a wide range of known and unknown organisms that are adapted to living in a quiescent state in soils when substrate is not available. How the organisms are distributed in the soil matrix, particularly at depth, is unknown, but they tend to be ubiquitous and multiply rapidly in the presence of substrate. The soil microbial biomass represents between 1% to 5% of the soil carbon, and its activities are controlled by the availability of substrate, temperature, and water regime. Between about 10°C and 30°C biological activity doubles for each 10°C rise in temperature providing water is not limiting or in excess.

The rate of decomposition decreases rapidly below 10°C and above 35°C. Decomposition and CO_2 production are most rapid at low soil water potentials (pores > 100 μm diameter are filled with air) and slow down as water becomes more tightly bound in soils, or under waterlogged conditions.

The influence of temperature is based largely on laboratory data with simple substrates. However, field experiments using ^{14}C-labeled plant materials have shown that the rate of decomposition doubled for each 9°C rise in mean annual temperature (Ladd et al. 1985).

Generally, data on temperature and water regimes can be obtained relatively easily and various models simulating carbon cycling in soils appear to handle these external variables well.

THE SOIL MATRIX

General Protection

When the decomposers utilize substrate in the soil, microbial biomass along with a range of metabolites results. The more recalcitrant biological chemicals are concentrated in the residual substrate. These processes take place in the highly polydispersed pore system of soils that normally offer extensive surface for adhesion of both microorganisms and their metabolic products. The soil environment mediates decomposition processes, and the rates of decay of biopolymers in the soil matrix are at least an order of magnitude less than the rates of decay in the absence of soil. For example, the half-lives of a range of biopolymers are measured in months, with phenols and waxes listed as the most recalcitrant chemicals (Paul and van Veen 1978). Similar biopolymers produced in soils become part of a system where half-lives are measured in decades. In some soils, part of the carbon is retained for thousands of years.

Intensive chemical extractions and fractionations have shown that old and new organic materials in soil tend to have similar chemical compositions. It must be the position in the soil matrix and depth within the profile that constrains further decomposition, not the chemistry of the biopolymers. Whether this applies to the very old carbon in soil is not known. Clays generally contain polymethylene components with low C:N ratio. These materials may be very recalcitrant (Derrene et al. 1993).

Texture and Clay Contents

Soil texture refers to the particle size distribution of soil after disaggregation. A relatively small clay content will control the chemical, physical, and biological properties of soils because of the very high surface area. Generally, there is a positive correlation between organic carbon content and clay content (Parton et al. 1987). However, it is not clear that clay is the direct causal agent since most clay soils are more fertile than sandy soils and thus have larger photosynthetic inputs.

A series of laboratory incubations have demonstrated that clay soils retain more carbon than do sandy soils (Sorensen 1975; Amato and Ladd 1992). After incubating ^{14}C-glucose in soils for 44 weeks, Amato and Ladd showed that residual ^{14}C was positively related to clay content expressed as cation exchange capacity (a measure of total negative charge associated with clay, or of surface area). Interestingly, the increased retention with clay content was all in the form of microbial biomass. The effect of the clay was greatest with 0% to 10% clay. It seems that a small amount of clay can offer as much adhesive surface to adsorb materials during biologic reactions as can higher amounts of clay. The retention of additional microbial biomass was related to pore spaces less than 6 μm diameter.

Consideration of pore spaces as habitats for microorganisms (Postma and van Veen 1990) leads to speculation that clays have considerably more surface area than ever reacts with by-products from organic matter decomposition. During a given decomposition event, only a very small portion of the clay surface ($\ll 1\%$) is involved in adsorption of biopolymers or microorganisms because organisms are actually clumped in close proximity to the substrate. It may be that clays managed appropriately offer considerable potential for sequestration of carbon. We have conceptual frameworks for the microstructure of clays; however, little is known about the juxtaposition of biopolymers, microorganisms, and clay microstructure.

Cation Bridges

Both clay microstructure and the association of organic materials (which possess functional groups such as carboxyl and hydroxyl) with negatively charged clay surfaces are very much influenced by cations with multiple charges. Quantitatively, the most important cations in soils are Ca^{2+} in neutral to alkaline soils and Al^{3+} in acid soils (Oades 1988).

Aluminum is associated with protons in acid soils, and hydrolyzed Al species play a major role in the amphoteric surfaces associated with variable charge in tropical soils and in the soils developed on volcanic ash (Andosols). There is convincing field and laboratory evidence to indicate that organic matter in Andosols has a significantly longer turnover time than that in soils developed on sedimentary rocks (Oades et al. 1989).

There is substantial evidence that Ca can act as a bridge between clay plates, between functional groups on biopolymers, and between biopolymers, bacteria, and clay surfaces (Oades 1989). Thus neutral to alkaline soils with Ca as a dominant exchangeable cation may retain organic matter to a greater extent than soils that are slightly acidic and not saturated with Ca. The role of Ca in promoting retention of C in soils requires further validation (Baldock et al. 1994).

Aggregation

Examination of the soil matrix at a scale of millimeters rather than micro- and nanometers reveals the existence of structural hierarchy, where smaller aggregates of clay particles, microorganisms, and biopolymers are clustered together in a larger aggregate that may be several millimeters in diameter. Within such aggregates, a range of partly decomposed plant remains are occluded. This entrapment within aggregates offers protection against rapid decomposition. There is good evidence to indicate that this occluded material is the organic matter which is lost during decades of cultivation. It may also be accreted during decades of growth of root systems of grasses.

There is no clear understanding of the role of soil structure in carbon cycling. The dynamics of soil structure is not understood other than the loss of larger aggregates and pores in cultivated soils, which is accompanied by a simultaneous loss of organic carbon. The inert or passive pool of carbon may well be strongly associated with clay surfaces, but so is modern carbon, which makes the inert pool impossible to isolate. One recent approach to define the inert or old carbon in soils involves oxidation of organic matter on the outside of aggregates using high-energy UV oxidation. The carbon inside the small aggregates was shown to be older than that on outer surfaces (Skjemstad et al. 1993).

INTEGRATION OF INFORMATION

The various factors discussed have been integrated in a range of models that simulate the dynamics of carbon in soils. Based on data currently available, the models consist of inputs and three major, largely conceptual pools, each of which has a different turnover rate. The most popular models at the moment (Jenkinson 1990; Parton et al. 1993) both have an active pool, a slow pool, and an inert or passive pool with turnover times of one year, decades, and hundreds to thousands of years respectively. Inputs have been divided into above and below ground and are characterized chemically into easily decomposed and resistant components. The rate constants for the various components of input and soil "pools" are all modified by temperature, water, regime, and clay content.

The models are being tested and modified continuously as they are applied to new situations. There is a need to match the conceptual pools in the models with actual pools in soils, which can be determined precisely. When the dynamics of carbon in the various pools can be measured, then the models can be validated rigorously, which will enhance their predictive capabilities.

SUMMARY

Substantial progress has been made in understanding carbon cycling in soils over the last twenty years; however, several areas still require quantitation and further development of concepts:

- Rhizodeposition: amounts, distribution, dynamics, and turnover.
- The inorganic matrix: the importance and dynamics of soil structure at all scales, including the role of cation bridges and clay microstructure on C cycling.
- The inert pool: Does it exist? How big is it? Is it charcoal or persistent biomacromolecules, or a clay organic interaction?

REFERENCES

Amato, M., and J.N. Ladd. 1992. Decomposition of ^{14}C-labeled glucose and legume material in soils: Properties influencing the accumulation of organic residue C and microbial biomass C. *Soil Biol. Biochem.* **24**:455–464.

Baldock, J.A., M. Aoyama, J.M. Oades, Susanto, and C.D. Grant. 1994. Structural amelioration of a South Australian Red brown earth using calcium and organic amendments. *Aust. J. Soil Res.* **32**:571–594.

Baldock, J.A., J.M. Oades, A.G. Waters, X. Peng, A.M. Vassallo, and M.A. Wilson. 1992. Aspects of the chemical structure of soil organic materials as revealed by solid-state ^{13}C NMR spectroscopy. *Biogeochemistry* **16**:1–42.

Balesdent, J., and M. Balabane. 1992. Maize root derived soil organic carbon estimated by natural ^{13}C abundance. *Soil Biol. Biochem.* **24**:97–101.

Bolin, B. 1983. The carbon cycle. In: The Major Biogeochemical Cycles and Their Interactions, ed B. Bolin and R.B. Cook, pp. 41–45. Chichester: Wiley.

Coupland, R.T., ed. 1979. Grassland Ecosystems of the World: Analysis of Grasslands and Their Uses. Cambridge: Cambridge Univ. Press.

Derrene, S., C. Largeau, and F. Taulelle. 1993. Occurrence of non-hydrolyzable amides in the macromolecular constituent *of Scenedesmus quadricauda* cell wall as revealed by ^{15}N NMR of *n*-alkylnitriles in pyrolysates of ultralaminae containing kerogens. *Geochim. Cosmochim. Acta* **57**: 851–858.

Hansson, A., and E. Steen. 1984. Methods of calculating root production and nitrogen uptake in an annual crop. *Swedish J. Agr. Res.* **14**:191–200.

Jenkinson, D.S. 1990. The turnover of organic carbon and nitrogen in soil. *Phil. Trans. R. Soc. Lond. B* **329**:361–368.

Johansson, G. 1992. Release of organic carbon from growing roots of meadow fescue (*Festuca pratensis* I.). *Soil Biol. Biochem.* **24**:427–433.

Keith, H., J.M. Oades, and J.K. Martin. 1986. Inputs of carbon to soil from wheat plants. *Soil Biol. Biochem.* **18**:445–449.

Ladd, J.N., J.M. Oades, and M. Amato. 1985. Decomposition of plant material in Australian soils. III. Residual organic and microbial biomass C and N from isotope labeled plant material and organic matter decomposing under field conditions. *Aust. J. Soil Res.* **23**:603–611.

Melillo, J.M., J.D. Aber, and J.F. Muratore. 1982. Nitrogen and lignin control of hardwood leaf litter decomposition dynamics. *Ecology* **63**: 621–626.

Nadelhoffer, K.J., and J.W. Raich. 1992. Fine root production estimates and below ground carbon allocation in forest ecosystems. *Ecology* **73**:1139–1147.

Oades, J.M. 1988. The retention of organic matter in soils. *Biogeochemistry* **5**:35–70.

Oades, J.M. 1989. An introduction to organic matter in mineral soils. In: Minerals in Soil Environments, ed. J.B. Dixon and S.B. Weed, pp. 89–160. Madison: Publ. Am. Soc. Soil Sci.

Oades, J.M., G.P. Gillman, and G. Uehara. 1989. Interactions of soil organic matter and variable charge clays. In: Dynamics of Soil Organic Matter in Tropical Ecosystems, ed. D.C. Coleman, J.M. Oades, and G. Uehara. Honolulu: Univ. of Hawaii Press.

Parton, W.J., D.S. Schimel, C.V. Cole, and D.S. Ojima. 1987. Analysis of factors controlling soil organic matter levels in Great Plains grasslands. *Soil Sci. Soc. Am. J.* **51**:1173–1179.

Parton, W.J., D.S. Schimel, D.S. Ojima, and C.V. Cole. 1994. A general model for soil organic matter dynamics: Sensitivity to litter chemistry, texture and management. *Soil Sci. Soc. Am. J.*, in press.

Paul, E.A., and H. van Veen. 1978. The Use of Tracers to Determine the Dynamics Nature of Organic Matter, vol. 3, pp. 61–102. Edmonton: 11th Congress Inst. Soc. Soil Science.

Paustian, K., O. Andrew, M. Clarholm, A.C. Hansson, G. Johansson, J. Lageri, T. Lindberg, R. Pettersson, and B. Sohlenius. 1990. Carbon and nitrogen budgets of four agroecosystems with annual and perrenial crops, with and without N fertilization. *J. Appl. Ecol.* **27**:60–84.

Paustian, K., J.P. Parton, and J. Persson. 1992. Modeling soil organic matter in organic amended and nitrogen fertilized long-term plots. *Soil Sci. Soc. Am. J.* **56**:476–488.

Postma, J., and J.A. van Veen. 1990. Habitable pore space and survival *of Rhizobium leguminosarum* biovar *trifoli* introduced into soil. *Microb. Ecol.* **19**:149–161.

Raich, J.W., and K.J. Nadelhoffer. 1989. Below ground carbon allocation in forest ecosystems: Global trends. *Ecology* **70**:1346–1354.

Skjemstad, J.O., L.J. Janik, M.J. Head, and S.G. McClure. 1993. High energy ultra-violet photo-oxidation: A novel technique for studying physically protected organic matter in clay and silt-sized aggregates. *J. Soil Sci.* **44**:485–499.

Sorensen, L.H. 1975. The influence of clay on the rate of decay of amino acid metabolites synthesised in soils during decomposition of cellulose. *Soil Biol. Biochem.* **13**:313–321.

Standing, left to right:
Cindy Lee, Richard Zepp, Claude Largeau, Ken Mopper, Kay-Christian Emeis, Göran Ågren, Otto Daniel
Seated, left to right:
David Schimel, Bill Reeburgh, Eric Davidson, Malcolm Oades

18

Group Report: What Are the Physical, Chemical, and Biological Processes that Control the Formation and Degradation of Nonliving Organic Matter?

E. DAVIDSON, Rapporteur
G. ÅGREN, O. DANIEL, K.-C. EMEIS, C. LARGEAU,
C. LEE, K. MOPPER, J.M. OADES, W.S. REEBURGH,
D.S. SCHIMEL, R.G. ZEPP

INTRODUCTION

Synthesis of organic compounds by primary producers and modification of these compounds by consumers and detritivores produces a broad spectrum of biomolecules with varying chemical and physical properties and with a wide range of residence times in the environment. We can estimate the global mean residence time of nonliving organic matter (NLOM) by dividing an estimate of total stocks by an estimate of total production. Yet this would reveal little about the nature of NLOM, because it lumps together types of NLOM that have different properties and that are formed and degraded at time scales ranging from fractions of a year to millions of years. Following a reductionist approach, we can subdivide the global estimates into smaller pools that are distinct from one another, and the new categories define less heterogeneous pools. If our classification system were successful, the range of mean residence times of compounds found within each class would be narrow.

Figure 18.1 illustrates one attempt to define pools of NLOM that are spatially or chemically distinguished from each other according to some of the

Role of Nonliving Organic Matter in the Earth's Carbon Cycle
Edited by R.G. Zepp and Ch. Sonntag © 1995 John Wiley & Sons Ltd.

sampling, analytical, and conceptual approaches commonly used in the marine and terrestrial sciences. Beside the fact that many of the estimates are no more than our best-educated guesses, the figure shows that the range of mean residence times attributed to many of the pools is very wide. Other classification schemes have been devised, but none is any more successful in overall characterization of classes of NLOM with meaningful residence times. In some cases, wide ranges in the estimates of Figure 18.1 are the result of inadequate information about those pools. In others, the wide ranges indicate that the identified pools are not well defined, i.e., these pools still consist of a diverse group of NLOM compounds that are formed and degraded at different rates.

The classification scheme of Figure 18.1 is inadequate because we do not understand the forms and alteration processes of NLOM well enough to identify pools that adequately characterize their rates of cycling. For example, ^{14}C data show that soil organic matter (SOM) adsorbed to clay surfaces cycles on time scales of centuries to millennia; however, if that identified pool is actually contaminated by some other pool, such as ancient organic matter inherited from a previous sedimentary cycle, the estimate of the rate of cycling of the more modern SOM would be in error. A similar problem in the marine sciences is the identification of pools and turnover times of dissolved organic carbon (DOC).

It may not be necessary to reduce further each pool into many smaller ones. New ways of defining NLOM pools, based on an understanding of how they are formed and degraded, may prove more fruitful. In the example given above, distinguishing between inherited SOM versus SOM formed *in situ* may be necessary to characterize the pool adsorbed to clay. Improving our understanding of the processes of formation, decomposition, and stabilization of NLOM in ecosystems may help us to improve definitions of appropriate pools of NLOM, determine their rates of cycling, and study how they might change as the human impact on biogeochemical cycles grows.

DEFINITION OF TERMS AND FOCI

In our effort to define pools of NLOM, we encountered problems of terminology as a plethora of terms (e.g., active, passive, fast, slow, recalcitrant, refractory, inert, labile) have been used to describe the persistence of organic materials in soils, sediments, and water. Across the various disciplines, the term "persistence" is used with no particular or specific meaning. It implies only time, and can be quantified as turnover times, half-lives, or residence times. NLOM may persist because it has a chemical structure that renders it resistant to degradation. On the other hand, NLOM that would normally have a transient existence in the environment may persist because of interaction with the inorganic matrix or because of an environmental factor such as low tem-

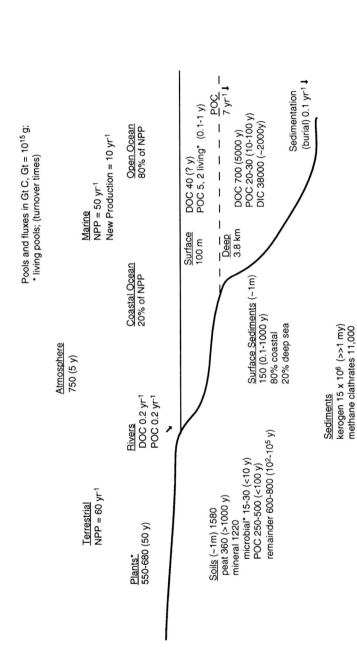

Figure 18.1 Global carbon reservoirs, fluxes, and turnover times (estimates). Most estimates are from the indicated literature sources; although in some cases, where published estimates are lacking the best guesses of the members of our group are offered (Hedges 1992; Eswaran et al. 1993; Siegenthaler and Sarmiento 1993; Schimel et al. 1994).

perature, anoxia, or lack of water. Here we will use "transient" to refer to NLOM that has turnover times of less than a century (further division of this category into "annual" and "decadal" pools may be helpful), and "persistent" to refer to NLOM with turnover times from centuries to millennia.

Our group focused most of its discussion on identifying the main sources and processes of the formation of persistent NLOM. With a couple of exceptions, we did not address NLOM that persists in the environment on time scales longer than millennia. We addressed transient NLOM, primarily with respect to the effects of this cycling on the formation of the persistent fraction.

The potentially important sources of persistent NLOM are:

1. complex compounds produced by condensation reactions of more simple decay products or simple primary compounds,
2. products of primary producers that are highly resistant to decomposition, and
3. physical and chemical complexes of decay products with minerals.

The first source has been the subject of active debate in both soil and marine sciences, regarding the pathways of humus formation, and was reviewed at a previous Dahlem Workshop (Hedges 1988). Briefly, complex macromolecules may be decomposed into successively smaller molecules, with some of the decay products containing highly resistant structures and functional groups. Alternatively, decay to very simple small structures may be followed by condensation reactions that create new, larger, more complex, and more resistant molecules. These pathways need not be mutually exclusive. In marine sediments, we conventionally view humic substances as being formed in sediment pore waters via the degradation-condensation pathway (e.g., high molecular weight compounds are degraded to low molecular weight compounds and then fulvic acids and humic acids are formed from condensation reactions). Although this view is widely accepted, it lacks strong empirical evidence for any of the proposed pathways. Pathways of formation of humic substances in surface waters are more uncertain than in marine sediments. Proposed pathways include the cross-linking of polyunsaturated fatty acids, browning-type reactions (e.g., between amino acids and sugars), cross-linking of phenolic compounds derived from algae and terrestrial sources, and fluxes out of sediments. Photochemical processes may play a role in the cross-linking reactions. As in the sediments, the relative importance of these pathways depends upon environmental conditions, such as proximity to sediments and rivers, light fields, biochemical composition of algal inputs, and microbial decomposition rates.

Most of our group's discussion focused on the second and third sources of persistent NLOM listed above: production of resistant compounds by primary producers and protection of decay products in organo-mineral complexes. To understand these sources of NLOM, we must explore the decomposition and stabilization processes of NLOM. We examined the effects of substrate quality, the matrix in which decomposition occurs, redox conditions, water, nutrient availability, acidity, faunal activity, and UV radiation.

SUBSTRATE QUALITY

The chemical composition of NLOM affects the rate of decomposition. At each phase of decomposition, the molecular structure and functional groups of an organic molecule influence the ability of decomposers to utilize, and thus to further decompose, the substrate. Some molecules, such as glucose and several amino acids, appear to be labile in most environments. At the other end of the spectrum, a class of compounds such as those described by Largeau et al. (this volume) are resistant to decomposition even on geologic time scales. Most compounds lie between these two extremes, however, and their degree of resistance is influenced not only by their chemical composition but also by the conditions under which they are transformed. Hence, for most NLOM, we cannot consider the concept of substrate quality without simultaneously considering the matrix in which decomposition occurs.

The quantitative significance of the highly resistant compounds in ecosystems is unknown. The preliminary evidence of Largeau et al. (this volume) shows that some freshwater and marine microalgae and some bacteria produce biomacromolecules that are highly aliphatic, nonhydrolyzable, and highly resistant to decomposition. It appears that marine diatoms and coccolithophores do not produce these compounds. Analogues in terrestrial vegetation include walls of pollen grains and macromolecular constituents of some seed coats, leaf cuticles, and outer bark tissues. Generally, we regard another class of higher plant compounds, including suberin, cutin, and waxes, as being highly resistant to decomposition; however, even these compounds can decompose in the soil environment. On the other hand, the very old NLOM associated with the clay fraction has many of the same properties (aliphatic groups, nonhydrolyzable) as the compounds described by Largeau et al. It is too early to judge whether the production of resistant biomacromolecules by plants is quantitatively significant relative to other processes of formation and stabilization of NLOM in persistent pools, but the identification, quantification, and fate of these resistant compounds deserve further study.

Another substrate generally regarded as highly resistant is charcoal. Fire is both a natural phenomenon and a human intervention in terrestrial systems that converts biomass C to charcoal, CO_2, and CO. If charcoal is persistent, in time (e.g., 1000 yrs) it will become a significant pool of soil C. It is also transferred by wind and water to the oceans. Particles of alleged charcoal are often observed in soils and sediments, but the amounts are seldom quanti-

E. Davidson et al.

Table 18.1 Persistence of selected compounds in seawater, sediments, and soils.

Authors	Compound	Media	Conditions	Turnover Time
Henrichs and Doyle (1986)	alanine	porewater	from glacially derived marine sediments (60 m depth)	0.12–0.45 days
	glutamic acid	porewater	from glacially derived marine sediments (60 m depth)	0.54–1.7 days
Lee (1992)	amino acids	seawater	oxic	0.03–6 days
			anoxic	0.5–28 days
	putrescine	seawater	oxic	2.1–31 days
			anoxic	12–160 days
	acetate	seawater	oxic	0.9–16 days
			anoxic	4–13 days
	formaldehyde	seawater	oxic	0.5–13 days
			anoxic	1–99 days
	trimethyl-amine	seawater	oxic	10 to >4000 days
			anoxic	14 to >4000 days
	urea	seawater	oxic	2–3.4 days
			anoxic	1.4 days
Mopper and Kieber (1991)	glucose	seawater	oxic (10 m depth)	1.7 days
			anoxic (200m)	346 days
			anoxic (1800 m)	177 days
	pyruvate	seawater	oxic (10 m depth)	1.8 days
			anoxic (200 m)	3303 days
			anoxic (1800 m)	153 days
	formaldehyde	seawater	oxic (10 m depth)	3.1 days
			anoxic (200 m)	1109 days
			anoxic (1800 m)	274 days
Sun et al. (1993)	chlorophyll-a	marine sediments	bound, oxic	0.10 days
			bound, anoxic	0.15 days
			free, oxic	0.044 days
			free, anoxic	0.3 days

(continued)

fied. Charcoal is not represented in soil organic C determined by wet oxidation procedures but is included in soil C estimates derived from other analytical procedures, so its representation in global data bases is not clear. Charcoal is chemically complex and may contain polycyclic aromatics and graphite. It is highly reactive as an absorbent for organic compounds, but its resistance to biological degradation is unknown. Charcoal has been equated with an "inert

Table 18.1   (*continued*)

Authors	Compound	Media/conditions	Half life
Sorensen (1972) Sorensen (1981)	amino acid labeled C	^{14}C labeled, in soil	2190–2555 days
		4.1% clay, field conditions	72 days
		11.4% clay, field conditions	72 days
		17.7% clay, field conditions	72 days
		34.4% clay, field conditions	180 days
	labeled amino acid	4.1% clay, field conditions	146 days
		11.4% clay, field conditions	146 days
		17.7% clay, field conditions	146 days
		34.4% clay, field conditions	438 days
Shields et al. (1973)	glucose	field conditions	4 days
	soil organic matter	^{14}C labeled	180 days

			Percent remaining after 12 weeks
Kassim et al. (1981)	glucose	^{14}C labeled, in soil	4.6–5.2%
	cellulose	^{14}C labeled, in soil	6.1–7.2%
	glucosamine	^{14}C labeled, in soil	3.5–6.4%
	wheat straw	^{14}C labeled, in soil	4.0–4.9%
	L. dextranicus polysaccharide	^{14}C labeled, in soil	3.6–5.0%
	H. holstii polysaccharide	^{14}C labeled, in soil	3.6–3.8%
	Anabaena flox aquae (whole cells)	^{14}C labeled, in soil	3.3–4.5%
	A. flox aqual (cell wall)	^{14}C labeled, in soil	0.8–4.2%
	A. flox aqaue (cytoplasm)	^{14}C labeled, in soil	3.3–5.0%

pool" in simulation models of C cycling in soils; however, this is more a result of the need for such an inert pool to fit empirical data on soil ^{14}C data than it is based on information about actual degradation of charcoal (Jenkinson 1990). There are speculative associations of charcoal aromatic compounds and soils formed under frequently burned grasslands.

Variation in substrate quality is particularly useful for the comparison of rates of decay of specific compounds or groups of compounds. Rate constants of the decay of specific amino acids and other compounds have been measured in both marine and terrestrial ecosystems (Table 18.1). In terrestrial ecosystems, indices based on ratios of lignin-to-nitrogen, lignin-to-cellulose, and structural-to-soluble components have been successful in predicting rates of decomposition of plant litter (Melillo et al. 1982). It should be noted, however, that these studies provide more information about the initial decay of these compounds rather than the eventual stabilization of residues of decomposition into persistent fractions of NLOM. Models of decomposition that use multiple-rate constants for a variety of substrate classes to describe observed rates of decomposition (e.g., the first-order decay model of Berner 1980; the CENTURY model of Parton et al. 1993; and the Rothamsted model of Jenkinson 1990) are empirical fits and do not represent mechanistically the physicochemical processes by which decay products are stabilized. Other modeling approaches have also been used to describe observed decomposition (e.g., Bosatta and Ågren 1991), but the important point is that processes of stabilization and decomposition need to be understood. The interaction of the decompositional products with their environment influences the rate of decay of the products as much as or more than does the original structure of the compounds produced by plants.

CHEMICAL AND PHYSICAL PROTECTION
WITHIN THE MATRIX

Soils and sediments are comprised of organo-mineral complexes of varying size, structure, and organization. As a result of this heterogeneity, which occurs on scales of microns to millimeters within soils and sediments, NLOM may be exposed to or protected from the microbial agents of decomposition. Similarly, concentrations of oxidants and nutrients vary among the heterogeneous microsites of soils and sediments, as does acidity and water content.

Cultivation exposes soil particulate organic carbon (POC) that is protected within soil aggregates (Cambardella and Elliot 1992). The soil POC is comprised of partially decomposed plant fragments that are larger than 20 microns. The aggregates typically range in size from tenths to tens of millimeters, are held together by microbial polysaccharides and fungal hyphae, are depleted in oxygen and nutrients in their interiors, and are disrupted when the soil is tilled. Exposing the NLOM from the interiors of these aggregates to oxygen and nutrients permits microbial decomposition to proceed. About 20% to 40% of soil organic matter within the A horizon is typically lost within a few years or decades after tilling previously uncultivated soils (Davidson and Ackerman 1993), and this loss emanates primarily from the POC fraction that has a mean residence time of decades (Figure 18.1). Similarly, bioturbation of marine sediments and bottom currents can, in some cases, expose organic matter previously protected within the sediment matrix.

The 60% or more of the soil organic matter that remains after decades of cultivation is thought to be the fraction adsorbed to clay surfaces. When other factors like vegetation type, mean annual temperature, and drainage class are

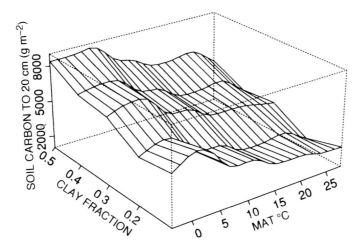

Figure 18.2 Model prediction of organic carbon content of surface soils as a function of clay content and mean annual temperature (Schimel et al.). This aspect of the CENTURY model is based on empirical relations described by Burke et al. (1989) and Parton et al. (1994).

relatively constant or are accounted for, clay content is positively correlated with soil organic matter content (Burke et al. 1989; Figure 18.2). Clay particles are less than 2 microns, they have a high surface-to-volume ratio, and certain types of clays (2:1 type minerals and allophane) are highly reactive with many NLOM compounds. Certain functional groups, such as aromatic rings and carboxylic acids, and the presence of cations facilitate adsorption and formation of microaggregates of clay particles at the scale of microns in soils (Oades, this volume) and in sediments. The compounds that are found adsorbed to clays are often aliphatic, although the aliphatic portions of the molecules are generally not reactive with clay. Hence, it appears that the persistent NLOM associated with the clay fraction of soils and sediments has two important characteristics: aliphatic regions that are resistant to decomposition and functional groups that facilitate adsorption to the clay.

In marine environments, clay minerals and silt-sized particles are important in the flocculation and aggregation of organic matter, forming organo-mineral complexes that have sufficient mass to sink to deep waters and to sediments. The sinking mineral particles are largely coated with organic matter of algal and bacterial origin. Adsorption of the NLOM to the mineral component of the sinking aggregate may help to stabilize the aggregate in the water column and once it reaches the sediment. Sedimentation rates are positively correlated with carbon concentrations of open marine oxic sediments (Figure 18.3).

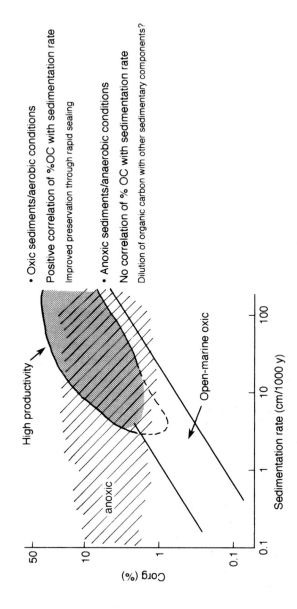

Figure 18.3 Preservation of organic matter in sediments. Relation between sedimentation rate and organic carbon content of sediments in oxic and anoxic marine sediments (Stein 1986).

In contrast to clay in sediments, clay within the soil matrix has only a small fraction of its surface coated by organic matter because only a small fraction of the surface area of soil particles is exposed to inputs of organic matter from plant and microbial sources (Oades, this volume). These observations regarding the roles of clays in sediments and soils raise several questions regarding potential sequestration of NLOM: Could soils sequester more C if increased allocation of C below ground caused fine roots and mycorrhizae to explore a greater fraction of the soil volume and come into contact with more clay surfaces? Would increased inputs of dust from terrestrial ecosystems to marine ecosystems significantly increase the rate of sedimentation of organic-coated mineral particles and/or reduce mineralization of NLOM once it is adsorbed to these particles?

Dating with ^{14}C indicates that the average age of soil C increases with depth (Figure 18.4). This observation suggests that either the processes of stabilization of NLOM must vary with depth, that a major source of deep NLOM is DOC derived from old NLOM and transported from the surface, that a larger fraction of the NLOM in deep soils is inherited from parent material, or that some combination of these processes occurs. The source and structure of persistent NLOM in deep soils are unknown, as is the potential for deep soils to sequester more NLOM in response to increased inputs.

SUPPLY OF OXIDANTS

Without an oxidant, NLOM cannot be mineralized. For this reason, it was once widely believed that a lack of oxygen was the primary reason that NLOM was preserved in anoxic marine sediments. Over the past several years, however, empirical observations and experimental evidence have led to a revision of the concept that low or zero O_2 concentrations in marine bottom waters necessarily enhance the preservation of NLOM (Pedersen and Calvert, this volume). For many biological compounds, the rates of microbial degradation are similar in aerobic and anaerobic environments (Henrichs and Reeburgh 1987; Lee 1992). Microbial efficiencies (growth rate per unit of substrate consumed) decline when O_2 is not present, but alternate oxidizing compounds, particularly sulfate and manganese oxides in marine environments, provide sufficient oxidizing capacity to promote decomposition in the absence of oxygen. Consortia of microorganisms are often required to break down many NLOM compounds in both aerobic and anaerobic environments; however, they may be especially important in anaerobic environments because anaerobes often possess a less versatile suite of enzymes than do aerobes. Hence, resistance to degradation under anaerobic conditions varies among substrates and environments.

E. Davidson et al.

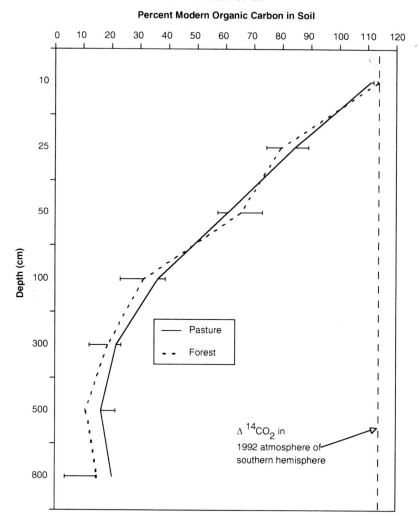

Figure 18.4 Relation between soil depth and ^{14}C content of soil organic carbon in deep Oxisols of the eastern Amazon forests of Brazil (Trumbore and Davidson, unpublished data). Values of 100% modern are equal to 0 per mil ^{14}C, which is the reference value for the atmospheric 1950 prior to atmospheric testing of nuclear weapons. Values > 100% modern result from a significant fraction of the organic-C being derived from C fixed by plants since atmospheric weapons testing in the early 1960s. A value of 0% modern carbon is equal to −1000 per mil and indicates that the soil-C is sufficiently old for ^{14}C decay to be essentially complete. Carbon at all depths is a mixture of varying proportions of young and old, but the fraction that is young appears to decline with depth.

In terrestrial wetlands, particularly nutrient-poor bogs, oxides of N, Mn, Fe, and S are not sufficient to promote anaerobic decomposition. Methanogenesis often becomes an important pathway of anaerobic respiration, which results in very slow rates of decomposition. Many of the compounds that persist for centuries to millennia in anaerobic peats would be readily decomposed in an aerobic environment, and they often are when peatlands are drained.

NUTRIENTS

Nutrient availability interacts with many of the processes already described. It is part of the lignin:nitrogen ratio that has been useful for describing initial decay rates of litter of varying substrate quality in terrestrial ecosystems. Rates of methanogenesis in wetlands, and hence rates of decomposition, are positively correlated with rates of net primary productivity (NPP), which in turn are largely determined by availability of N and P. Fens receive inputs of nutrients from inflowing water and they generally have higher rates of methanogenesis and decomposition than do bogs, which are nutrient poor. When nutrients are limiting, as in bogs, the plants adapted to those conditions make efficient use of their nutrients and lose little N and P in their litter (Valentine et al. 1994). A positive feedback develops whereby low inputs of nutrients to the bog from vegetation further reinforce the low rates of decomposition and the accumulation of peat. This type of localized feedback between NPP and nutrients in the soil–sediment matrix is unlikely to occur in open ocean waters because the primary production in the surface waters is not tightly coupled with the remineralization of nutrients from sediments on the ocean floor. A long and indirect path of upwelling and ocean circulation separates the two processes. Within oceanic surface waters, however, NPP and nutrient cycling are tightly coupled as they are in terrestrial ecosystems.

Living biomass in terrestrial systems includes structural components such as wood with very high ratios of carbon to nutrients. In general, these ratios decrease as CO_2 is released during decomposition of terrestrial NLOM. In some cases, nutrients are added, as in biological N fixation within decaying logs, or are transported between loci of different nutrient content by fungal mycelium. The equilibrium C:N ratio of most mineral soils is about 10 to 20. The equilibrium C:N ratio commonly found in marine sediments is also about 10. In marine systems, however, the primary producers lack analogues of structural wood, and most living biomass has a C:N ratio of about 7. The mineralization of proteins, which occurs as NLOM sinks to the sediments and is reworked in the sediments, causes the C:N ratio to increase to about 10. It is interesting that in both terrestrial and marine systems, the equilibrium C:N ratio is only somewhat higher than the common range for microorganisms (6 to 10). This observation suggests that microbial decay plays a major role in the decomposition of NLOM and in the formation of NLOM components that become stabilized into persistent pools. Microbial biomass and its associated microbial products may be protected from grazing and from further decomposition in microaggregates on the submicron scale. This is not to say that processes such as photolytic decomposition and condensation reactions are not important; however, the common C:N ratios observed in widely different ecosystems suggest a dominant role for microbial processes in determining these final ratios of elements.

WATER

Water is important as a medium through which nutrients and organic substrates move in soils and sediments. Mass flow carries DOC down the soil profile, and diffusion through water films is the primary means by which soil microorganisms obtain nutrients and organic substrates. The water content of a soil varies widely on time scales from minutes to years. Availability of soil water to plants is a key factor determining NPP and allocation of carbon below ground in many ecosystems. The water content of soil pore spaces affects diffusion of gases, including the availability of O_2 to decomposers. Wetting and drying events and freezing and thawing events in soils stimulate turnover of microbial biomass and mineralization of carbon and nitrogen. Marine sediments are continuously wet, but compaction can alter the pore volume and hence the rate of diffusion of substrates through the sediments. Although pore water affects decomposition in both soils and sediments, the wide range of variability of water content in soils indicates that the effect of water as a limitation to rates of decomposition is more important in soils than in sediments.

ACIDITY

Many organisms adapt to acidic environments. In general, increasing acidity results in lower rates of decomposition. Acidification of soils also may mobilize aluminum, which is toxic to many plants. Changes in acidity also may affect the solubility of substrates used by microorganisms and the rate of enzymatic activity. Many factors determine the acidity of soils, sediments, and aquatic systems. Among them is the production of organic acids. As in the feedback described above, relating nutrient-poor vegetation to low rates of decomposition in peat bogs, accumulation of organic acids can also slow rates of decomposition.

FAUNA

The fauna of soils and sediments, ranging in size from protozoa to ground hogs, influence NLOM decomposition in four ways:

1. They stimulate microbial activity by grazing and by changing the microorganisms' habitat (redox, pH, nutrients, moisture).
2. They alter the size of NLOM particles.
3. They mix the soil matrix, which exposes NLOM that may have been protected from decomposition, forms new aggregates that may provide greater protection from decomposition, and redistributes NLOM vertically in soils and sediments to areas of higher or lower rates of decomposition.
4. They use NLOM for their own metabolism and thereby change its chemical composition.

Microcosm and lysimeter experiments have shown that all four mechanisms occur in both terrestrial and marine ecosystems, but their relative importance is generally unknown. In most cases, however, it is believed that the effects of mixing the matrix and of stimulating activity of microbial populations by feeding on them is of greater importance than the actual chemical decomposition carried out by the fauna themselves.

Protozoans and other grazers of bacteria and fungi may affect rates of decomposition by stimulating new growth of the grazed populations. In aquatic ecosystems, zooplankton grazing can affect both decomposition and primary production. Grazing can also influence the chemical quality of molecules synthesized by the vegetation to discourage grazing. The importance of these secondary plant products on microbial decomposition of plant tissues is not well known.

Like microfauna, macrofauna also affect the fate of NLOM through their actions as grazers and mixers. Macrofauna also affect the hydrologic cycle in terrestrial ecosystems by creating macropores through which water and gases move. As we have already discussed, soil aeration and transport of water and nutrients affects rates of decomposition. Earthworms process large amounts of C in some ecosystems, such as meadows (Daniel 1992); leaf cutter ants, termites, and a variety of other organisms perform similar processing roles in other terrestrial ecosystems (Lavelle et al. 1993). Termites and ruminants divert significant amounts of NLOM from aerobic decomposition pathways to anaerobic decomposition involving methane production. Polychaetes are important agents of bioturbation in marine sediments (Aller 1982). Indeed, geologists have inferred the abundance of benthic fauna from the degree of lamination found in sediments (Savdra and Bottjer 1991).

ULTRAVIOLET RADIATION

Global changes in atmospheric composition, especially in total ozone and cloud cover, have important effects on the amount of solar ultraviolet radiation that reaches the Earth's surface. The most energetic part of this radiation, the ultraviolet-B (UV-B) component (280–315 nm), may have important effects on the production and degradation of NLOM. Responses of photosynthetic organisms to UV-B exposure include possible reduction in plant growth shifts in species distributions, increased production of protective flavonoids, and changes in litter composition resulting in lower decomposition rates.

Penetration of solar radiation into soils is confined to the upper 10 microns, but UV radiation may burn down into the litter layer as degradation proceeds. In aquatic ecosystems, solar UV radiation penetrates to depths ranging from a few centimeters in wetlands to tens of meters in open ocean water. Most of the NLOM in the upper layer of the sea may be exposed to UV because of mixing of surface waters (Najjar, this volume). Degradation of NLOM by UV radiation typically results in fading (loss of NLOM light absorption) in the ultraviolet spectral region. Because NLOM is an important UV light-screening substance (Blough, this volume), the fading process may result in deeper penetration of solar UV radiation into aquatic ecosystems. Thus, photodegradation of NLOM in surface waters could permit the inhibitory effect of UV radiation on photosynthesis to extend deeper into the water column.

Photooxidation of NLOM in aquatic environments is accompanied by the production of a variety of reactive transients which in turn impact other processes (Mopper et al. 1991). For example, photoreactions have been shown to convert biologically unavailable forms of trace metal nutrients (iron, manganese) to bioactive forms. These various degradation processes thus may have important effects on productivity of nutrient-limited ecosystems. Photodegradation of NLOM also rapidly degrades its fluorescent components, thus changing relationships between aquatic DOC and fluorescence. The light-absorbing components of NLOM that are susceptible to UV photodegradation include aromatic and unsaturated moieties. Saturated organic parts of the NLOM are more resistant to direct degradation but may be indirectly degraded by secondary photochemical reactions.

Various recent laboratory, field, and modeling studies have demonstrated that solar UV radiation increases decomposition rates of NLOM (Valentine and Zepp 1993). Most of these studies have focused on NLOM in marine or freshwater ecosystems, including wetlands, although there is evidence that lignin and other litter components are degraded by solar UV. Photodegradation also converts complex high molecular weight NLOM to lower molecular weight species that are more readily consumed by microbiota. This is probably an important mechanism by which riverine DOC is removed from ocean water and old (millennial residence time) DOC from deep ocean waters is broken down when it is recirculated through surface waters.

CONCLUSIONS

The most rewarding aspect of our group's discussions was the opportunity for scientists from marine and terrestrial disciplines to compare and contrast processes of NLOM formation and decomposition in the environments they have studied. We each learned about the other discipline, and the new things we learned about other systems often yielded insights about what is important in the systems each of us knows best. Below are some of the comparisons we found particularly useful:

In both marine and terrestrial systems, the amount of C in living organic matter is very small relative to stocks of NLOM, but the living microbial biomass looms large as a processor of NLOM, and microbial processes are important pathways of stabilization of NLOM into much larger persistent NLOM pools.

A significant fraction of the total NLOM is older than the cycling time of the matrix in both systems, e.g., some of the DOC is older than the cycling time of the ocean water and some of the SOM adsorbed to clay surfaces is older than the soil profile.

The matrix provides physical and chemical protection of NLOM from decomposition processes in both systems, and this matrix effect is a key factor in stabilizing NLOM into persistent pools (i.e., pools that cycle on time scales of centuries to millennia). The presence of functional groups in NLOM decay products facilitates adsorption to clay surfaces, and aggregation of particles at spatial scales from microns to millimeters provides further protection from microbial decomposition and grazing.

Substrate quality has been characterized by similar methods in both systems, and these methods are useful in describing the early stages of decomposition of a wide variety of compounds. In neither system, however, does the initial chemical composition of the NLOM alone determine rates of formation of persistent NLOM.

Fauna of all sizes play important roles as mixers of matrix material and as grazers in both systems. Faunal activity usually enhances decompositional processes, although faunal activity can also stabilize NLOM by helping to form new microsites of physical protection from microbial decomposition.

Oxidants are required for the decomposition of NLOM in both marine and terrestrial systems, and the dynamics of use of O_2 and oxides of N, Mn, Fe, and S are similar. The abundance of sulfate in marine systems and its low concentrations in many terrestrial wetlands, however, makes accumulation of NLOM due to the absence of oxidants more common in terrestrial wetlands than in marine systems.

Water is a medium for the transfer of organic substrate and nutrients in both systems, but many soils experience large variations in water content, and these variations strongly influence the formation and decomposition of NLOM in soils.

Terrestrial plants can respond to atmospheric concentrations of CO_2 by altering the C:N ratios of the tissues they synthesize, which, in turn, can affect rates of decomposition of plant litter. An analogous response by primary producers of the oceans to transients in atmospheric CO_2 does not occur.

Solar UV radiation decomposes exposed surface litter in terrestrial ecosystems and probably has important effects on the degradation of NLOM in fresh waters and the upper ocean.

The interactions of nutrient supply to primary producers and the mineralization of nutrients from depositional stocks of nutrients are more closely coupled in terrestrial ecosystems than in marine systems. Primary production in the open ocean is far removed from the remineralization of nutrients from ocean floor sediments (although it is tightly coupled with remineralization within surface waters), whereas land plants obtain a significant fraction of their nutrients from the mineralization that occurs within soils.

The C:N ratio of terrestrial NLOM decreases as decomposition progresses, whereas the C:N ratio of marine NLOM increases as it decomposes. In both systems, however, the C:N ratio converges at a value somewhat higher than the C:N ratio of microbial cells.

REFERENCES

Aller, R.C. 1982. The effects of macrobenthos on chemical properties of marine sediments and overlying water. In: Animal–Sediment Relations, ed. P.L. McCall and M.J.S. Tevesz, pp. 53–102. New York: Plenum.

Berner, R.A. 1980. A rate model for organic matter decomposition during bacterial sulfate reduction in marine sediments. In: Biogeochemistry of Organic Matter at the Sediment–Water Interface. *Colloq. Intl. C.N.R.S.* **293**:35–44.

Bosatta, E., and G.I. Ågren. 1991. Dynamics of carbon and nitrogen in the organic matter of the soil: A generic theory. *Am. Nat.* **138**:227–245.

Burke, C., C.M. Yonker, W.J. Parton, C.V. Cole, K. Flach, and D.S. Schimel. 1989. Texture, climate, and cultivation effects on soil organic matter content in U.S. grassland soils. *Soil Sci. Soc. Am. J.* **53**:800–805.

Cambardella, C.A., and E.T. Elliot. 1992. Particulate soil organic-matter changes across a grassland cultivation sequence. *Soil Sci. Soc. Am. J.* **56**:777–783.

Daniel, O. 1992. Population dynamics of *Lumbricus Terrestris* L. (*Oligochaeta: Lumbricidae*) in a meadow. *Soil Biol. Biochem.* **24(12)**:1425–1431.

Davidson, E.A., and I.L. Ackerman. 1993. Changes in soil carbon inventories following cultivation of previously untilled soils. *Biogeochemistry* **20**:161–193.

Eswaran, H., E. Van Den Berg, and P. Reich. 1993. Organic carbon in soils of the world. *Soil Sci. Soc. Am. J.* **57**:192–194.

Hedges, J.I. 1988. Polymerization of humic substances in natural environments. In: Humic Substances and Their Role in the Environment, ed. F.H. Frimmel and R.F. Christman, pp. 45–48. Dahlem Workshop Report LS 41. Chichester: Wiley.

Hedges, J.I. 1992. Global biogeochemical cycles: Progresses and problems. *Marine Chem.* **39**:67–93.

Henrichs, S.M., and A.P. Doyle. 1986. Decomposition of ^{14}C-labeled organic substances in marine sediments. *Limnol. Oceanogr.* **31(4)**:765–778.

Henrichs, S.M., and W.S. Reeburgh. 1987. Anaerobic mineralization of marine sediment organic matter: Notes and the role of anaerobic processes in the oceanic carbon economy. *Geomicro. J.* **5**:191–237.

Jenkinson, D.S. 1990. The turnover of organic carbon and nitrogen in soil. *Phil. Trans. R. Soc. Lond. B* **329**:361–368.

Kassim, G., J.P. Martin, and K.Haider. 1981. Incorporation of a wide variety of organic substrate carbons into soil biomass as estimated by the fumigation procedure. *Soil Sci. Soc. of Am. J.* **45**:1106–1112.

Lavelle, P., E. Blanchart, A. Martin, and S. Martin. 1993. A hierarchical model for decomposition in terrestrial ecosystems: Application to soils of the humid tropics. *Biotropica* **25**:2:130–150.

Lee, C. 1992. Controls on organic carbon preservation: The use of stratified water bodies to compare intrinsic sites of decomposition in oxic and anoxic systems. *Geochim. Cosmochim. Acta* **56**:3323–3335.

Melillo, J.M., J.D. Aber, and J.F. Muratore. 1982. Nitrogen and lignin control of hardwood leaf litter decomposition dynamics. *Ecology* **63**:621–626.

Mopper, K., and D.J. Kieber. 1991. Distribution and biological turnover of dissolved organic compounds in the water column of the Black Sea. *Deep-Sea Res.* **38(2)**:S1021–S1047.

Mopper, K., X. Zhou, R.J. Kieber, D.J Kieber, R.J. Sikorski, and R.D. Jones. 1991. Photochemical degradation of dissolved organic carbon and its impact on the oceanic carbon cycle. *Nature* **353**:60.

Parton, W.J., D.S. Schimel, D.S. Ojima, and C.V. Cole. 1994. A general model for soil organic matter dynamics: Sensitivity to litter chemistry, texture and management. *Soil Sci. Soc. Am. J.*, in press.

Parton, W.J., J.M.O. Scurlock, D.S. Ojima, T.G. Gilmanov, R.S. Scholes, D.S. Schimel, T. Kirchner, J.C. Menaut, T. Seastedt, E. Garcia Moya, A. Kamnalrut, and J.L. Kinyamario. 1993. Observations and modeling of biomass and soil organic matter dynamics for the grassland biome worldwide. *Glob. Biogeochem. Cyc.* **7**:785–809.

Savdra, C.E., and D.J. Bottjer. 1991 Oxygen-related biofacies in marine strata: An overview and update. In: Modern and Ancient Continental Shelf Anoxia, ed. R.V. Tyson and T.H. Pearson, pp. 201–219. Geol. Soc. Spec. Publ. 58. London: Geol. Soc.

Schimel, D.S., B.H. Braswell, E.A. Holland, R. McKeown, D.S. Ojima, T.H. Painter, W.J. Parton, and A.R. Townsend. 1994. Climatic, edaphic, and biotic controls over storage and turnover of carbon in soils. *Glob Biogeochem. Cyc.* **8**:279–293.

Shields, J.A., E.A. Paul, and W.E. Lowe. 1973. Factors influencing the stability of labelled microbial materials in soils. *Soil Biol. Biochem.* **6**:31–37.

Siegenthaler, U., and J.L. Sarmiento. 1993. Atmospheric carbon dioxide and the ocean. *Nature* **365**:119–125.

Sorensen, L.H. 1972. Stabilization of newly formed amino acid metabolites in soil by clay minerals. *Soil Sci.* **114**:5–11.

Sorensen, L.H. 1981. Carbon-nitrogen relationships during the humification of cellulose in soils containing different amounts of clay. *Soil Biol. Biochem.* **13**:313–321.

Stein, R. 1986. Organic carbon and sedimentation rate: Further evidence for anoxic deep-water conditons in the Cenomanian/Turonian Atlantic Ocean. *Mar. Geol.* **72**:199–209.

Sun, M.-Y., C. Lee, and R.C. Aller. 1993. Laboratory studies of oxic and anoxic degradation of chlorophyll-*a* in Long Island Sound sediments. *Geochim. Cosmochim. Acta* **57**:147–157.

Valentine, D., E.A. Holland, and D.S. Schimel. 1994. Ecosystem and physiological controls over methane production in northern wetlands. *J. Geophys. Res.* **99**:1563–1571.

Valentine, R., and R.G. Zepp. 1993. Formation of carbon monoxide from photodegradation of terrestrial organic matter. *Environ. Sci. Technol.* **27**:409–412.

Author Index

Subject Index

Subject index compiled by Colin Will

Dahlem Konferenzen Workshop Reports

Life Sciences Research Reports
(LS)

Physical, Chemical and Earth Science Research Reports (PC)

Environmental Sciences Research Reports (ES)

ES 16   The Role of Nonliving Organic Matter in the Earth's Carbon Cycle
Editors: R.G. Zepp and Ch. Sonntag
0 471 85463 2   358pp   1995

ES 17   Aerosol Forcing of Climate
Editors: R.J. Charlson and J. Heintzenberg
0 471 95693 7   approx 300pp   due July 1995

Prices on application from the Publisher